Universitext

Universitext

Universitext is a series of textbooks that presents material from a wide variety of mathematical disciplines at master's level and beyond. The books, often well class-tested by their author, may have an informal, personal even experimental approach to their subject matter. Some of the most successful and established books in the series have evolved through several editions, always following the evolution of teaching curricula, to very polished texts.

Thus as research topics trickle down into graduate-level teaching, first textbooks written for new, cutting-edge courses may make their way into *Universitext*.

More information about this series at http://www.springer.com/series/223

Bruno Bouchard • Jean-François Chassagneux

Fundamentals and Advanced Techniques in Derivatives Hedging

 Springer

Bruno Bouchard
Université Paris Dauphine
Paris, France

Jean-François Chassagneux
Université Paris Diderot
Paris, France

Translation from the French language edition:
Valorisation des produits dérivés by Bruno Bouchard and Jean-François Chassagneux
Copyright © Economica 2014
All Rights Reserved.

ISSN 0172-5939   ISSN 2191-6675   (electronic)
Universitext
ISBN 978-3-319-38988-2   ISBN 978-3-319-38990-5   (eBook)
DOI 10.1007/978-3-319-38990-5

Library of Congress Control Number: 2016943430

Mathematics Subject Classification (2010): 91G20, 91G10, 91G80, 49L25, 60H07

Printed on acid-free paper

This Springer imprint is published by Springer Nature
The registered company is Springer International Publishing AG Switzerland

Foreword

This book is dedicated to master students or young PhD students in applied mathematics who wish to learn mathematical finance. It uses a top-down approach, starting from the fundamentals and going little by little towards applications, while covering a large spectrum of concepts and mathematical tools. This is complemented by various (corrected) exercises in which many examples of practical interest will be provided. However, it is not our aim to detail all practitioners' models. We rather try to keep a general viewpoint, show various classes of techniques and try to adopt as much as possible a rigorous mathematical approach. Very good complementary readings are [7, 18, 33, 46, 49].

We first study the so-called *fundamental theorems of asset pricing* that are the basis of the theory of pricing and hedging of financial risks. This part is of abstract nature but is the basis of any realistic model and should be understood before going to practical considerations.

Indeed, we explain how the absence of arbitrage opportunities is related to the existence of risk-neutral measures and the central role the latter play in the pricing of derivatives and in the theory of portfolio management. We first consider discrete time models in which the analysis can be carried out in full generality and with rather simple mathematical tools. Continuous time models are studied in a subsequent chapter. In both cases, we shall see that the absence of arbitrage opportunities does not permit, in general, to define a unique possible price for a given financial derivative product, unless the market is complete and transactions are not restricted. This only provides an interval of *viable prices*, i.e. that do not create new arbitrage opportunities. Its upper bound is the so-called *super-hedging price*: the minimal price at which an option should be sold to ensure that it can be covered without risk.

To solve portfolio management issues, we follow the classical approach that consists in modelling preferences by utility functions. We explain how the so-called *dual formulation* approach leads to an explicit characterisation of the optimal portfolios in complete markets and how it can be used in incomplete markets to select viable prices of financial derivatives.

In the second part, we go closer to applications and confine ourselves to Markovian diffusion models, which are the ones used in practice. Using the fundamentals of the first part, we provide general tools for the pricing and hedging of risk in practical situations.

We first study complete markets and explain why, in this context, the super-hedging price of a derivative can be characterised as the solution of linear second-order parabolic equation, from which a perfect hedging strategy can be inferred as the gradient of the option price with respect to the values of the underlying assets. This is the *delta hedging* strategy. We consider different types of payoffs: plain vanilla options, barrier options and American options. As a first step, we will assume that the pricing functional is smooth and then appeal to the notion of viscosity solutions in the case where it may be irregular. We also present various techniques allowing one to compute the hedging strategy: tangent process and Malliavin calculus approaches and characterisations by partial differential equations.

We then consider *imperfect markets*: either incomplete or with constraints on the hedging positions. The super-hedging price functional is now related to a non-linear parabolic equation. When the imperfection is due to a constraint on the hedging positions, this equation can be replaced by a linear one after modifying the option payoff; this is the so-called *face-lift*. When the imperfection is due to a risk that cannot be hedged, it cannot be simplified in general but still provides some information on how the option can be covered without taking any risk. Unfortunately, this typically leads to hedging costs that are much too high for practical purposes, as can be observed in the classical example of stochastic volatility models when the volatility cannot be hedged.

This leads us to the notion of *shortfall price*. Instead of trying to super-hedge a risk, we look for a minimal price such that a given loss criteria can be satisfied by the P&L of a suitable (partial-)hedging strategy. In complete markets, this price can be determined by the same *dual formulation* approach as the one used in the analysis of portfolio management issues. For incomplete markets, we use the recent theory of *stochastic targets* that provides a characterisation of the *shortfall price* as the solution of a non-linear parabolic equation of Hamilton-Jacobi-Bellman type that can be solved numerically.

The last part of this book goes one step further towards applications: we provide a more detailed description of local and stochastic volatility models, in particular how they can be calibrated. We also discuss the effect of discrete time portfolio rebalancing, the use of semi-static hedging policies and the issues related to the recalibration procedures.

Paris, France   Bruno Bouchard
Paris, France   Jean-François Chassagneux
April 2016

Contents

General Notations

We list here general notations that will be used all over this book. Other notations will be introduced progressively; the list can be found in the Index of Notations.

An element $x = (x^1, \ldots, x^d)$ of $\mathbb{R}^d$ is identified to a column vector, with transpose x'. We let $\langle x, y \rangle$ denote the scalar product of x and y and $\|x\|$ be the Euclidean norm of x. The notation diag $[x]$ is used for the diagonal matrix whose i-th diagonal element is x^i. The collection of $d \times n$ matrices is $\mathbb{M}^{d,n}$, M' is the transpose of M, Tr $[M]$ denotes its trace, and $\|M\|$ is its Euclidean norm when viewed as the element of $\mathbb{R}^{d \times n}$ obtained by stacking the columns of M.

We shall always work on a probability space $(\Omega, \mathcal{F}, \mathbb{P})$. If $\mathcal{G} \subset \mathcal{F}$ and $\mathbb{Q} \ll \mathbb{P}$, we denote by $\mathbb{L}^p(\mathbb{Q}, \mathcal{G})$ the space of $\mathcal{G}$-measurable random variables that admit a moment of order p under $\mathbb{Q}$, $0 \leq p < \infty$. For $p = \infty$, this means that it is essentially bounded. We write $\mathbb{L}^p(B, \mathbb{Q}, \mathcal{G})$ if we want to insist on the fact that the random variables take values in the set B. The corresponding strong norm is $\| \cdot \|_{\mathbb{L}^p(\mathbb{Q})}$. The arguments $\mathcal{G}$ and/or $\mathbb{Q}$ are omitted if they are clearly identified by the context, in particular if $\mathcal{G} = \mathcal{F}$ or $\mathbb{Q} \sim \mathbb{P}$ and $p = \infty$ or $p = 0$. By default, all relations involving random variables are understood in the $\mathbb{P} - $ a.s. sense; relations between event sets are up to a null set. We denote by $\mathbb{E}[\cdot]$ the expectation under $\mathbb{P}$ and by $\mathbb{E}^\mathbb{Q}[\cdot]$ the expectation under $\mathbb{Q} \ll \mathbb{P}$. We use the same rule for conditional expectations. The random variable $\mathbf{1}_A$ takes the value 1 on the event set A, 0 on its complement A^c.

Given $(t, x) \in [0, T] \times B \mapsto f(t, x) \in \mathbb{R}$, $B \subset \mathbb{R}^d$, we denote by $\partial_t f$ its partial derivative with respect to its first component t and by ∂f (resp. ∇f) the line (resp. column) vector collecting its partial derivatives with respect to the components of the second argument x; it is the Jacobian (resp. the gradient) with respect to x. The Hessian matrix with respect to x is $\partial^2 f$. The partial derivative (resp. crossed second partial derivative) with respect to x^i (resp. (x^i, x^j)) is $\partial^2_{x^i} f$ (resp. $\partial^2_{x^i x^j} f$). If f takes values in $\mathbb{R}^m$, for some $m > 1$, then ∂f is the Jacobian matrix of f. A map f is $C^{p,q}(\tilde{B})$ if it is continuous, p-times differentiable with respect to its first argument t

and q-times differentiable with respect to its second argument x, on $\tilde{B} \subset [0, T] \times B$. We write $f \in C_b^{p,q}(\tilde{B})$ if the map and its derivatives up to order p in t and up to order q in x are bounded. When $\tilde{B}$ is clearly identified by the context, we omit it.

Given a set $\mathcal{O}$ of $\mathbb{R}^d$, we denote by $\bar{\mathcal{O}}$ its closure, $\partial\mathcal{O}$ its boundary and $\mathcal{O}^c$ its complement.

Part I
Fundamental Theorems

Chapter 1
Discrete Time Models

This first chapter is dedicated to discrete time markets. We first relate the *absence of arbitrage opportunities* to the existence of equivalent *martingale measures*, i.e. of equivalent probability measures that turn discounted asset prices into martingales. These measures are the basis of the whole pricing theory. They define the price intervals for derivatives products that are acceptable for the market. When the market is complete, meaning that any source of risk can be hedged perfectly by trading liquid assets, these intervals are reduced to one single point. This unique price allows one to hedge the corresponding derivative perfectly. However, in general, these intervals are not reduced to a singleton, and only their upper-bound, the so-called *super-hedging price*, permits to offset all risks by using a suitable dynamic hedging strategy. We shall study in details both European and American options. At the end of the chapter, the impact of portfolio constraints will also be discussed.

1.1 Financial Assets and Portfolio Strategies

We shall call *perfect market*[1] a financial market on which one can buy and sell the different assets freely, without restriction nor price impact or taxes. From now on, the market will be considered to be perfect.

From the mathematical point of view, we work on a complete probability space $(\Omega, \mathcal{F}, \mathbb{P})$ endowed with a filtration $\mathbb{F} = (\mathcal{F}_t)_{t \in \mathbb{T}}$ where $\mathbb{T} := \{0, \ldots, T\}$, $T > 0$. We assume that $\mathcal{F}_0 = \{\Omega, \emptyset\}$ and that $\mathcal{F}_T = \mathcal{F}$. We shall make use of the convention $\mathcal{F}_{-1} = \mathcal{F}_0$. The filtration $\mathbb{F}$ should be interpreted as modeling the flow of information.

---

[1]Extensions will be considered later on, in particular in Sect. 1.6 below.

© Springer International Publishing Switzerland 2016
B. Bouchard, J.-F. Chassagneux, *Fundamentals and Advanced Techniques in Derivatives Hedging*, Universitext, DOI 10.1007/978-3-319-38990-5_1

We follow the traditional approach which consists in considering two classes of assets:

- The *risk-free asset*: its return is known in advance, its value at time $t + 1$ is known at time t. This notion typically refers to a sovereign short term interest rate.
- The *risky assets*: their returns are not known in advance. It can be stocks, bonds, swaps, etc. Any financial product that is sufficiently liquid to be traded dynamically without transaction costs and restrictions.[2]

More precisely, we assume that there exists a financial product whose return is $r_t \geq 0$ on the time period $[t, t+1]$. By investing 1 at t, we obtain $1 + r_t$ at $t + 1$. Let us denote

$$\beta_t := \left(\prod_{s=0}^{t-1} (1 + r_s) \right)^{-1} \mathbf{1}_{\{t \geq 1\}} + \mathbf{1}_{\{t=0\}}$$

the corresponding discounted value at time 0 of 1 paid at t. The process $r = (r_t)_{t < T}$ is *adapted*, i.e. r_t is $\mathcal{F}_t$-measurable for all $t < T$. It is called the *risk-free interest rate*, this label comes from the fact that r_t is known at t, although it can be random.

Note that $\beta = (\beta_t)_{t \in \mathbb{T}}$ is *predictable* as β_t is $\mathcal{F}_{t-1}$-measurable for all $t \in \mathbb{T}$. It is called *discount factor*.[3]

The *risky assets* are modeled by a d-dimensional stochastic process $S = (S_t^1, \ldots, S_t^d)_{t \in \mathbb{T}}$, each component S_t^i representing the value at time t of the asset i. The process S is only *adapted*, meaning that S_t is $\mathcal{F}_t$-measurable. Hence, its return on the time period $[t - 1, t]$ is not known at $t - 1$, it is a risky investment.

With these different assets, one can build a *portfolio strategy* which consists in buying/selling dynamically the risky assets and borrowing or lending money at the risk-free interest rate. Mathematically, a *portfolio strategy* is modeled by an adapted stochastic process (α, ϕ) with values in $\mathbb{R} \times \mathbb{R}^d$. The quantity α_t stands for the amount of money lent (borrowed if negative) on the time interval $[t, t+1]$, while ϕ_t^i denotes the number of units of the i-th risky asset held in the portfolio on the same time interval.

Since $(\alpha_{t-1}, \phi_{t-1})$ is the position between $t - 1$ and t, and since the market is perfect, the value of the corresponding *financial portfolio* (or *wealth process*) at t is

$$V_t = \alpha_{t-1}(1 + r_{t-1}) + \langle \phi_{t-1}, S_t \rangle .$$

A financial strategy is said to be *self-financing* when this value V_t is enough at (each) time t to build the new position (α_t, ϕ_t) for the next time period $[t, t + 1]$. Otherwise

---

[2]In practice, transaction costs and restrictions are always present, but they are neglected here, and so are they in practice whenever they are sufficiently small.

[3]One must be careful with this terminology. The time t value β_t of the discount factor may refer, in a part of the financial literature, to the value at 0 of 1 paid at t. Here it is not the case, unless β_t is deterministic. In general, β_t is only known at $t - 1$.

stated, if at any t it satisfies

$$V_t = \alpha_{t-1}(1 + r_{t-1}) + \langle \phi_{t-1}, S_t \rangle = \alpha_t + \langle \phi_t, S_t \rangle . \tag{1.1}$$

Then, the *discounted value* of the portfolio is given by

$$\tilde{V}_t := \beta_t V_t = \beta_t \alpha_t + \langle \phi_t, \tilde{S}_t \rangle$$

in which

$$\tilde{S}_t := \beta_t S_t ,$$

is the vector of discounted values of the risky assets. In particular,

$$\alpha_t = B_t \left(\tilde{V}_t - \langle \phi_t, \tilde{S}_t \rangle \right) \quad \text{where} \quad B_t := \beta_t^{-1} .$$

Given (1.1), this implies

$$\tilde{V}_{t+1} = \beta_t \alpha_t + \langle \phi_t, \tilde{S}_{t+1} \rangle = \tilde{V}_t + \langle \phi_t, \tilde{S}_{t+1} - \tilde{S}_t \rangle, \tag{1.2}$$

or equivalently

$$\tilde{V}_t = V_0 + \sum_{s=0}^{t-1} \langle \phi_s, \tilde{S}_{s+1} - \tilde{S}_s \rangle, \tag{1.3}$$

i.e.

$$V_t = B_t \left(V_0 + \sum_{s=0}^{t-1} \langle \phi_s, \tilde{S}_{s+1} - \tilde{S}_s \rangle \right) . \tag{1.4}$$

As can be seen in (1.4), under the *self-financing* condition, the portfolio value V only depends on ϕ, the amount α of lent/borrowed money is uniquely determined by (V, ϕ). The collection of financial strategies $\mathcal{A}$ satisfying the self-financing condition can thus be identified to the set of adapted processes ϕ with values in $\mathbb{R}^d$.

In the following, we denote by $V^{v,\phi}$ the value of the financial portfolio induced by the strategy ϕ and the initial value v, at 0, and by $\tilde{V}^{v,\phi} := \beta V^{v,\phi}$ its discounted value.

Remark 1.1 It follows from (1.3) that

$$\tilde{V}^{y,\phi} = \tilde{V}^{x,\phi} + y - x ,$$

which implies

$$V^{y,\phi} = V^{x,\phi} + \beta^{-1}(y - x) .$$

Remark 1.2 As will be seen in the subsequent sections, only discounted values are important in our analysis. This explain why authors often assume $r \equiv 0$. This means that they work directly in discounted terms. In many exercises, we shall make the same assumption to ease notations and computations.

1.2   No-Arbitrage and Martingale Measures

This section is dedicated to the study of the no-arbitrage condition and its link with the existence of a martingale measure (also called risk neutral measure), i.e. a $\mathbb{P}$-equivalent probability measure that turns the discounted risky-assets price process into a martingale.

1.2.1   Definition of the Absence of Arbitrage Opportunities

An *arbitrage opportunity* is a financial strategy ϕ such that

$$V_T^{0,\phi} \geq 0 \quad \mathbb{P} - \text{a.s.} \quad \text{and} \quad \mathbb{P}\left[V_T^{0,\phi} > 0\right] > 0 .$$

The above means that, starting from 0, the strategy ϕ leads to a possible gain $(\mathbb{P}\left[V_T^{0,\phi} > 0\right] > 0)$ without taking any risk $(V_T^{0,\phi} \geq 0 \ \mathbb{P}-\text{a.s.})$. It can be equivalently defined in terms of discounted values, since the discount factor β is positive:

$$\tilde{V}_T^{0,\phi} \geq 0 \ \mathbb{P} - \text{a.s.} \quad \text{and} \quad \mathbb{P}\left[\tilde{V}_T^{0,\phi} > 0\right] > 0 .$$

Note that an arbitrage is not a situation where one makes a gain with certainty, in general.

The fundamental concept of the *absence of arbitrage opportunities* rules out the existence of such strategies.

Definition 1.1 (NA) We say that there is no arbitrage opportunity, if the following holds:

$$(\text{NA}) \ : \ \tilde{V}_T^{0,\phi} \geq 0 \ \mathbb{P} - \text{a.s.} \quad \Rightarrow \quad \tilde{V}_T^{0,\phi} = 0 \ \mathbb{P} - \text{a.s.}, \quad \text{for all } \phi \in \mathcal{A}.$$

Remark 1.3 The definition of (NA) only depends on $\mathbb{P}$ though its null sets, i.e. the sets of measure zero for $\mathbb{P}$. The exact weights associated to the measure $\mathbb{P}$ therefore play no role, the only question is whether an event has positive or zero measure for $\mathbb{P}$, is feasible or not. The definition does not change if stated under a $\mathbb{P}$-equivalent probability measure.

1.2.2   First Fundamental Theorem

In this section, we will prove the *first fundamental theorem of asset pricing*, see Theorem 1.1 below, which is the corner stone of the whole theory of derivatives pricing. It says that the *absence of arbitrage opportunities* and the existence of *risk neutral measures*, as defined below, are equivalent.

Definition 1.2 (Risk neutral/martingale measure) The set of *risk neutral measures*, or *martingale measures*, denoted by $\mathcal{M}(\tilde{S})$, is the collection of measures $\mathbb{Q} \sim \mathbb{P}$ such that $\tilde{S}$ is a $\mathbb{Q}$-*local martingale*, i.e.

1. $\tilde{S}_{t+1} \in \mathbb{L}^1(\mathbb{Q})$ for all $t < T$;
2. $\mathbb{E}^{\mathbb{Q}}\left[\tilde{S}_{t+1} \mid \mathcal{F}_t\right] = \tilde{S}_t$ $\mathbb{Q}$-a.s. for all $t < T$.

The terminology *martingale measure* is natural given the definition. The label *risk neutral* is more ambiguous. It refers to the foundations of the asset pricing theory in economy. We will comment on this later in Remark 1.8 below.

We shall show that the condition (NA) implies the existence of an element of $\mathcal{M}(\tilde{S})$ with density in $\mathbb{L}^\infty$ (with respect to $\mathbb{P}$). In the following, let us denote by $\mathcal{M}_b(\tilde{S})$ the subset of elements in $\mathcal{M}(\tilde{S})$ with density in $\mathbb{L}^\infty$.

In order to slightly simplify the proofs, we use the traditional approach that consists in reducing to a one step model. We therefore define the notion of *local no arbitrage condition*:

$$(\text{NA}_t) \; : \; \langle \xi, \tilde{S}_{t+1} - \tilde{S}_t \rangle \geq 0 \; \mathbb{P} - \text{a.s.} \; \Rightarrow \; \langle \xi, \tilde{S}_{t+1} - \tilde{S}_t \rangle = 0 \; \mathbb{P} - \text{a.s.},$$

$$\forall \, \xi \in \mathbb{L}^0(\mathbb{R}^d, \mathcal{F}_t) \, .$$

The condition (NA_t) means that one cannot make a profit, with non-zero probability and without taking any risk, by trading only on $[t, t+1]$.

Theorem 1.1 (First Fundamental Theorem of Asset Pricing) *The following are equivalent:*

  (i) (NA) *holds.*
 (ii) (NA_t) *holds for* $t = 0, \ldots, T - 1$.
(iii) $\mathcal{M}_b(\tilde{S}) \neq \emptyset$.

From now on, we assume that

$$\tilde{S}_t \in L^1(\mathbb{P}) \quad \text{for all } t \in \mathbb{T} \, . \tag{1.5}$$

This is without loss of generality. Indeed, if the above does not hold, then one can replace $\mathbb{P}$ by the equivalent measure with density

$$e^{-\sum_{t\in T} \|\tilde{S}_t\|} / \mathbb{E}\left[e^{-\sum_{t\in T} \|\tilde{S}_t\|}\right] .$$

It has the same null-sets as $\mathbb{P}$ and therefore the condition (NA) is not affected, see Remark 1.3.

1.2.3   Proof of the First Fundamental Theorem

The following subsections are dedicated to the proof of Theorem 1.1. Given its length, it is split in different intermediate results.

1.2.3.1   (NA) Implies (NA$_t$)

This is clear. Indeed, if there exists $t \leq T-1$ and $\xi \in \mathbb{L}^0(\mathbb{R}^d, \mathcal{F}_t)$ such that $\langle \xi, \tilde{S}_{t+1} - \tilde{S}_t \rangle \geq 0 \; \mathbb{P} - \text{a.s.}$ and $\langle \xi, \tilde{S}_{t+1} - \tilde{S}_t \rangle \neq 0$ then (NA) is violated by the strategy $\phi \in \mathcal{A}$ defined by $\phi_s = \xi_t \mathbf{1}_{s=t}$ for $s \in \mathbb{T}$.

The converse implication will be proved at the end of the proof of Theorem 1.1.

1.2.3.2   A Closure Property Under (NA$_t$)

Let us first prove that

$$\tilde{\Gamma}_t := \{ \langle \xi, \tilde{S}_{t+1} - \tilde{S}_t \rangle - \rho \; : \; \xi \in \mathbb{L}^0(\mathbb{R}^d, \mathcal{F}_t), \; \rho \in \mathbb{L}^0(\mathbb{R}_+) \}$$

is closed in probability under (NA$_t$), or equivalently under

$$\tilde{\Gamma}_t \cap \mathbb{L}^0(\mathbb{R}_+) = \{0\} . \tag{1.6}$$

Once proved, this will be combined with a Hahn-Banach separation type argument to deduce the existence of martingale measures, see Sect. 1.2.3.3.

To show the above claim, we shall appeal to the following technical lemma.

Lemma 1.1 *Let $(\xi_n)_{n\geq 1}$ be a sequence in $\mathbb{L}^0(\mathbb{R}^d, \mathcal{F}_t)$. If*

$$\eta^0 := \liminf_{n\to\infty} \|\xi_n\| < \infty \; \mathbb{P} - \text{a.s.}$$

then, there exists a sequence $(k_n)_{n\geq 1} \in \mathbb{L}^0(\mathbb{N}, \mathcal{F}_t)$ such that $k_n \to \infty \; \mathbb{P} - \text{a.s.}$ and $\xi_{k_n} \to \xi_* \; \mathbb{P} - \text{a.s.}$ for some random variable $\xi_* \in \mathbb{L}^0(\mathbb{R}^d, \mathcal{F}_t)$.

Proof Let us define $(k_n^0)_{n\geq 0}$ by $k_0^0 = 0$ and $k_{n+1}^0 := \inf\{j > k_n^0 : | \, \|\xi_j\| - \eta^0| \leq n^{-1}\}$. Clearly, $(k_n^0)_{n\geq 0}$ is $\mathcal{F}_t$-measurable and $\sup_{n\geq 1} \|\xi_{k_n^0}\| < \infty$. This implies that $\eta^1 := \liminf_{n\to\infty} |\xi_n^1| < \infty \; \mathbb{P}$ – a.s. Let $\epsilon^1 \in \mathbb{L}^0(\mathcal{F}_t)$ be defined by $\epsilon^1 = 1$ when $\text{card}\{n \in \mathbb{N} : \exists \, j \geq 1 \text{ s.t. } |\xi_{k_j^0}^1 - \eta^1| \leq n^{-1}\} = \infty$ and $\epsilon^1 = -1$ otherwise. We then introduce the sequence $(k_n^1)_{n\geq 0}$ as $k_0^1 = 0$ and $k_{n+1}^1 := \inf\{j > k_n^1 : |\xi_{k_j^0}^1 - \epsilon^1\eta^1| \leq n^{-1}\}$. Clearly, $(k_n^1)_{n\geq 0}$ is $\mathcal{F}_t$-measurable and $\xi_{k_n}^1 \to \epsilon^1\eta^1 \; \mathbb{P}$ – a.s. We can then define $\xi_*^1 := \epsilon^1\eta^1$. We go on and now construct $(k_n^2)_{n\geq 0}$ in a similar way from $(\xi_{k_n^1})_{n\geq 1}$ so that $(\xi_{k_n^2}^2)_{n\geq 1}$ converges $\mathbb{P}$ – a.s. By iterating this procedure, we obtain a sequence $(k_n^d)_{n\geq 0}$ which satisfies the requirements of our lemma. $\qquad\square$

Proposition 1.1 *If* (NA_t) *holds, then* $\tilde{\Gamma}_t$ *is closed in probability.*

Proof Since the convergence in probability implies the a.s. convergence, up to a subsequence, we only have to show that if a sequence $(\xi_n, \rho_n)_{n\geq 1}$ of $\mathbb{L}^0(\mathbb{R}^d, \mathcal{F}_t) \times L^0(\mathbb{R}_+)$ satisfies

$$G_n := \langle \xi_n, \tilde{S}_{t+1} - \tilde{S}_t \rangle - \rho_n \to G_* \; \mathbb{P} - \text{a.s.}$$

then we can find $(\xi_*, \rho_*) \in \mathbb{L}^0(\mathbb{R}^d, \mathcal{F}_t) \times L^0(\mathbb{R}_+)$ such that

$$\langle \xi_*, \tilde{S}_{t+1} - \tilde{S}_t \rangle - \rho_* = G_* . \tag{1.7}$$

For sake of simplicity, we shall only consider the one dimensional case $d = 1$. The general case is handled similarly, essentially by iterating on the dimension, see [39].

1. Set $A := \{\liminf_{n\to\infty} \|\xi_n\| = \infty\}$, and observe that $A \in \mathcal{F}_t$. The sequence $(\bar{\xi}_n)_{n\geq 1}$ of $\mathbb{L}^0(\mathbb{R}^d, \mathcal{F}_t)$ defined by $\bar{\xi}_n = \xi_n \mathbf{1}_A$ satisfies

$$\langle \bar{\xi}_n, \tilde{S}_{t+1} - \tilde{S}_t \rangle - \rho_n \mathbf{1}_A \to G_* \mathbf{1}_A \; \mathbb{P} - \text{a.s.}$$

and therefore

$$\langle \hat{\xi}_n, \tilde{S}_{t+1} - \tilde{S}_t \rangle - \hat{\rho}_n \to 0 \; \mathbb{P} - \text{a.s.}$$

in which $(\hat{\xi}_n, \hat{\rho}_n) = (\xi_n, \rho_n)\mathbf{1}_A/(1 + \|\xi_n\|)$. Since $\liminf_{n\to\infty} \|\hat{\xi}_n\| = 1 < \infty$ on $A \; \mathbb{P}$ – a.s. by construction, Lemma 1.1 implies that we can find a sequence $(k_n)_{n\geq 1}$ in $\mathbb{L}^0(\mathbb{N}, \mathcal{F}_t)$ such that $\langle \hat{\xi}_{k_n}, \tilde{S}_{t+1} - \tilde{S}_t \rangle \to \langle \hat{\xi}_*, \tilde{S}_{t+1} - \tilde{S}_t \rangle$, in which $\hat{\xi}_* \in \mathbb{L}^0(\mathbb{R}, \mathcal{F}_t)$. Since $\hat{\rho}_{k_n} \geq 0 \; \mathbb{P}$ – a.s., it must hold that $\langle \hat{\xi}_*, \tilde{S}_{t+1} - \tilde{S}_t \rangle \geq 0$. By our assumption (NA_t), recall its equivalent formulation (1.6), this implies that $\langle \hat{\xi}_*, \tilde{S}_{t+1} - \tilde{S}_t \rangle = 0$ and $\hat{\rho}_{k_n} \to 0 \; \mathbb{P}$ – a.s. Since $|\hat{\xi}_*| = 1$ on A by construction, this shows that $\tilde{S}_{t+1} - \tilde{S}_t = 0$ on A (recall that $d = 1$ for this proof). One can then replace the initial sequence $(\xi_n, \rho_n)_{n\geq 0}$ by $(\check{\xi}_n, \check{\rho}_n)_{n\geq 0}$ defined by $(\check{\xi}_n, \check{\rho}_n) = (\xi_n, \rho_n)\mathbf{1}_{A^c}$ and still have that

$$\langle \breve{\xi}_n, \tilde{S}_{t+1} - \tilde{S}_t \rangle - \breve{\rho}_n \to G_* \quad \mathbb{P} - \text{a.s.}$$

in which $\liminf_{n \to \infty} \|\breve{\xi}_n\| < \infty \ \mathbb{P} - \text{a.s.}$

2. In view of 1., one can now assume that $\liminf_{n \to \infty} \|\xi_n\| < \infty \ \mathbb{P}$–a.s. Lemma 1.1 then implies that there exists a sequence $(k_n)_{n \geq 1}$ in $\mathbb{L}^0(\mathbb{N}, \mathcal{F}_t)$ such that $G_{k_n} \to G_*$ and $\langle \xi_{k_n}, \tilde{S}_{t+1} - \tilde{S}_t \rangle \to \langle \xi_*, \tilde{S}_{t+1} - \tilde{S}_t \rangle$, in which $\xi_* \in \mathbb{L}^0(\mathbb{R}, \mathcal{F}_t)$. From this, one deduces that $(\rho_{k_n})_n$ converges to some $\rho_* \in \mathbb{L}^0(\mathbb{R}_+)$, a.s., for which (1.7) holds.

$\square$

1.2.3.3   Existence of a Martingale Measure Under (NA)$_t$ for $t < T$

We first show that (NA)$_t$ implies the existence of an equivalent measure that turns $\tilde{S}$ into a martingale on $[t, t+1]$.

To this purpose, we combine the closure property of Proposition 1.1 and the *Hahn-Banach separation theorem*, see for instance [15, Theorem 1.7], which we recall below.

Theorem 1.2 (Hahn-Banach separation theorem) *Let E be a normed vector space, A be a convex compact subset of E, and B be a closed convex subset of E. If $A \cap B = \emptyset$, then there exists a linear map φ on E such that*

$$\sup_{b \in B} \varphi(b) < \inf_{a \in A} \varphi(a).$$

Proposition 1.2 *If (NA$_t$) holds, then there exists $H_t \in \mathbb{L}^\infty$ such that*

$$\mathbb{E}\left[H_t(\tilde{S}_{t+1} - \tilde{S}_t) \mid \mathcal{F}_t\right] = 0 \ \text{and} \ H_t > 0 \ \mathbb{P} - \text{a.s.}$$

Proof [4]

1. Since $\tilde{\Gamma}_t$ is closed for the convergence in probability, see Proposition 1.1, $\tilde{\Gamma}_t \cap \mathbb{L}^1$ is closed in $\mathbb{L}^1$. Moreover, (NA$_t$) implies that $\mathbf{1}_A \notin \tilde{\Gamma}_t$ for all $A \in \mathcal{F}_t$ such that $\mathbb{P}[A] \neq 0$. Hence, the set $\tilde{\Gamma}_t$ being convex, we deduce from Theorem 1.2, that there exists a bounded random variable Y such that

$$\sup_{G \in \tilde{\Gamma}_t \cap \mathbb{L}^1} \mathbb{E}[YG] < \mathbb{E}[Y\mathbf{1}_A].$$

---

[4]The proof is a version of Kreps-Yan theorem, see Exercise 1.7 below.

Since $\tilde{\Gamma}_t$ is a cone (which contains 0), this implies that $\mathbb{E}\left[Y\mathbf{1}_A\right] > 0$ and

$$\sup_{G \in \tilde{\Gamma}_t \cap \mathbb{L}^1} \mathbb{E}\left[YG\right] \leq 0 . \tag{1.8}$$

Furthermore, since $-\mathbf{1}_{\{Y<0\}} \in \tilde{\Gamma}_t \cap \mathbb{L}^1$, it must also hold that

$$Y \geq 0 \quad \mathbb{P} - \text{a.s.}$$

2. Let $\mathcal{G}$ denote the set of non-zero random variables Y in $\mathbb{L}^\infty(\mathbb{R}_+)$ satisfying (1.8). We set $s(Y) := \{Y \neq 0\}$, this is the support of Y. One can then find $\hat{Y} \in \mathcal{G}$ such that

$$\mathbb{P}\left[s(\hat{Y})\right] = \max\{\mathbb{P}\left[s(Y)\right] \; : \; Y \in \mathcal{G}\} .$$

Indeed, if $(Y_n)_{n\geq 1}$ is a maximizing sequence for $\{\mathbb{P}\left[s(Y)\right] \; : \; Y \in \mathcal{G}\}$, then one can find[5] a sequence $(a_n)_{n\geq 1}$ of positive real numbers such that

$$\hat{Y} := \sum_{n\geq 1} a_n Y_n$$

is well-defined and has the support

$$s(\hat{Y}) = \bigcup_{n\geq 1} s(Y_n) .$$

Then,

$$\mathbb{P}\left[\bigcup_{n\leq N} s(Y_n)\right] \geq \mathbb{P}\left[s(Y_N)\right] \to \sup\{\mathbb{P}\left[s(Y)\right] \; : \; Y \in \mathcal{G}\} .$$

3. We now prove that $\mathbb{P}\left[s(\hat{Y})\right] = 1$. If this was not the case, then the arguments used in step 1. would allow one to find $\bar{Y} \in \mathcal{G}$ such that

$$\sup_{G \in \tilde{\Gamma}_t \cap \mathbb{L}^1} \mathbb{E}\left[\bar{Y}G\right] \leq 0 < \mathbb{E}\left[\bar{Y}\mathbf{1}_{s(\hat{Y})^c}\right] .$$

This would imply $\{\bar{Y} \neq 0\} \cap s(\hat{Y})^c \neq \emptyset$, thus contradicting the very definition of $\hat{Y}$ since $\bar{Y} + \hat{Y}$ satisfies (1.8) but has a support strictly bigger than the support of $\hat{Y}$.

---

[5] Take for instance $a_n := 2^{-n}(1 + \|Y_n\|_{L^\infty})^{-1}$.

4. One can now conclude by noticing that, for any $i \leq d$,

$$-\mathbf{1}_{\{\mathbb{E}\left[H_t(\tilde{S}^i_{t+1}-\tilde{S}^i_t)\mid \mathcal{F}_t\right]<0\}} \text{ and } \mathbf{1}_{\{\mathbb{E}\left[H_t(\tilde{S}^i_{t+1}-\tilde{S}^i_t)\mid \mathcal{F}_t\right]>0\}}$$

belong to $\mathbb{L}^0(\mathcal{F}_t)$, and by using (1.8).

$\square$

We are now in position to prove the existence of a martingale measure under (NA).

Proposition 1.3 *If* (NA) *holds, then there exists* $\mathbb{Q} \sim \mathbb{P}$ *such that* $\tilde{S}$ *is a* $\mathbb{Q}$-*martingale. Moreover,* $d\mathbb{Q}/d\mathbb{P}$ *is essentially bounded.*

Proof For all $t \in \mathbb{T}$, we can find H_t satisfying the requirements of Proposition 1.2. One can then define

$$H := \prod_{t \leq T-1} \left(\mathbb{E}\left[H_t \mid \mathcal{F}_{t+1}\right]/\mathbb{E}\left[H_t \mid \mathcal{F}_t\right]\right),$$

and the corresponding probability measure $\mathbb{Q}$ through $d\mathbb{Q}/d\mathbb{P} = H$. Saying that $\tilde{S}$ is a $\mathbb{Q}$-martingale is equivalent to saying that $\hat{H}\tilde{S}$ is a $\mathbb{P}$-martingale, in which $\hat{H}_t := \mathbb{E}\left[H \mid \mathcal{F}_t\right]$, i.e.

$$\mathbb{E}^{\mathbb{Q}}[\tilde{S}_{t+1} \mid \mathcal{F}_t] = \mathbb{E}\left[H\tilde{S}_{t+1} \mid \mathcal{F}_t\right]/\mathbb{E}\left[H \mid \mathcal{F}_t\right]$$

$$= \mathbb{E}\left[\hat{H}_{t+1}\tilde{S}_{t+1} \mid \mathcal{F}_t\right]/\hat{H}_t$$

$$= \tilde{S}_t.$$

This is easily checked after noticing that

$$\hat{H}_t := \prod_{j \leq t-1} \left(\mathbb{E}\left[H_j \mid \mathcal{F}_{j+1}\right]/\mathbb{E}\left[H_j \mid \mathcal{F}_j\right]\right),$$

and by appealing to Proposition 1.2.

$\square$

1.2.3.4 Absence of Arbitrage Under $\mathcal{M}(\tilde{S}) \neq \emptyset$:

We now conclude the proof of Theorem 1.1 by proving that (iii) implies (i).

Recall that a process ξ is a $\mathbb{Q}$-*local martingale* if $\mathbb{E}^{\mathbb{Q}}[\xi_{t+1} \mid \mathcal{F}_t]$ is well-defined and $\mathbb{E}^{\mathbb{Q}}\left[\xi_{t+1} \mid \mathcal{F}_t\right] = \xi_t$ for all $t < T$. We denote by $\mathbb{L}^0_b$ the set of random variables that are essentially bounded from below.

Proposition 1.4 *Assume that* $\mathbb{Q} \in \mathcal{M}(\tilde{S})$ *and fix* $\phi \in \mathcal{A}$. *Then,*

(i) $\tilde{V}^{0,\phi}$ *is a* $\mathbb{Q}$-*local martingale.*

(ii) *If* $\tilde{V}_T^{0,\phi} \geq G\ \mathbb{P} - a.s.$ *for some* $G \in \mathbb{L}_b^0$, *then* $\tilde{V}_t^{0,\phi} \geq \mathbb{E}^{\mathbb{Q}}\left[G \mid \mathcal{F}_t\right]\ \mathbb{P} - a.s.$ *for all* $t \in \mathbb{T}$.

(iii) *If* (ii) *holds with* $G \geq 0\ \mathbb{P} - a.s.$ *and* $\mathbb{P}\left[G > 0\right] > 0$, *then* $V_0^{0,\phi} > 0$.

Proof Let $t \leq T - 1$. Since ϕ_t is $\mathcal{F}_t$-measurable, we have

$$\mathbb{E}^{\mathbb{Q}}\left[\langle \phi_t, \tilde{S}_{t+1} - \tilde{S}_t \rangle \mid \mathcal{F}_t\right] = \langle \phi_t, \mathbb{E}^{\mathbb{Q}}\left[\tilde{S}_{t+1} - \tilde{S}_t \mid \mathcal{F}_t\right]\rangle ,$$

which implies (i), recall (1.2). This in turns implies that $\tilde{V}_t^{0,\phi} \geq \mathbb{E}^{\mathbb{Q}}\left[\tilde{V}_{t+1}^{0,\phi} \mid \mathcal{F}_t\right]$, for all $t \leq T - 1$. Hence, $\tilde{V}_{t+1}^{0,\phi} \geq \mathbb{E}^{\mathbb{Q}}\left[G \mid \mathcal{F}_{t+1}\right]\ \mathbb{P} - a.s.$ implies that $\tilde{V}_t^{0,\phi} \geq \mathbb{E}^{\mathbb{Q}}\left[G \mid \mathcal{F}_t\right]\ \mathbb{P} - a.s.$ by taking the conditional expectation. Since $\tilde{V}_T^{0,\phi} \geq G = \mathbb{E}^{\mathbb{Q}}\left[G \mid \mathcal{F}_T\right]\ \mathbb{P} - a.s.$, (ii) follows by induction. The assertion (iii) is an immediate consequence of (ii) since $G \geq 0\ \mathbb{P} - a.s.$ and $\mathbb{P}\left[G > 0\right] > 0$ is equivalent to $G \geq 0\ \mathbb{P} - a.s.$ and $\mathbb{Q}\left[G > 0\right] > 0$, as $\mathbb{Q} \sim \mathbb{P}$, which implies $\mathbb{E}^{\mathbb{Q}}\left[G\right] > 0$. $\qquad \square$

Remark 1.4 It follows from the above that $\tilde{V}^{0,\phi}$ is bounded from below by the $\mathbb{Q}$-martingale X defined as $X_t := \mathbb{E}^{\mathbb{Q}}\left[G \mid \mathcal{F}_t\right]$, whenever $\tilde{V}_T^{0,\phi} \geq G\ \mathbb{P} - a.s.$ with $G \in \mathbb{L}^1(\mathbb{Q})$. One can then define $\mathbb{E}^{\mathbb{Q}}\left[\tilde{V}_t^{0,\phi}\right]$ for all $t \in \mathbb{T}$, up to assigning to it the value $+\infty$ if the positive part $\tilde{V}_t^{0,\phi}$ is not $\mathbb{Q}$-integrable. Moreover, a simple induction based on the local martingale property implies that $\tilde{V}_0^{0,\phi} \geq \mathbb{E}^{\mathbb{Q}}\left[\tilde{V}_t^{0,\phi}\right]$ for all $t \in \mathbb{T}$. In particular, $\tilde{V}_t^{0,\phi} \in \mathbb{L}^1(\mathbb{Q})$ for all $t \in \mathbb{T}$, otherwise stated, $\tilde{V}^{0,\phi}$ is a (true) $\mathbb{Q}$-martingale.

1.3 European Option Pricing

This section is dedicated to the foundations of the pricing and hedging of *European options*, i.e. financial derivatives that pay a random amount $G \in \mathbb{L}^0$ at time T, called *maturity*.

1.3.1 Various Notions of Price

The question we try to answer here is: given that the seller of the option will deliver the random amount (called *payoff*) G at T to the buyer, which price, also called *premium*, should be paid at time 0 by the buyer in exchange.

The following notions of price are natural.

Definition 1.3 (Viable/super-hedging price)

1. A price p is a *viable price* for the option of payoff G if buying or selling this
   option at this price does not create an arbitrage:

$$\nexists \phi \in \mathcal{A} \text{ et } \epsilon \in \{-1, 1\} \text{ s.t. } V_T^{\epsilon p, \phi} - \epsilon \, G \in \mathbb{L}^0(\mathbb{R}_+) \setminus \{0\} \,. \qquad (1.9)$$

2. The price p allows the seller to *super-hedge* the payoff G if

$$\exists \, \phi \in \mathcal{A} \text{ s.t. } V_T^{p, \phi} - G \in \mathbb{L}^0(\mathbb{R}_+) \,. \qquad (1.10)$$

3. The *super-hedging price* of G is

$$p(G) := \inf\{p \in \mathbb{R} \ : \ \exists \, \phi \in \mathcal{A} \text{ s.t. } V_T^{p, \phi} - G \in \mathbb{L}^0(\mathbb{R}_+)\} \,. \qquad (1.11)$$

1.3.2 Dual Description of the Payoffs That Can Be Super-Hedged

We first provide a description of the collection of payoffs of European options that
can be super-hedged from a zero initial wealth. For technical reasons, we restrict to
payoffs $G \in \mathbb{L}^0$ such that $\beta_T G^-$ belongs to the class L_S of random variables $Y \in \mathbb{L}^0$
satisfying

$$|Y| \le c_Y^0 + \sum_{i=1}^d c_Y^i \tilde{S}_T^i \ \mathbb{P} - \text{a.s.} \qquad (1.12)$$

for some constants $c_Y^0, \ldots, c_Y^d$ that may depend on Y.

Theorem 1.3 (Dual formulation) *Assume that $\mathcal{M}(\tilde{S}) \ne \emptyset$, then the following sets*

$$\Gamma := \left\{ G \in \mathbb{L}^0 \ : \ \beta_T G^- \in L_S \,, \ \exists \, \phi \in \mathcal{A} \text{ s.t. } V_T^{0, \phi} \ge G \right\}$$

$$\Theta := \left\{ G \in \mathbb{L}^0 \ : \ \beta_T G^- \in L_S \,, \ \sup_{Q \in \mathcal{M}(\tilde{S})} \mathbb{E}^Q \left[\beta_T G \right] \le 0 \right\}$$

are equal. Moreover,

$$\left\{ G \in \mathbb{L}^0 \ : \ \exists \, \phi \in \mathcal{A} \text{ s.t. } V_T^{0, \phi} \ge G \right\}$$

is closed in probability.

Remark 1.5 We deduce from Remark 1.1 and the above result that $p(G-\beta_T^{-1}v) \le 0$ implies: for all $\varepsilon > 0$, there exists $\phi \in \mathcal{A}$ such that $V_T^{\varepsilon,\phi} - (G - \beta_T^{-1}v) \in \mathbb{L}^0(\mathbb{R}_+)$ or equivalently $\tilde{V}_T^{\varepsilon,\phi} - \beta_T G + v \in \mathbb{L}^0(\mathbb{R}_+)$, i.e. $\tilde{V}_T^{v+\varepsilon,\phi} - \beta_T G \in \mathbb{L}^0(\mathbb{R}_+)$. So that $p(G) \le v$. Describing Γ is thus equivalent to describing the set of European options that can be super-hedged from an arbitrary initial wealth x. Moreover, since Γ is closed in probability, the infimum in the definition of $p(G)$ is achieved. Indeed, $G - p(G) - \varepsilon \in \Gamma$ for all $\varepsilon > 0$ implies $G - p(G) \in \Gamma$.

Remark 1.6 The above is easily generalised to the case where we start from a random initial $\mathcal{F}_t$-measurable wealth at time $t > 0$: Fix $\xi \in \mathbb{L}^0(\mathbb{R}, \mathcal{F}_t)$ and $G \in \mathbb{L}^0$ such that $\beta_T G^- \in L_S$, then there exists $\phi \in \mathcal{A}$ such that

$$\xi + \tilde{V}_T^{0,\phi} - \tilde{V}_t^{0,\phi} \ge \beta_T G$$

if and only if

$$\mathbb{E}^{\mathbb{Q}}[\beta_T G \mid \mathcal{F}_t] \le \xi \quad \mathbb{P}-\text{a.s.}, \quad \forall\, \mathbb{Q} \in \mathcal{M}(\tilde{S}).$$

In order to prove the above theorem, we shall use the following key lemma.

Lemma 1.2 *Assume that* (NA) *holds. If* $\phi \in \mathcal{A}$ *satisfies*

$$\sum_{s=t}^{T-1} \langle \phi_s, \tilde{S}_{s+1} - \tilde{S}_s \rangle = 0 \quad \mathbb{P}-a.s.,$$

then

$$\langle \phi_t, \tilde{S}_{t+1} - \tilde{S}_t \rangle = 0 \quad \mathbb{P}-a.s.$$

Proof On $A := \{\langle \phi_t, \tilde{S}_{t+1} - \tilde{S}_t \rangle > 0\}$, we have

$$\sum_{s=t+1}^{T-1} \langle \phi_s, \tilde{S}_{s+1} - \tilde{S}_s \rangle = -\langle \phi_t, \tilde{S}_{t+1} - \tilde{S}_t \rangle < 0.$$

Since $A \in \mathcal{F}_{t+1}$, $\hat{\phi}$ defined by $\hat{\phi}_s = -\phi_s 1_A 1_{s \ge t+1}$ belongs to $\mathcal{A}$. On the other hand,

$$\tilde{V}_T^{0,\hat{\phi}} \ge 0, \quad \mathbb{P}\left[\tilde{V}_T^{0,\hat{\phi}} > 0\right] = \mathbb{P}[A],$$

which implies $\mathbb{P}[A] = 0$ under (NA). We show similarly that $\{\langle \phi_t, \tilde{S}_{t+1} - \tilde{S}_t \rangle < 0\}$ is a null-set. $\qquad\square$

Proof of Theorem 1.3 1. We first use an induction to show that Γ is closed in probability. For $s < t$, let us define $\Gamma_{s,t}$ as the set of random variables $G \in \mathbb{L}^0$

such that

$$\exists \phi \in \mathcal{A} \text{ s.t. } \phi = 0 \text{ on } \mathbb{T} \setminus [s, t[\text{ and } V_T^{0,\phi} - G \in \mathbb{L}^0(\mathbb{R}_+) .$$

By Proposition 1.1 and Theorem 1.1, $\Gamma_{T-1,T}$ is closed in probability. Let us assume that $\Gamma_{t,T}$ is closed in probability for $1 \le t \le T - 1$ and deduce that this implies that $\Gamma_{t-1,T}$ is closed as well. Let $(G_n)_n$ be a sequence of $\Gamma_{t-1,T}$ that converges to G_* a.s., after possibly considering a subsequence. Let $(\phi^n, \rho_n)_n$ be a sequence in $\mathcal{A} \times \mathbb{L}^0(\mathbb{R}_+)$ such that $\phi_s^n = 0$ for $s < t - 1$ and

$$\tilde{V}_T^{0,\phi^n} - \rho_n = \beta_T G_n \quad \mathbb{P} - \text{a.s.}$$

For sake of simplicity, we now restrict to the case $d = 1$. We use similar arguments as in the proof of Proposition 1.1.

Set $A := \{\liminf_{n\to\infty} \|\phi_{t-1}^n\| = \infty\}$ and

$$(\hat{\phi}^n, \hat{\rho}_n, \hat{G}_n) := ((\phi^n, \rho_n, G_n)\mathbf{1}_A/(1 + \|\phi_{t-1}^n\|)_n.$$

After possibly passing to a random subsequence, we can assume that $(\hat{\phi}_{t-1}^n)_n$ converges a.s. to some element $\hat{\phi}_{t-1}^*$ of $\mathbb{L}^0(\mathcal{F}_t)$, see Lemma 1.1. This implies that $\hat{G}_n - \langle \hat{\phi}_{t-1}^n, \tilde{S}_t - \tilde{S}_{t-1} \rangle$ converges in probability to $-\langle \hat{\phi}_{t-1}^*, \tilde{S}_t - \tilde{S}_{t-1} \rangle$. Since

$$\sum_{s=t}^{T-1} \langle \hat{\phi}_s^n, \tilde{S}_{s+1} - \tilde{S}_s \rangle - \hat{\rho}_n = \beta_T \hat{G}_n - \langle \hat{\phi}_{t-1}^n, \tilde{S}_t - \tilde{S}_{t-1} \rangle \quad \mathbb{P} - \text{a.s.}$$

and since $\Gamma_{t,T}$ is closed by assumption, we deduce the existence of $\check{\phi}^* \in \mathcal{A}$ and $\rho_* \in \mathbb{L}^0(\mathbb{R}_+)$ such that

$$\sum_{s=t}^{T-1} \langle \check{\phi}_s^*, \tilde{S}_{s+1} - \tilde{S}_s \rangle - \hat{\rho}_* = -\langle \hat{\phi}_{t-1}^*, \tilde{S}_t - \tilde{S}_{t-1} \rangle \quad \mathbb{P} - \text{a.s.}$$

Since (NA) holds, by Theorem 1.1, we must have $\rho_* = 0 \, \mathbb{P} - \text{a.s.}$ and therefore

$$\sum_{s=t}^{T-1} \langle \check{\phi}_s^*, \tilde{S}_{s+1} - \tilde{S}_s \rangle = -\langle \hat{\phi}_{t-1}^*, \tilde{S}_t - \tilde{S}_{t-1} \rangle \quad \mathbb{P} - \text{a.s.}$$

By Lemma 1.2, this leads to $\langle \hat{\phi}_{t-1}^*, \tilde{S}_t - \tilde{S}_{t-1} \rangle = 0 \, \mathbb{P} - \text{a.s.}$ Then, by arguing as in the proof of Proposition 1.1, we can reduce to the situation where $\mathbb{P}[A] = 0$. The fact that $\Gamma_{t-1,T}$ is closed is then a consequence of Lemma 1.1 and of the fact that $\Gamma_{t,T}$ is closed.

2. The fact that $\Gamma \subset \Theta$ is a consequence of Proposition 1.4, see Remark 1.4. It remains to prove that the converse inclusion holds. Assume to the contrary

that we can find $\hat{G} \in \Theta \setminus \Gamma$. Then, $\hat{G} \in \Theta \setminus \Gamma_{0,T}$. By (1.12), $\beta_T \hat{G} \in \mathbb{L}^1(\mathbb{Q})$ for some $\mathbb{Q} \in \mathcal{M}(\tilde{S})$. Moreover, since $\Gamma_{0,T}$ is closed, the restriction $\tilde{\Gamma}_1$ of $\{\beta_T G \; : \; G \in \Gamma_{0,T}\}$ to $\mathbb{L}^1(\mathbb{Q})$ is closed in $\mathbb{L}^1(\mathbb{Q})$. Since $\Gamma_{0,T}$ is convex, we can appeal to Theorem 1.2 to deduce the existence of $Y \in \mathbb{L}^\infty$ satisfying

$$\sup_{G \in \tilde{\Gamma}_1} \mathbb{E}^{\mathbb{Q}}[YG] < \mathbb{E}^{\mathbb{Q}}\left[Y\beta_T \hat{G}\right] .$$

Let H be the density $\mathbb{Q}$ with respect to $\mathbb{P}$, and note that

$$\sup_{G \in \tilde{\Gamma}_1} \mathbb{E}[HYG] < \mathbb{E}\left[HY\beta_T \hat{G}\right] .$$

By the same arguments as in the proof of Proposition 1.2, one easily checks that $\hat{H} := HY/\mathbb{E}[HY]$ is the density of a measure $\hat{\mathbb{Q}}$ which is absolutely continuous with respect to $\mathbb{P}$ and which turns $\tilde{S}$ into a martingale. We now observe that, for any $\varepsilon \in (0, 1)$, the density $H^\varepsilon := (\varepsilon H + (1 - \varepsilon)\hat{H})$ is associated to an equivalent measure $\mathbb{Q}^\varepsilon \in \mathcal{M}(\tilde{S})$. It then follows from the above inequality and the inclusion $\Gamma \subset \Theta$ that we can choose $\varepsilon \in (0, 1)$ in such a way that

$$\sup_{G \in \tilde{\Gamma}_1} \mathbb{E}[H^\varepsilon G] < \mathbb{E}\left[H^\varepsilon \beta_T \hat{G}\right] .$$

Since $0 \in \tilde{\Gamma}_1$, this contradicts the fact that $\hat{G} \in \Theta$. $\qquad\square$

1.3.3 Dual Formulation for the Super-Hedging Price and the Set of Viable Prices

As an immediate consequence of the preceding result, we can now provide a *dual formulation* for the super-hedging price and for the set of viable prices associated to an European option.

Theorem 1.4 (Dual formulation) *Let $G \in \mathbb{L}^0$ be such that $\beta_T G^- \in L_S$. If $\mathcal{M}(\tilde{S}) \neq \emptyset$, then*

$$p(G) = \sup_{\mathbb{Q} \in \mathcal{M}(\tilde{S})} \mathbb{E}^{\mathbb{Q}}[\beta_T G] . \tag{1.13}$$

If in addition $\beta_T |G| \in L_S$ *then:*

(i) *the collection of viable prices for* G *is the relative interior*[6] *of the set* $[-p(-G), p(G)]$.
(ii) $p(G) = -p(-G) \Leftrightarrow \exists \phi \in \mathcal{A}$ *s.t.* $V_T^{p(G),\phi} = G \, \mathbb{P} - a.s.$

Remark 1.7 When there exists $(p, \phi) \in \mathbb{R} \times \mathcal{A}$ s.t. $V_T^{p,\phi} = G \, \mathbb{P} -$ a.s., we say that the payoff G is *replicable*, or that it can be hedged perfectly.

Remark 1.8 One can easily deduce from the previous theorem and the fact that $\mathcal{M}(\tilde{S})$ is convex that a price is viable if and only if it is of the form $\mathbb{E}^{\mathbb{Q}}[\beta_T G]$ for some $\mathbb{Q} \in \mathcal{M}(\tilde{S})$. It must be the expectation under some *risk neutral measure* of the discounted payoff. In economy, fixing the value of a contingent claim by simply taking an expectation refers to an absence of risk aversion. This explains the terminology *risk neutral*: the price is defined as if the seller was risk neutral when considering suitable probability weights for the future. Obviously, this is just an interpretation. When (ii) of Theorem 1.4 holds, this price allows him to hedge his risk completely. He therefore does not take any risk.

Proof of Theorem 1.4 The first assertion is an immediate consequence of Theorem 1.3, Remark 1.5 and of Remark 1.4. The assertion (i) stems form (ii) and of (1.13). The implication $p(G) = -p(-G) \Leftarrow \exists \phi \in \mathcal{A}$ s.t. $V_T^{p(G),\phi} = G \, \mathbb{P} -$ a.s. is obvious. Let us prove that the reverse implication holds as well. Let us assume that $p := p(G) = -p(-G)$. Then, there exist ϕ_+ and ϕ_- in $\mathcal{A}$ such that $V_T^{p,\phi_+} \geq G$ and $V_T^{-p,\phi_-} \geq -G$. If one of these inequalities is strict on a non nul set A, then $V_T^{p,\phi_+} + V_T^{-p,\phi_-} \in \mathbb{L}^0(\mathbb{R}_+) \setminus \{0\}$. Since $V_T^{p,\phi_+} + V_T^{-p,\phi_-} = V_T^{0,\phi_+ + \phi_-}$, this would contradict (NA). $\qquad\square$

1.3.4 Characterisation of the Hedging Strategy of a Replicable Payoff

We conclude this section by a characterisation of the hedging strategy in the case $p(G) = -p(-G)$, i.e. when the payoff can be replicated.

Proposition 1.5 *Assume that* $\mathcal{M}(\tilde{S}) \neq \emptyset$, $\tilde{S}_t \in \mathbb{L}^2(\mathbb{P})$ *for all* $t \in \mathbb{T}$, *that* $G \in \mathbb{L}^0$ *satisfies* $|G| \in L_S$ *and that* $p(G) = -p(-G)$. *Let* $\mathbb{Q}$ *be an element of* $\mathcal{M}(\tilde{S})$ *whose density is bounded. Then,* $V_T^{p(G),\hat{\phi}} = G \, \mathbb{P} - a.s.$ *for all* $\hat{\phi}$ *satisfying the following backward induction*

$$\tilde{V}_t := \mathbb{E}^{\mathbb{Q}}\left[\tilde{V}_{t+1} \mid \mathcal{F}_t\right], \quad \hat{\phi}_t \in \mathcal{S}_t(\tilde{V}, \mathbb{Q}) \quad t \leq T - 1$$

---

[6]I.e. $] - p(-G), p(G)[$ if $-p(-G) \neq p(G)$, $\{p(G)\}$ otherwise.

with $\tilde{V}_T = \beta_T G$ and $\mathcal{S}_t(\tilde{V}, \mathbb{Q})$ *defined as the set of* $\mathcal{F}_t$-*measurable random variables such that*

$$\langle \hat{\phi}_t, \mathbb{E}^{\mathbb{Q}}\left[(\tilde{S}_{t+1} - \tilde{S}_t)(\tilde{S}^i_{t+1} - \tilde{S}^i_t) \mid \mathcal{F}_t\right]\rangle = \mathbb{E}^{\mathbb{Q}}\left[\tilde{V}_{t+1}(\tilde{S}^i_{t+1} - \tilde{S}^i_t) \mid \mathcal{F}_t\right]$$

for $i = 1, \ldots, d$.

Proof By Theorem 1.4, we can find $\hat{\phi} \in \mathcal{A}$ such that $V^{p(G),\hat{\phi}}_T = G$. Set $\tilde{V} = \beta V^{p(G),\hat{\phi}}$. Then, $\tilde{V}_T = \beta_T G$ and Remark 1.4 implies that the induction on $\tilde{V}$ is indeed satisfied. Moreover, since $\tilde{S}$ is a $\mathbb{Q}$-martingale,

$$\mathbb{E}^{\mathbb{Q}}\left[\tilde{V}_{t+1}(\tilde{S}^i_{t+1} - \tilde{S}^i_t) \mid \mathcal{F}_t\right]$$

$$= \mathbb{E}^{\mathbb{Q}}\left[\tilde{V}_t(\tilde{S}^i_{t+1} - \tilde{S}^i_t) + \langle \hat{\phi}_t, \tilde{S}_{t+1} - \tilde{S}_t\rangle(\tilde{S}^i_{t+1} - \tilde{S}^i_t) \mid \mathcal{F}_t\right]$$

$$= \langle \hat{\phi}_t, \mathbb{E}^{\mathbb{Q}}\left[(\tilde{S}_{t+1} - \tilde{S}_t)(\tilde{S}^i_{t+1} - \tilde{S}^i_t) \mid \mathcal{F}_t\right]\rangle$$

for all $i = 1, \ldots, d$. If $\bar{\phi}$ is another element of $\mathcal{A}$ such that $\bar{\phi}_t \in \mathcal{S}_t(\tilde{V}, \mathbb{Q})$ for all $t \leq T - 1$, we also have, for $t \leq T - 1$,

$$\mathbb{E}^{\mathbb{Q}}\left[\left|\langle \hat{\phi}_t - \bar{\phi}_t, \tilde{S}_{t+1} - \tilde{S}_t\rangle\right|^2 \mid \mathcal{F}_t\right]$$

$$= (\hat{\phi}_t - \bar{\phi}_t)' \mathbb{E}^{\mathbb{Q}}\left[(\tilde{S}_{t+1} - \tilde{S}_t)(\tilde{S}_{t+1} - \tilde{S}_t)' \mid \mathcal{F}_t\right](\hat{\phi}_t - \bar{\phi}_t)$$

$$= 0$$

since

$$\hat{\phi}'_t \mathbb{E}^{\mathbb{Q}}\left[(\tilde{S}_{t+1} - \tilde{S}_t)(\tilde{S}_{t+1} - \tilde{S}_t)' \mid \mathcal{F}_t\right]$$

$$= \bar{\phi}'_t \mathbb{E}^{\mathbb{Q}}\left[(\tilde{S}_{t+1} - \tilde{S}_t)(\tilde{S}_{t+1} - \tilde{S}_t)' \mid \mathcal{F}_t\right]$$

$$= \mathbb{E}^{\mathbb{Q}}\left[\tilde{V}_{t+1}(\tilde{S}_{t+1} - \tilde{S}_t)' \mid \mathcal{F}_t\right].$$

This implies that $\langle \hat{\phi}_t - \bar{\phi}_t, \tilde{S}_{t+1} - \tilde{S}_t\rangle = 0$ $\mathbb{P}$–a.s. We deduce that $V^{p(G),\hat{\phi}} = V^{p(G),\bar{\phi}}$ $\mathbb{P}$– a.s. $\square$

1.4   Complete Markets

Intuitively, a market is said to be complete when the number of tradable assets is enough to hedge perfectly any source of risk. In this case, there is only one possible viable price and the corresponding hedging strategy is clearly identified. This is a desirable property in practice.

1.4.1 Definition and Characterisation

Definition 1.4 (Complete/incomplete markets) The market is said to be a *complete market* if, for all $G \in \mathbb{L}^0$ such that $|G| \in L_S$, there exists $(p, \phi) \in \mathbb{R} \times \mathcal{A}$ satisfying $V_T^{p,\phi} = G \ \mathbb{P}$–a.s. If this is not the case, we say that we have an *incomplete market*.

Theorem 1.4 provides a precise description of complete markets in terms of the cardinality of $\mathcal{M}(\tilde{S})$.

Theorem 1.5 (*Second fundamental theorem of asset pricing*) *Assume that* $\mathcal{M}(\tilde{S}) \neq \emptyset$. *Then, the market is complete if and only if* $\mathcal{M}(\tilde{S})$ *is a singleton.*

Proof If $\mathcal{M}(\tilde{S})$ is reduced to a singleton, the fact that the market is complete follows from (1.13) and (ii) of Theorem 1.4. Let us now assume that the market is complete. By (1.13) and (ii) of Theorem 1.4, we have

$$\sup_{\mathbb{Q} \in \mathcal{M}(\tilde{S})} \mathbb{E}^{\mathbb{Q}}[\beta_T G] = \inf_{\mathbb{Q} \in \mathcal{M}(\tilde{S})} \mathbb{E}^{\mathbb{Q}}[\beta_T G] \quad \forall G \in \mathbb{L}^0 \text{ s.t. } |G| \in L_S. \tag{1.14}$$

Let us assume now that $\mathcal{M}(\tilde{S})$ contains two different elements with densities H^1 and H^2. Then, $G := \mathbf{1}_{\{H^1 > H^2\}}$ is non identically null and

$$\mathbb{E}\left[H^1 \beta_T G\right] > \mathbb{E}\left[H^2 \beta_T G\right],$$

which contradicts (1.14). $\square$

Remark 1.9 By Theorem 1.4 and the above, the unique viable price in a complete market is given by $\mathbb{E}^{\mathbb{Q}}[\beta_T G]$ in which $\mathbb{Q}$ is the unique equivalent measure that turns $\tilde{S}$ into a martingale. The hedging strategy can be computed by using the algorithm of Proposition 1.5.

We conclude this section by showing that, if the market is complete, then, for all $t < T$, the conditional law of S_{t+1} given $\mathcal{F}_t$ is concentrated on at most $d + 1$ points (they are random but $\mathcal{F}_t$-measurable). In particular, if there is only one risky asset, then it can take only two different values a $t + 1$, given its value at t. This is rather limiting, continuous time models will prove to be much more rich from this perspective.

Proposition 1.6 *Assume that* $\mathcal{M}(\tilde{S}) = \{\mathbb{Q}\}$, *then, for all* $t < T$, *there exist* $d + 1$ *random vectors* $\xi_1, \ldots, \xi_{d+1} \in \mathbb{L}^0(\mathbb{R}^d; \mathcal{F}_t)$ *such that*

$$\sum_{j=1}^{d+1} \mathbb{P}\left[\tilde{S}_{t+1} - \tilde{S}_t = \xi_j \mid \mathcal{F}_t\right] = 1, \ \mathbb{P}-a.s.$$

Proof Assume that our claim is not true for some $t < T$. Then, we can find a set $A \in \mathcal{F}_t$ such that $\mathbb{Q}[A] > 0$ and $d + 2$ random variables $\zeta_1, \ldots, \zeta_{d+2} \in \mathbb{L}^0(\mathbb{R}^d; \mathcal{F}_t)$, and $\eta \in \mathbb{L}^0((0,\infty); \mathcal{F}_t)$ satisfying

$$\mathbb{P}\left[B_j \mid \mathcal{F}_t\right] > 0 \quad \forall\, j \le d+2 \text{ on } A,$$

where $B_j := \{\|(\tilde{S}_{t+1} - \tilde{S}_t) - \zeta_j\| \le \eta\} \cap A$, and the B_j are disjoint. Then, the random vectors $(\mathbb{Q}[B_j \mid \mathcal{F}_t], \mathbb{E}^{\mathbb{Q}}\left[(\tilde{S}_{t+1} - \tilde{S}_t)' \mathbf{1}_{B_j} \mid \mathcal{F}_t\right]), j = 1, \ldots, d+2$, cannot be linearly independent (because $d + 1 < d + 2$), and there exist $d + 2$ random variables $a_1, \ldots, a_{d+2}$ of $\mathbb{L}^0(\mathbb{R}; \mathcal{F}_t)$, not identically equal to 0 on A, such that

$$\sum_{j=1}^{d+2} a_j \left(\mathbb{Q}[B_j \mid \mathcal{F}_t], \ \mathbb{E}^{\mathbb{Q}}\left[(\tilde{S}_{t+1} - \tilde{S}_t)' \mathbf{1}_{B_j} \mid \mathcal{F}_t\right]\right) = 0 \quad \mathbb{Q}\text{-a.s.}$$

Without loss of generality we can assume that each a_j is bounded by $1/(2d + 4)$ $\mathbb{Q}$-a.s. Let us then define

$$\bar{H} := 1 + \sum_{j=1}^{d+2} a_j \mathbf{1}_{B_j},$$

so that $\mathbb{E}^{\mathbb{Q}}\left[\bar{H} \mid \mathcal{F}_t\right] = 1, \bar{H} > 0$ $\mathbb{Q}$-a.s. and, since $\tilde{S}$ is a $\mathbb{Q}$-martingale,

$$\mathbb{E}^{\mathbb{Q}}\left[\bar{H}(\tilde{S}_{t+1} - \tilde{S}_t) \mid \mathcal{F}_t\right] = 0.$$

This shows that the measure $\bar{\mathbb{Q}}$ of density $\bar{H}$ with respect to $\mathbb{Q}$ is another element of $\mathcal{M}(\tilde{S})$, which is a contradiction. $\qquad\qquad\square$

1.4.2 Martingale Representation Theorem and European Option Hedging

As a corollary of Theorem 1.5, we now obtain the following *martingales representation theorem*.

Corollary 1.1 (Martingales representation theorem) *Assume that $\mathcal{M}(\tilde{S}) = \{\mathbb{Q}\}$. Then, any $\mathbb{Q}$-martingale M can be written in the form*

$$M_t = M_0 + \sum_{s=0}^{t-1} \langle \phi_s, \tilde{S}_{s+1} - \tilde{S}_s \rangle,$$

for some $\phi \in \mathcal{A}$.

Proof Let us first assume that $|M_T| \leq n$, $n \geq 1$, so that $M_T \in L_S$. By Theorem 1.5, we can find $\phi \in \mathcal{A}$ and $x \in \mathbb{R}$ such that

$$M_T = x + \sum_{s=0}^{T-1} \langle \phi_s, \tilde{S}_{s+1} - \tilde{S}_s \rangle .$$

Next assume that

$$M_{t+1} = x + \sum_{s=0}^{t} \langle \phi_s, \tilde{S}_{s+1} - \tilde{S}_s \rangle$$

for some $t < T$. Then, by taking the $\mathcal{F}_t$-conditional expectation under $\mathbb{Q}$, we obtain

$$M_t = x + \sum_{s=0}^{t-1} \langle \phi_s, \tilde{S}_{s+1} - \tilde{S}_s \rangle + \langle \phi_t, \mathbb{E}^{\mathbb{Q}} \left[\tilde{S}_{t+1} - \tilde{S}_t \mid \mathcal{F}_t \right] \rangle$$

$$= x + \sum_{s=0}^{t-1} \langle \phi_s, \tilde{S}_{s+1} - \tilde{S}_s \rangle .$$

A simple induction proves that the representation holds at any time. In particular, $M_0 = x$.

Let us now consider the general case. Given M, we set $M_t^n := \mathbb{E}^{\mathbb{Q}}[M_T \vee (-n) \wedge n | \mathcal{F}_t]$, $n \geq 1$. By the above, we can find $\phi^n \in \mathcal{A}$ such that $\tilde{V}_T^{M_0^n, \phi^n} = M_T^n$. Since $M_T \in \mathbb{L}^1(\mathbb{Q})$, the dominated convergence theorem implies that $M_0^n \to M_0$ as $n \to \infty$. In particular, $M_T^n - M_0^n \in \beta_T \Gamma$. Since the latter is closed in probability (Theorem 1.3) and since $M_T^n - M_0^n \to M_T - M_0$ as $n \to \infty$, we deduce that $M_T - M_0 \in \beta_T \Gamma$. Similarly $-(M_T^n - M_0^n) \in \beta_T \Gamma$, so that $-(M_T - M_0) \in \beta_T \Gamma$. To conclude, it suffices to repeat the arguments used in the proof of Theorem 1.4.   □

Corollary 1.2 (Representation theorem) *Assume that $\{\mathbb{Q}\} = \mathcal{M}(\tilde{S})$. Then, any $G \in \mathbb{L}^1(\mathbb{Q})$ admits a representation of the form*

$$G = \mathbb{E}^{\mathbb{Q}} [G] + \sum_{s=0}^{T-1} \langle \phi_s, \tilde{S}_{s+1} - \tilde{S}_s \rangle ,$$

for some $\phi \in \mathcal{A}$.

Proof It suffices to apply Corollary 1.1 to the martingale $(\mathbb{E}^{\mathbb{Q}} [G \mid \mathcal{F}_t])_{t \in \mathbb{T}}$.   □

1.5 American Options

An *American option* is a financial derivative that gives the right to its buyer to receive an amount G_t of money at a time t that she can choose, the *exercise date*, before a time T fixed in advance, the *maturity* of the option.

From the mathematical point of view, the option is described by an adapted process $G = (G_t)_{t\in\mathbb{T}}$. In this section, we shall fix it and assume that it is bounded:

$$\|G_t\|_{\mathbb{L}^\infty} < \infty \ \forall \ t \in \mathbb{T}. \tag{1.15}$$

Since the American option can be exercised at any time before its maturity, the price at which it should be sold in order to offset all risk for the buyer, its *super-hedging price*, is defined as

$$p^{US}(G) := \inf\{p \in \mathbb{R} \ : \ \exists \ \phi \in \mathcal{A} \text{ s.t. } V_t^{p,\phi} - G_t \in \mathbb{L}^0(\mathbb{R}_+) \ \forall \ t \in \mathbb{T}\} \ . \tag{1.16}$$

In this section, we shall characterise this super-hedging price under the no-arbitrage condition, i.e. $\mathcal{M}(\tilde{S}) \neq \emptyset$. For this, we shall use the notion of *Snell envelope*.

1.5.1 Supermartingale and Snell Envelope

Let us first recall what a supermartingale is.

Definition 1.5 (Supermartingale) An $\mathbb{F}$-adapted process ξ is a $\mathbb{Q}$-*supermartingale* if $\xi_{t+1}^- \in \mathbb{L}^1(\mathbb{Q})$ and $\mathbb{E}^\mathbb{Q}[\xi_{t+1} \mid \mathcal{F}_t] \leq \xi_t$ $\mathbb{Q}$-a.s. for all $t < T$.

More generally, it is possible to define the notion of $\mathcal{M}(\tilde{S})$-supermartingale (resp. martingale) ξ by imposing that ξ is a $\mathbb{Q}$-supermartingale (resp. martingale) for all $\mathbb{Q} \in \mathcal{M}(\tilde{S})$.

Remark 1.10 If ξ is a $\mathbb{Q}$-martingale, then a simple induction implies that

$$\mathbb{E}^\mathbb{Q}[\xi_t \mid \mathcal{F}_s] \leq \xi_s \ \mathbb{Q}\text{-a.s. for all } s \leq t \leq T \ .$$

We will first show that a bounded $\mathcal{M}(\tilde{S})$-supermartingale is always dominated by a bounded $\mathcal{M}(\tilde{S})$-martingale.

Theorem 1.6 *Let Y be a bounded $\mathcal{M}(\tilde{S})$-supermartingale, then there exists a $\mathcal{M}(\tilde{S})$-martingale M such that $M_0 = Y_0$ and $M \geq Y$ on $[0, T]$. Moreover, there exists $\phi \in \mathcal{A}$ such that $M = \tilde{V}^{Y_0,\phi}$.*

Proof Let $t < T$ be fixed. The $\mathcal{M}(\tilde{S})$-supermartingale property implies that

$$\mathbb{E}^\mathbb{Q}[Y_{t+1} \mid \mathcal{F}_t] \leq Y_t \ , \ \mathbb{P} - \text{a.s.} \ \forall \ \mathbb{Q} \in \mathcal{M}(\tilde{S}).$$

By Remark 1.6, we can then find $\phi^t \in \mathcal{A}$ such that $Y_t + \tilde{V}_T^{0,\phi^t} - \tilde{V}_t^{0,\phi^t} \geq Y_{t+1}$. In view of Remark 1.4, $\tilde{V}_{\cdot \vee t}^{0,\phi^t} - \tilde{V}_t^{0,\phi^t}$ is a $\mathbb{Q}$-supermatingale for all $\mathbb{Q} \in \mathcal{M}(\tilde{S})$. In particular, $Y_t + \langle \phi_t^t, \tilde{S}_{t+1} - \tilde{S}_t \rangle \geq Y_{t+1}$ by taking the conditional expectation under any $\mathbb{Q} \in \mathcal{M}(\tilde{S})$. Let us then consider ϕ defined by $\phi_t := \phi_t^t$. From the above, $M := \tilde{V}^{Y_0,\phi} \geq Y$. $\qquad\qquad\qquad\qquad\qquad\qquad\qquad\qquad\qquad\qquad\qquad\qquad\qquad\qquad\qquad\square$

Remark 1.11 (Doob-Meyer decomposition) When the family $\mathcal{M}(\tilde{S})$ is reduced to a singleton, one can obtain a more precise decomposition, called *Doob-Meyer* decomposition: if Y is a $\mathbb{Q}$-supermartingale, then there exists a $\mathbb{Q}$-martingale $M^{\mathbb{Q}}$ and a predictable non-decreasing process $A^{\mathbb{Q}}$, taking the value 0 at 0, such that $Y = M^{\mathbb{Q}} - A^{\mathbb{Q}}$ $\mathbb{Q}$-a.s. This decomposition is unique.
To see this, it suffices to define : $(M_0^{\mathbb{Q}}, A_0^{\mathbb{Q}}) = (Y_0, 0)$ and then

$$M_{t+1}^{\mathbb{Q}} = M_t^{\mathbb{Q}} + Y_{t+1} - \mathbb{E}^{\mathbb{Q}}[Y_{t+1} \mid \mathcal{F}_t],$$

$$A_{t+1}^{\mathbb{Q}} = A_t^{\mathbb{Q}} + Y_t - \mathbb{E}^{\mathbb{Q}}[Y_{t+1} \mid \mathcal{F}_t]$$

by forward induction.

In the following, we show that the super-hedging price of an American option G is given by the value at time 0 of the $\mathcal{M}(\tilde{S})$-Snell envelope of the discounted payoff process

$$\tilde{G} := (\beta_t G_t)_{t \in \mathbb{T}}.$$

Definition 1.6 (Snell envelope) The $\mathcal{M}(\tilde{S})$-*Snell envelope* of $\tilde{G}$ is the smallest $\mathcal{M}(\tilde{S})$-supermartingale Y such that $Y \geq \tilde{G}$ $\mathbb{P}$-a.s.

In the above definition, the assertion *smallest* means that any other $\mathcal{M}(\tilde{S})$-supermartingale Y' satisfying $Y' \geq \tilde{G}$ $\mathbb{P}$-a.s. is such that $Y' \geq Y$ $\mathbb{P}$-a.s. The so-called *dynamic programming* algorithm described in Proposition 1.7 below provides an explicit characterisation of the $\mathcal{M}(\tilde{S})$-Snell envelope, and thereby shows that it exists. It makes use of the notion of essential-supremum, which can be understood as the counter-part of the supremum of a family of real numbers but for a family of random variables.

Definition 1.7 (Essential supremum) Let $\mathcal{J}$ be a family of real valued random variables. The *essential supremum* of $\mathcal{J}$, which we write $\text{esssup}\,\mathcal{J}$, is the smallest random variable ξ satisfying $\xi \geq J$ $\mathbb{P}$ – a.s. for all $J \in \mathcal{J}$.

Remark 1.12 One can obviously define the supremum of a family of random variables ω by ω by considering it as a family of real numbers for ω fixed. However, this supremum is in general not measurable, i.e. the result is not a random variable. The notion of essential supremum takes care of this issue. We refer to [47] for more details.

Proposition 1.7 *The $\mathcal{M}(\tilde{S})$-Snell envelope of $\tilde{G}$ is the unique solution of the following backward dynamic programming algorithm:*

$$Y_t = \max\left\{\tilde{G}_t \,, \underset{\mathbb{Q}\in\mathcal{M}(\tilde{S})}{\text{ess sup}}\; \mathbb{E}^{\mathbb{Q}}\left[Y_{t+1}\mid\mathcal{F}_t\right]\right\} \quad t < T\,, \tag{1.17}$$

with the terminal condition

$$Y_T = \tilde{G}_T\,. \tag{1.18}$$

Proof In view of (1.17), we have $Y_t \geq \mathbb{E}^{\mathbb{Q}}[Y_{t+1}\mid\mathcal{F}_t]$ for all $\mathbb{Q}\in\mathcal{M}(\tilde{S})$ and $t < T$. Moreover, a simple induction using the fact that $\tilde{G}$ is bounded shows that Y is bounded as well. Hence, Y is a $\mathbb{Q}$-supermartingale for all $\mathbb{Q}\in\mathcal{M}(\tilde{S})$. Moreover, (1.17) and (1.18) implies that $Y \geq \tilde{G}$ $\mathbb{Q}$-a.s. By definition of the Snell envelope $\bar{Y}$ of $\tilde{G}$, we thus have $Y \geq \bar{Y}$ $\mathbb{P}$-a.s. We now prove the reverse inequality. First, $Y_T = \tilde{G}_T \leq \bar{Y}_T$ $\mathbb{P}$-a.s. Let us now assume that $Y_{t+1} \leq \bar{Y}_{t+1}$ $\mathbb{P}$-a.s. for some $t \leq T - 1$. By (1.17),

$$\begin{aligned}
Y_t &= \max\left\{\tilde{G}_t \,, \underset{\mathbb{Q}\in\mathcal{M}(\tilde{S})}{\text{ess sup}}\; \mathbb{E}^{\mathbb{Q}}\left[Y_{t+1}\mid\mathcal{F}_t\right]\right\} \\
&\leq \max\left\{\bar{Y}_t \,, \underset{\mathbb{Q}\in\mathcal{M}(\tilde{S})}{\text{ess sup}}\; \mathbb{E}^{\mathbb{Q}}\left[\bar{Y}_{t+1}\mid\mathcal{F}_t\right]\right\} \\
&\leq \bar{Y}_t
\end{aligned}$$

since $\bar{Y}$ is a $\mathcal{M}(\tilde{S})$-supermartingale. An induction concludes the proof. $\qquad\square$

1.5.2   Super-Replication Price and $\mathcal{M}(\tilde{S})$-Snell Envelope

We establish a first characterisation of the super-replication price in terms of the $\mathcal{M}(\tilde{S})$-Snell envelope. We will use it later to provide a dual formulation in terms of expectations under the risk neutral measures of the discounted payoff stopped in an optimal way.

Theorem 1.7 $p^{US}(G) = Y_0$.

Proof

1. By definition of Y, we have $Y \geq \tilde{G}$ $\mathbb{P}$ – a.s. Moreover, Theorem 1.6 implies that there exists $\phi \in \mathcal{A}$ such that $\tilde{V}^{Y_0,\phi} \geq Y$. Hence, $Y_0 \geq p^{US}(G)$.
2. Let us now fix $\varepsilon > 0$ so that we can find $\phi \in \mathcal{A}$ for which $\tilde{V}^{p_\varepsilon,\phi} \geq \tilde{G}$ $\mathbb{P}$ – a.s., where $p_\varepsilon := p^{US}(G) + \varepsilon$. In view of Remark 1.4, $\tilde{V}^{p_\varepsilon,\phi}$ is a $\mathcal{M}(\tilde{S})$-

supermartingale. Since Y is the $\mathcal{M}(\tilde{S})$-Snell envelope, it follows that $\tilde{V}^{p_\varepsilon, \phi} \geq Y$. In particular, $p^{US}(G) + \varepsilon = p_\varepsilon \geq Y_0$, which leads to $p^{US}(G) \geq Y_0$ since $\varepsilon > 0$ is arbitrary.

$\square$

1.5.3 Super-Replication Price and Optimal Stopping

We now denote by $\mathcal{T}_t$ the collection of $(\mathbb{F}, \mathbb{T})$-stopping times with values in $[t, T]$, $t \in \mathbb{T}$. We recall that a $(\mathbb{F}, \mathbb{T})$-*stopping time* is a random variable τ with values in $\mathbb{T}$ such that $\{\tau = t\} \in \mathcal{F}_t$ for all $t \in \mathbb{T}$.

Remark 1.13 If $\tau \in \mathcal{T}_0$ then $\{\tau \leq t\} \in \mathcal{F}_t$ for all $t \in \mathbb{T}$.

Remark 1.14 Let ξ be a $\mathbb{Q}$-martingale and fix $\tau \in \mathcal{T}_0$. Then, the *stopped process* $\xi^\tau := (\xi_{t \wedge \tau})_{t \in \mathbb{T}}$ is a $\mathbb{Q}$-martingale. Indeed,

$$\xi^\tau_t = \xi_t \mathbf{1}_{\tau \geq t} + \sum_{s < t} \xi_s \mathbf{1}_{\tau = s} \quad t \in \mathbb{T} \,,$$

which shows that ξ^τ is integrable. Moreover, by Remark 1.13 combined with the martingale property of ξ, this also implies that

$$\mathbb{E}^{\mathbb{Q}} \left[\xi^\tau_{t+1} \mid \mathcal{F}_t \right] = \mathbb{E}^{\mathbb{Q}} \left[\xi_{t+1} \mathbf{1}_{\tau > t} + \sum_{s \leq t} \xi_s \mathbf{1}_{\tau = s} \mid \mathcal{F}_t \right]$$

$$= \mathbb{E}^{\mathbb{Q}} \left[\xi_{t+1} \mid \mathcal{F}_t \right] \mathbf{1}_{\tau > t} + \sum_{s \leq t} \xi_s \mathbf{1}_{\tau = s}$$

$$= \xi_t \mathbf{1}_{\tau > t} + \sum_{s \leq t} \xi_s \mathbf{1}_{\tau = s} = \xi^\tau_t \quad \forall \, t < T \,.$$

In particular, $\xi_0 = \xi^\tau_0 = \mathbb{E}^{\mathbb{Q}} [\xi_{t \wedge \tau}] = \mathbb{E}^{\mathbb{Q}} [\xi_{T \wedge \tau}] = \mathbb{E}^{\mathbb{Q}} [\xi_\tau]$ for all $t \in \mathbb{T}$.

Similarly, if ξ is a $\mathbb{Q}$-supermartingale, then $\xi^\tau := (\xi_{t \wedge \tau})_{t \in \mathbb{T}}$ is a $\mathbb{Q}$-supermartingale.

We now provide the characterisation of Y as the value of an *optimal stopping* problem. This is the *dual formulation* for American options.

Theorem 1.8 (Dual formulation) *Define*

$$\hat{\tau} := \inf\{t \in \mathbb{T} : Y_t = \tilde{G}_t\} \,. \tag{1.19}$$

Then,

$$p^{US}(G) = Y_0 = \sup_{\tau \in \mathcal{T}_0} \sup_{\mathbb{Q} \in \mathcal{M}(\tilde{S})} \mathbb{E}^{\mathbb{Q}} \left[\tilde{G}_\tau \right] = \sup_{\mathbb{Q} \in \mathcal{M}(\tilde{S})} \mathbb{E}^{\mathbb{Q}} \left[\tilde{G}_{\hat{\tau}} \right] .$$

Moreover, there exists $\phi \in \mathcal{A}$ such that $\tilde{V}^{p^{US}(G), \phi} \geq Y \geq \tilde{G}$.

The identity $p^{US}(G) = Y_0$ has already been mentioned in Theorem 1.7. The last assertion is a consequence of Theorem 1.6 and of the inequality $Y \geq \tilde{G}$, that holds by construction.

Let us split the rest of the proof in several steps.

Proposition 1.8 *The following holds:*

$$Y_0 \geq \sup_{\tau \in \mathcal{T}_0} \sup_{\mathbb{Q} \in \mathcal{M}(\tilde{S})} \mathbb{E}^{\mathbb{Q}} \left[\tilde{G}_\tau \right] \geq \sup_{\mathbb{Q} \in \mathcal{M}(\tilde{S})} \mathbb{E}^{\mathbb{Q}} \left[\tilde{G}_{\hat{\tau}} \right] .$$

Proof Since Y is a $\mathcal{M}(\tilde{S})$-supermartingale above $\tilde{G}$, Remark 1.14 leads to

$$Y_0 \geq \sup_{\tau \in \mathcal{T}_0} \sup_{\mathbb{Q} \in \mathcal{M}(\tilde{S})} \mathbb{E}^{\mathbb{Q}} [Y_\tau] \geq \sup_{\tau \in \mathcal{T}_0} \sup_{\mathbb{Q} \in \mathcal{M}(\tilde{S})} \mathbb{E}^{\mathbb{Q}} \left[\tilde{G}_\tau \right] .$$

Since $\hat{\tau} \in \mathcal{T}_0$, our claim is proved. □

It remains to prove the opposite inequalities. For this, we shall use the following important technical result, see [47].

Definition 1.8 (Directed family) Let $\mathcal{J}$ be a family of real valued random variables. We say that $\mathcal{J}$ is *directed upward* if, for any $J_1, J_2 \in \mathcal{J}$, we can find $J_3 \in \mathcal{J}$ such that $J_3 \geq \max\{J_1, J_2\}$ $\mathbb{P}$ − a.s.

Proposition 1.9 *Let $\mathcal{J}$ be a family of real valued random variables that is directed upward. Then, there exists a countable subfamily $(J_n)_{n \geq 1} \subset \mathcal{J}$ satisfying $J_n \uparrow$ esssup $\mathcal{J}$ $\mathbb{P}$ − a.s. as $n \to \infty$.*

Proposition 1.10 *The following holds:*

$$Y_0 \leq \sup_{\mathbb{Q} \in \mathcal{M}(\tilde{S})} \mathbb{E}^{\mathbb{Q}} \left[\tilde{G}_{\hat{\tau}} \right] .$$

Proof

1. Fix $t < T$ and $\mathbb{Q} \in \mathcal{M}(\tilde{S})$. Then,

$$Y_{t+1}^{\hat{\tau}} - Y_t^{\hat{\tau}} = (Y_{t+1} - Y_t) \mathbf{1}_{\hat{\tau} \geq t+1}$$

and

$$Y_t = \text{ess} \sup_{\mathbb{Q} \in \mathcal{M}(\tilde{S})} \mathbb{E}^{\mathbb{Q}} [Y_{t+1} \mid \mathcal{F}_t] \quad \text{on } \{\hat{\tau} \geq t+1\} \in \mathcal{F}_t$$

by definition of Y and $\hat{\tau}$. Combined with Remark 1.13, this leads to

$$Y_t^{\hat{\tau}} = \text{ess} \sup_{\mathbb{Q} \in \mathcal{M}(\tilde{S})} \mathbb{E}^{\mathbb{Q}} \left[Y_{t+1}^{\hat{\tau}} \mid \mathcal{F}_t \right]$$

for all $t < T$. To $\mathbb{Q} \in \mathcal{M}(\tilde{S})$, we can associate its density process $H^{\mathbb{Q}}$ with respect to $\mathbb{P}$ and define $H_t^{\mathbb{Q}} := \mathbb{E}\left[H^{\mathbb{Q}} \mid \mathcal{F}_t\right]$. Then,

$$\mathbb{E}^{\mathbb{Q}} \left[Y_{t+1}^{\hat{\tau}} \mid \mathcal{F}_t \right] = \mathbb{E} \left[H_{t+1}^{\mathbb{Q}} Y_{t+1}^{\hat{\tau}} \mid \mathcal{F}_t \right] / H_t^{\mathbb{Q}}.$$

Let us set $\mathcal{H}_t := \{H_{t+1}^{\mathbb{Q}}/H_t^{\mathbb{Q}}, \mathbb{Q} \in \mathcal{M}(\tilde{S})\}$, so that the previous equation reads

$$Y_t^{\hat{\tau}} = \text{ess} \sup_{H^t \in \mathcal{H}_t} \mathbb{E} \left[H^t Y_{t+1}^{\hat{\tau}} \mid \mathcal{F}_t \right].$$

2. Let $\mathcal{J}^t = \{J^{H^t}, H^t \in \mathcal{H}_t\}$ with $J^{H^t} := \mathbb{E}\left[H^t Y_{t+1}^{\hat{\tau}} \mid \mathcal{F}_t\right]$. This family is directed upward. Indeed, if $J_1, J_2 \in \mathcal{J}^t$ then there exist $H^{t,1}, H^{t,2} \in \mathcal{H}_t$ such that $J_i = \mathbb{E}\left[H^{t,i} Y_{t+1}^{\hat{\tau}} \mid \mathcal{F}_t\right]$, $i = 1, 2$. Set $H^{t,3} := H^{t,1} \mathbf{1}_{J^1 \geq J^2} + H^{t,2} \mathbf{1}_{J^1 < J^2}$. Since $\{J^1 \geq J^2\} \in \mathcal{F}_t$, we have $J_3 := J^{H^{t,3}} = \max\{J_1, J_2\}$. Let us now consider the measures $\mathbb{Q}^1, \mathbb{Q}^2 \in \mathcal{M}(\tilde{S})$ associated to the densities $H^{t,1}$ and $H^{t,2}$. One easily checks that $H^{t,3}$ is the density of the measure $\mathbb{Q}^3$ whose density with respect to $\mathbb{P}$ is

$$H^3 := H_t^{\mathbb{Q}^1} \left(\mathbf{1}_{J^1 \geq J^2} H_T^{\mathbb{Q}^1}/H_t^{\mathbb{Q}^1} + \mathbf{1}_{J^1 < J^2} H_T^{\mathbb{Q}^2}/H_t^{\mathbb{Q}^2} \right),$$

and that $\mathbb{Q}^3$ is an element of $\mathcal{M}(\tilde{S})$.

3. Given $t < T$, we now appeal to steps 1. and 2. and to Proposition 1.9 to find a countable sequence $(H^{t,n_t})_{n_t \geq 1} \subset \mathcal{H}_t$ such that

$$Y_t^{\hat{\tau}} = \lim_{n_t \to \infty} \uparrow \mathbb{E}\left[H^{t,n_t} Y_{t+1}^{\hat{\tau}} \mid \mathcal{F}_t\right].$$

We can then observe that, for all $H^{t-1} \in \mathcal{H}_{t-1}$, we have

$$\mathbb{E}\left[H^{t-1} Y_t^{\hat{\tau}} \mid \mathcal{F}_{t-1}\right] = \mathbb{E}\left[H^{t-1} \lim_{n_t \to \infty} \uparrow \mathbb{E}\left[H^{t,n_t} Y_{t+1}^{\hat{\tau}} \mid \mathcal{F}_t\right] \mid \mathcal{F}_{t-1}\right]$$

$$= \lim_{n_t \to \infty} \uparrow \mathbb{E}\left[H^{t-1} \mathbb{E}\left[H^{t,n_t} Y_{t+1}^{\hat{\tau}} \mid \mathcal{F}_t\right] \mid \mathcal{F}_{t-1}\right]$$

$$= \lim_{n_t \to \infty} \uparrow \mathbb{E}\left[H^{t-1} H^{t,n_t} Y_{t+1}^{\hat{\tau}} \mid \mathcal{F}_{t-1} \right]$$

by monotone convergence. Iterating this argument leads to

$$Y_0^{\hat{\tau}} = \lim_{n_0,\dots,n_{T-1} \to \infty} \uparrow \mathbb{E}\left[\prod_{i=0}^{T-1} H^{i,n_i} Y_T^{\hat{\tau}} \right].$$

Then, it suffices to check that the density $\prod_{i=0}^{T-1} H^{i,n_i}$ actually defines an element of $\mathcal{M}(\tilde{S})$ to conclude that

$$Y_0 = Y_0^{\hat{\tau}} \leq \sup_{\mathbb{Q} \in \mathcal{M}(\tilde{S})} \mathbb{E}^{\mathbb{Q}}\left[Y_T^{\hat{\tau}} \right] = \sup_{\mathbb{Q} \in \mathcal{M}(\tilde{S})} \mathbb{E}^{\mathbb{Q}}\left[\tilde{G}_{\hat{\tau}} \right],$$

since $Y_T^{\hat{\tau}} = Y_{\hat{\tau}} = \tilde{G}_{\hat{\tau}}$. □

1.5.4 Viable Prices

In this section, we characterise the collection of viable prices for an American option.

Definition 1.9 (Viable price) The price p is a *viable price* for the American option of payoff process G if

NA1. $\nexists \phi \in \mathcal{A}$ s.t. $V_\tau^{p,\phi} - G_\tau \in \mathbb{L}^0(\mathbb{R}_+) \setminus \{0\}$ for all $\tau \in \mathcal{T}_0$;
NA2. $\nexists (\phi, \tau) \in \mathcal{A} \times \mathcal{T}_0$ s.t. $V_\tau^{-p,\phi} + G_\tau \in \mathbb{L}^0(\mathbb{R}_+) \setminus \{0\}$.

The first assertion implies that this price does not create any arbitrage for the seller, the second that it is neither the case for the buyer. Given the absence of symmetry between the seller and the buyer, one needs to study the existence of arbitrage separately.

Theorem 1.9 *A price p satisfies the condition NA1 if and only if*

(i) $p < p^{US}(G)$ *if* $-p(-\tilde{G}_{\hat{\tau}}/\beta_T) \neq p(\tilde{G}_{\hat{\tau}}/\beta_T)$,
(ii) $p \leq p^{US}(G)$ *otherwise.*

Proof We only provide a proof for the first case. The second case is treated similarly by combining the arguments below with (ii) of Theorem 1.4.

1. Let us assume that $p \geq p^{US}(G)$. By Theorem 1.8, there exists $\phi \in \mathcal{A}$ such that $\tilde{V}^{p,\phi} \geq \tilde{G}$. To show that p is not viable, it suffices to prove that $\mathbb{P}\left[\tilde{V}_\tau^{p,\phi} > \tilde{G}_\tau \right] > 0$, for all $\tau \in \mathcal{T}_0$. Given $\tau \in \mathcal{T}_0$, we assume that $\tilde{V}_\tau^{p,\phi} = \tilde{G}_\tau$ $\mathbb{P}$ – a.s. and work

towards a contradiction. If the previous assertion was true, then Theorem 1.8 would imply that $\tau \geq \hat{\tau}$ and $\tilde{V}^{p,\phi}_\tau = Y_\tau$. The former being a $\mathcal{M}(\tilde{S})$-martingale and the latter a $\mathcal{M}(\tilde{S})$-supermatingale, this would imply that $\tilde{V}^{p,\phi}_{\hat{\tau}} = Y_{\hat{\tau}} = \tilde{G}_{\hat{\tau}}$, by taking the $\mathcal{F}_{\hat{\tau}}$-conditional expectation under any martingale measure. On the other hand, $-p(-\tilde{G}_{\hat{\tau}}/\beta_T) \neq p(\tilde{G}_{\hat{\tau}}/\beta_T)$. Therefore, $\tilde{G}_{\hat{\tau}}/\beta_T$ cannot be replicated, see Theorem 1.4. Hence, it cannot be that $\tilde{V}^{p,\phi}_{\hat{\tau}} = \tilde{G}_{\hat{\tau}}$.

2. If $p < p^{US}(G)$, there cannot be any arbitrage by definition $p^{US}(G)$.   □

To discuss the point of view of the buyer, let us introduce the auxiliary process $\check{Y}$ defined by

$$\check{Y}_t = \min\left\{-\tilde{G}_t \ , \ \underset{\mathbb{Q}\in\mathcal{M}(\tilde{S})}{\text{ess sup}} \ \mathbb{E}^{\mathbb{Q}}\left[\check{Y}_{t+1} \mid \mathcal{F}_t\right]\right\} \quad t < T \ ,$$

and the terminal condition

$$\check{Y}_T = -\tilde{G}_T \ .$$

Following line by line the arguments in the proof of Theorem 1.8, we obtain the following characterisation for $\check{Y}$.

Theorem 1.10 *Let*

$$\check{\tau} := \inf\{t \in \mathbb{T} \ : \ \check{Y}_t = -\tilde{G}_t\} \ .$$

Then,

$$\check{Y}_0 = \inf_{\tau\in\mathcal{T}_0} \ \sup_{\mathbb{Q}\in\mathcal{M}(\tilde{S})} \ \mathbb{E}^{\mathbb{Q}}\left[-\tilde{G}_\tau\right] = \sup_{\mathbb{Q}\in\mathcal{M}(\tilde{S})} \ \mathbb{E}^{\mathbb{Q}}\left[-\tilde{G}_{\check{\tau}}\right] \ .$$

Moreover, there exists $\phi \in \mathcal{A}$ such that $\tilde{V}^{\check{Y}_0,\phi}_{\check{\tau}} \geq -\tilde{G}_{\check{\tau}}$.

We can now characterise the fact that the price does not allow the buyer to realise an arbitrage.

Theorem 1.11 *The price p satisfies the condition NA2 if and only if*

(i) $-p < \check{Y}_0$ *if* $-p(-\tilde{G}_{\check{\tau}}/\beta_T) \neq p(\tilde{G}_{\check{\tau}}/\beta_T)$,
(ii) $-p \leq \check{Y}_0$ *otherwise.*

Proof Again, we only consider the situation $-p(-\tilde{G}_{\check{\tau}}/\beta_T) \neq p(\tilde{G}_{\check{\tau}}/\beta_T)$.

1. Assume that $-p \geq \check{Y}_0$. To show that NA2 is violated, it suffices to exhibit $\phi \in \mathcal{A}$ such that $\tilde{V}^{-p,\phi}_{\check{\tau}} + \tilde{G}_{\check{\tau}} \in \mathbb{L}^0(\mathbb{R}_+) \setminus \{0\}$. But, this is an immediate consequence of Theorem 1.4 and of the fact that $\check{Y}_0 = p(-\tilde{G}_{\check{\tau}}/\beta_T) \neq -p(\tilde{G}_{\check{\tau}}/\beta_T)$ by assumption and by Theorem 1.10.

2. Let us now assume that $-p < \check{Y}_0$. Then, Theorem 1.10 and Theorem 1.4 imply that $-p < p(-\tilde{G}_\tau/\beta_T)$ for all $\tau \in \mathcal{T}_0$. By the definition of $p(-\tilde{G}_\tau/\beta_T)$, NA2 is satisfied. □

Remark 1.15 If the market is complete, i.e. $\mathcal{M}(\tilde{S}) = \{\mathbb{Q}\}$, then the only viable price is

$$p^{US}(G) = \sup_{\tau \in \mathcal{T}_0} \mathbb{E}^{\mathbb{Q}} \left[\tilde{G}_\tau \right] = \mathbb{E}^{\mathbb{Q}} \left[\tilde{G}_{\check{\tau}} \right].$$

Indeed, in this case $-\check{Y}_0 = p^{US}(G)$, $\hat{\tau} = \check{\tau}$, and our claim follows from the above arguments.

The above formula is often interpreted as the maximisation of the value of the option by the buyer: given a fixed exercise strategy $\tau \in \mathcal{T}_0$ the price is $\mathbb{E}^{\mathbb{Q}} \left[\tilde{G}_\tau \right]$, the buyer maximises the value over the exercise strategies.

This is in a sense true, see the next section, but this is only an interpretation. Nobody is maximising anything in the definition (1.16).

1.5.5 Rational Exercise Strategy

Suppose that the price quoted on the market is P_t at time t. Clearly, the holder of the option should not exercise if $P > G$, it is much better to sell it.

Conversely, the holder should in principle not exercise his option after $\hat{\tau}$. Indeed, at the stopping time $\hat{\tau}$, the exercise provides $G_{\hat{\tau}} = Y_{\hat{\tau}}$. By similar arguments as the one used in the proof of Theorem 1.6, we can then show that there exists $\phi \in \mathcal{A}$ such that $Y_{\hat{\tau}} + \tilde{V}^{0,\phi}_{t \vee \hat{\tau}} - \tilde{V}^{0,\phi}_{\hat{\tau}} \geq Y_{t \vee \hat{\tau}}$ for all $t \in \mathbb{T}$. Since $Y_{\cdot \vee \hat{\tau}} \geq \tilde{G}_{\cdot \vee \hat{\tau}}$, this shows that the holder can invest $G_{\hat{\tau}}$ received at $\hat{\tau}$, if she exercises at this date, and follow a financial strategy such that the induced portfolio value is always greater than the exercise value of the option at each date t between $\hat{\tau}$ and the maturity T. In general, the portfolio value is even strictly greater with non-zero probability. The holder should therefore exercise his option at time $\hat{\tau}$, if not done before.

Note that, if P is a viable price process, then $P \leq \beta^{-1} Y$, by Theorem 1.9. The first time when $P \leq G$ must thus be before $\hat{\tau}$.

When the market is complete, the above arguments are much simpler. Indeed, the only viable price process is $\beta^{-1} Y$, see Theorem 1.9. Hence, the holder should exercise exactly at $\hat{\tau}$, This stopping time is called the *rational exercise time*.

The formula $p^{US}(G) = \mathbb{E}^{\mathbb{Q}} \left[\tilde{G}_{\hat{\tau}} \right]$ can then be interpreted as the value of the option of payoff $G_{\hat{\tau}}$ paid at the *rational exercise time* $\hat{\tau}$. Let us insist on the fact that this is a just an interpretation. The rationality of the buyer has nothing to do with the definition (1.16) of the super-hedging price.

1.6 Models with Portfolio Constraints

We now consider a market in which the financial assets can no more be traded freely, positions are now subject to restrictions. More precisely, we impose *portfolio constraints* of the form

$$\phi_t \in K \quad \mathbb{P} - \text{a.s.} \quad \text{for all } t \in \mathbb{T},$$

in which K is a closed convex cone of $\mathbb{R}^d$. We shall use the notation $\mathcal{A}_K$ to denote the collection of elements of $\mathcal{A}$ satisfying this condition.

Example 1.1

1. $K = \mathbb{R}^d$: there is no constraint.
2. $K = \mathbb{R}^d_+$: no asset can be short-sold.
3. $K = \mathbb{R}^{d-1} \times \{0\}$: the d-th asset can not be traded.

We refer to Chap. 5 below for more comments on this kind of market imperfection.

The arguments used in the preceding sections can be followed almost without any modification to characterise the absence of arbitrage opportunities in this context

Theorem 1.12 (First fundamental theorem of asset pricing) *The following are equivalent:*

(i) *(NA^K) holds.*
(ii) *(NA^K_t) holds for all $t = 0, \dots, T - 1$.*
(iii) *$\mathcal{H}^K_b(\tilde{S}) \neq \emptyset$.*

In the above, (NA^K) et (NA^K_t) are defined as (NA) and (NA$_t$) after replacing $\mathcal{A}$ by $\mathcal{A}_K$, and $\mathcal{H}^K_b(\tilde{S})$ is the set of positive random variables H of $\mathbb{L}^\infty$ such that $\mathbb{E}[H] = 1$ and

$$\sup_{\xi \in \mathbb{L}^\infty(K; \mathcal{F}_t)} \mathbb{E}\left[H\langle \xi, \tilde{S}_{t+1} - \tilde{S}_t\rangle\right] \leq 0 \quad \text{for all } t < T.$$

Example 1.2

1. $K = \mathbb{R}^d$: $\mathcal{H}^K_b(\tilde{S})$ is the set of densities of elements $\mathcal{M}_b(\tilde{S})$.
2. $K = \mathbb{R}^d_+$: $\mathcal{H}^K_b(\tilde{S})$ is the set of densities of the measures $\mathbb{Q} \sim \mathbb{P}$ such that $\tilde{S}$ is a $\mathbb{Q}$-supermartingale and $d\mathbb{Q}/d\mathbb{P} \in \mathbb{L}^\infty$.
3. $K = \mathbb{R}^{d-1} \times \{0\}$: $\mathcal{H}^K_b(\tilde{S})$ is the set of densities of the measures $\mathbb{Q} \sim \mathbb{P}$ such that $(\tilde{S}^1, \dots, \tilde{S}^{d-1})$ is a $\mathbb{Q}$-martingale and $d\mathbb{Q}/d\mathbb{P} \in \mathbb{L}^\infty$.

The results of Sect. 1.3 can be extended in a similar manner. The corresponding notion of super-replication price is

$$p_K(G) := \inf\{p \in \mathbb{R} : \exists\, \phi \in \mathcal{A}_K \text{ s.t. } V_T^{p,\phi} - G \in \mathbb{L}^0(\mathbb{R}_+)\},$$

and it can be characterised as follows.

Theorem 1.13 *Assume that* $\mathcal{H}_b^K(\tilde{S}) \neq \emptyset$. *Then, the following sets*

$$\Gamma_K := \{G \in \mathbb{L}^\infty : p_K(G) \leq 0\} ,$$

$$\Theta_K := \left\{ G \in \mathbb{L}^\infty : \sup_{H \in \mathcal{H}_b^K(\tilde{S})} \mathbb{E}\left[H\beta_T G\right] \leq 0 \right\}$$

are equal and

$$\{G \in \mathbb{L}^0 : p_K(G) \leq 0\}$$

is closed for the convergence in probability.

Corollary 1.3 *Fix* $G \in \mathbb{L}^\infty$. *If* $\mathcal{H}_b^K(\tilde{S}) \neq \emptyset$, *then*

$$p_K(G) = \sup_{H \in \mathcal{H}_b^K(\tilde{S})} \mathbb{E}\left[H\beta_T G\right] .$$

One can also define a notion of viable price. See Exercise 2.19 below.

1.7 Problems

1.1 (Cox-Ross-Rubinstein model) The model of Cox, Ross and Rubinstein is a simple binomial tree describing the dynamics of a single risky asset. At each period, its variations can take only two values. Although quite restrictive, it turns out to provide a good approximation of the Black and Scholes model in continuous time, which is very helpful for numerical computations.

Given two real numbers $d < u$ satisfying $ud = 1$, we consider the probability space $\Omega = \{\omega = (\omega^1, \ldots, \omega^T) \in \{d, u\}^T\}$ endowed with the probability measure $\mathbb{P}$ defined as

$$\mathbb{P}[\omega] = \pi^{N_T(\omega)}(1 - \pi)^{T - N_T(\omega)} , \quad \omega \in \Omega ,$$

in which $\pi \in (0, 1)$ and

$$N_t(\omega) = \sum_{j=1}^{t} \mathbf{1}_{\{\omega^j = u\}} , \quad 1 \leq t \leq T ,$$

so that $N_T(\omega)$ is the number of components of ω equal to u. We use the convention $N_0 = 0$.

We assume that the risk free interest rate is a constant $r > 0$. The risky asset S evolves according to

$$S_t(\omega) := S_0 e^{bt} u^{N_t(\omega)} d^{t-N_t(\omega)} \ , \ t \in \mathbb{T},$$

for some $b \in \mathbb{R}$.

1. Show that, in this model, the absence of arbitrage opportunities implies that

$$de^b < 1 + r < ue^b . \tag{1.20}$$

2. Show that the condition (1.20) implies the existence of a unique probability measure $\mathbb{Q}$ in $\mathcal{M}(\tilde{S})$ which is given by

$$\mathbb{Q}[\omega] = q^{N_T(\omega)}(1-q)^{T-N_T(\omega)} \ , \ \omega \in \Omega, \tag{1.21}$$

where

$$q := \frac{1 + r - e^b d}{e^b u - e^b d} . \tag{1.22}$$

3. Let us consider a European option of payoff $G \in \mathbb{L}^0(\mathcal{F}_T)$. Write down the backward dynamic programming algorithm that describes the price process of the option and the hedging strategy.
4. Answer to the above questions in the case of an American option $G = (G_t)_{t \in \mathbb{T}}$.

1.2 (A simple two-period model) Let us consider a simple model with only five possibilities $\Omega = \{\omega_1, \omega_2, \omega_3, \omega_4, \omega_5\}$, two periods and three assets: one non-risky with return equal to 0 $(r = 0)$ and two stocks S^1 and S^2 whose dynamics are given by the following table:

ω	$S_0^1(\omega)$	$S_1^1(\omega)$	$S_2^1(\omega)$
ω_1	6	5	3
ω_2	6	5	4
ω_3	6	5	8
ω_4	6	7	6
ω_5	6	7	8

ω	$S_0^2(\omega)$	$S_1^2(\omega)$	$S_2^2(\omega)$
ω_1	3.75	3	2
ω_2	3.75	3	3
ω_3	3.75	3	4
ω_4	3.75	4.5	4
ω_5	3.75	4.5	5

1. a. Describe the evolution of each asset in a tree.
   b. What does Ω represent?
   c. Describe the sets $\mathcal{F}_0$, $\mathcal{F}_1$ and $\mathcal{F}_2$ (i.e. the flow of information provided at the dates 0, 1 and 2). Show in particular that one can only know at time 1 that the realised path is either in $\{\omega_1$ or ω_2 or $\omega_3\}$ or in $\{\omega_4$ or $\omega_5\}$ without being able to distinguish these two sets.
2. Let us denote by ϕ_t^i the number of shares of S^i in the portfolio between t and $t+1$.

   a. Describe the wealth dynamics associated to (ϕ^1, ϕ^2).
   b. Recall the definition of the absence of arbitrage.

3. a. Show that a measure $\mathbb{Q}$ such that $\mathbb{E}^{\mathbb{Q}}[S_1^1] = 6$ and $\mathbb{E}^{\mathbb{Q}}[S_1^2] = 3.75$ necessarily
      satisfies $\mathbb{Q}[\omega_4 \text{ or } \omega_5] = \mathbb{Q}[\omega_1 \text{ or } \omega_2 \text{ or } \omega_3] = 1/2$.
   b. Show that a measure $\mathbb{Q}$ such that $\mathbb{E}^{\mathbb{Q}}[S_2^1 \mid \mathcal{F}_1] = S_1^1$ and $\mathbb{E}^{\mathbb{Q}}[S_2^2 \mid \mathcal{F}_1] = S_1^2$
      necessarily satisfies $\mathbb{Q}[\omega_4 \mid \omega_4 \text{ or } \omega_5] = \mathbb{Q}[\omega_5 \mid \omega_4 \text{ or } \omega_5] = 1/2$.
   c. What should satisfy $\mathbb{Q}[\omega_1 \mid \omega_1 \text{ or } \omega_2 \text{ or } \omega_3]$ and $\mathbb{Q}[\omega_2 \mid \omega_1 \text{ or } \omega_2 \text{ or } \omega_3]$ for S^1
      and S^2 to be martingales under $\mathbb{Q}$?
   d. Does the market admit an arbitrage?

4. Let us consider a European option whose payoff at time 1 equals $G_1 := [S_1^1 - S_1^2 - 2]^+$.

   a. Describe in a table the payoff depending on the value of ω.
   b. Compute the price at 0 of the option.
   c. Compute the hedging strategy.

5. Let us consider a European option whose payoff at time 2 equals $G_2 := [S_2^1 - S_2^2 - 2]^+$.

   a. Describe in a table the payoff depending on the value of ω.
   b. Compute the price at 1 of the value of the option (as a function of ω), and then
      compute its price at time 0.
   c. Compute the hedging strategy.

6. Let us now consider an American option with payoff $G_t := [S_t^1 - S_t^2 - 2]^+$ if
   exercised at t.

   a. Compute the price at 1 of the value of the option (as a function of ω), and then
      compute its price at time 0.
   b. Compute the hedging strategy.
   c. When should it be exercised?

1.3 (Short-sell constraints) We consider a one period model with one risky asset.
We denote by $R = 1 + r$ the return of the risk-free asset, in which the risk free rate is
a constant r. The risky asset evolves according to a binomial tree on $\Omega := \{\omega_1, \omega_2\}$,
each element having a positive probability:

$$\frac{S_1(\omega_1)}{S_0} = u \text{ and } \frac{S_1(\omega_2)}{S_0} = d,$$

in which $u > d > 0$ are given.
   We consider the following condition, hereafter called NA$_+$;

$$\theta(S_1 - RS_0) \in \mathbb{L}^0(\mathbb{R}_+) \implies \theta(S_1 - RS_0) = 0, \text{ for all } \theta \geq 0.$$

1. How should NA$_+$ be interpreted?
2. Show that $d < R$ if NA$_+$ holds.

3. Deduce that NA$_+$ implies the existence of a real number π such that

$$0 \ < \ \pi \ < \ 1 \ \text{and} \ \ \pi u \ + \ (1 - \pi) \, d \ \leq \ R \, . \tag{1.23}$$

4. How can we relate π to the notion of risk neutral probability measure?
5. Conversely, let us now assume that there exists π satisfying (1.23). Show that NA$_+$ is then satisfied.
6. Provide an example of a financial market in which NA$_+$ holds but the condition NA of Definition 1.1 does not.

1.4 (Proportional transaction costs) Let us consider a market consisting of only one risky asset S and in which the risk free rate is set to zero, $r \equiv 0$. We assume that trading the risky asset leads to paying a proportional transaction cost equal to λ times the amount of money traded, with $\lambda > 0$. The wealth is now described as a two dimensional process $V = (V^0, V^1)$, V^0 (resp. V^1) stands for the amount of numéraire/cash (resp. risky asset) in the portfolio. We denote by M_t (resp. L_t) the cumulated amount of bought (resp. sold) shares of S between 0 and t. They should be $\mathbb{F}$-adapted. In the following, we assume that S takes positive values.

1. Write the number of shares of S held between t and $t+1$ in terms of V_t^1 and S_t.
2. Deduce the dynamics of V in terms of M and L.
3. Let K denote the set of values of portfolios $x = (x^1, x^2)$ such that the position x^2 in risky asset can be cleared so as to obtain a new portfolio of the form $(y, 0)$ such that $y \geq 0$ (this is the so-called solvability region). Provide the explicit form of K in terms of λ.
4. Let $\mathcal{Z}$ be the collection of $\mathbb{F}$-adapted processes $Z = (Z^1, Z^2)$ that take values in $(0, \infty)^2$ and such that $(Z^1, Z^2 S)$ is a $(\mathbb{F}, \mathbb{P})$-martingale with values in $K^o = \{(z^1, z^2) : x^1 z^1 + x^2 z^2 \geq 0 \ \forall \ (x^1, x^2) \in K\}$ (this is the positive polar of K in the sense of convex analysis).
   Show that $\langle Z, V \rangle$ is a $(\mathbb{F}, \mathbb{P})$-supermartingale for any $Z \in \mathcal{Z}$.
5. Assuming that $\mathcal{Z} \neq \emptyset$, show that we cannot have $V_{0-} = 0$ (values before the first trade at 0) and $V_T \in \mathbb{R}_+^d$ with $\mathbb{P}[V_T \neq 0] > 0$.

1.5 (Fixed transaction cost) We consider a one period market with risk free rate equal to r. There is only one risky asset. At 0, it is worth $S_0 = 1$ and can take two values at time 1, uS_0 or dS_0, each with probability 0.5. Here, $u > d$ are given.

A strategy consists in buying $\theta \in \mathbb{R}$ shares of the risky asset at 0. The position is cleared at time 1. If $\theta \geq 0$, this is a buy at 0 and therefore a sell at 1. Conversely if $\theta \leq 0$. Each transaction, buy or sell, is subject to a fixed fee $c > 0$, i.e. each time we buy or sell one has to pay immediately c, whatever the amount of the transaction is. We denote by $V_1^{x,\theta}$ the value at time 1, after having cleared the position, given that we use the strategy θ and that the initial endowment in cash is x, before the first transaction (no initial position in the risky asset is assumed before the first transaction at 0).

In this model, we say that there is no arbitrage opportunities if we cannot find $\theta \in \mathbb{R}$ such that $\mathbb{P}\left[V_1^{0,\theta} \geq 0\right] = 1$ and $\mathbb{P}\left[V_1^{0,\theta} > 0\right] > 0$.

1. Show that $V_1^{x,\theta} = \theta S_1 + (x - \theta S_0 - c\mathbf{1}_{\{\theta \neq 0\}})(1 + r) - c\mathbf{1}_{\{\theta \neq 0\}}$.
2. Show that there is no arbitrage opportunities if $1 + r \in [d, u]$.
3. We now want to prove the reciprocal, namely that $1 + r \in [d, u]$ rules out arbitrage opportunities.

   a. Assume that $1 + r < d$. Show that we can find $\theta > 0$ such that $d - 1 - r > (2 + r)c/\theta$. Deduce an arbitrage.
   b. Assume that $1 + r > u$. Show that we can find $\theta < 0$ such that $\theta(u - 1 - r) > (2 + r)c$. Deduce an arbitrage.

4. Provide a necessary and sufficient condition on u, d and r for the absence of arbitrage opportunities.

   From now on, we assume that $1 + r \in [d, u]$.
5. We look for a probability measure $\mathbb{Q}$ such that $\mathbb{E}^{\mathbb{Q}}[V_1^{0,\theta}/(1 + r)] \leq 0$ for all $\theta \in \mathbb{R}$.

   a. By using the notation $q := \mathbb{Q}(S_1 = uS_0) = 1 - \mathbb{Q}(S_1 = dS_0)$, write down the expression of $\mathbb{E}^{\mathbb{Q}}[V_1^{0,\theta}]$ for $\theta \in \mathbb{R}$.
   b. Deduce that $\theta S_0(qu + (1 - q)d - (1 + r)) \leq (2 + r)c\mathbf{1}_{\{\theta \neq 0\}}$ for all $\theta \in \mathbb{R}$.
   c. Deduce that $qu + (1 - q)d = 1 + r$.
   d. Deduce that there exists a unique probability measure $\mathbb{Q}$ such that $\mathbb{E}^{\mathbb{Q}}[V_1^{0,\theta}/(1 + r)] \leq 0$ for $\theta \in \mathbb{R}$. Provide its explicit expression.
   e. Under which condition on u, d and r is this measure $\mathbb{Q}$ equivalent to $\mathbb{P}$?

6. Let us consider an European option of payoff G defined by $G = G(u) \in \mathbb{R}$ if $S_1 = uS_0$ and $G = G(d) \in \mathbb{R}$ if $S_1 = dS_0$. We assume that $G(u) \neq G(d)$.

   a. Set $p^+ := \mathbb{E}^{\mathbb{Q}}[G/(1 + r)] + c(1 + 1/(1 + r))$. Show that there exists $\theta \in \mathbb{R}$ such that $\mathbb{P}\left[V_1^{p^+,\theta} = G\right] = 1$.
   b. Set $p^- := \mathbb{E}^{\mathbb{Q}}[-G/(1 + r)] + c(1 + 1/(1 + r))$. Show that there exists $\theta \in \mathbb{R}$ such that $\mathbb{P}\left[V_1^{p^-,\theta} = -G\right] = 1$.
   c. What is the set of viable prices for G in this model?

1.6 (Ho and Lee model) We consider a finite probability space $(\Omega, \mathcal{F}, \mathbb{P})$ in which exchanges take place at times $0, 1, 2, \ldots, T^*$. Here T^* is a positive integer. We set $\mathbb{T} := \{0, 1, \ldots, T^*\}$ and we endow $(\Omega, \mathcal{F})$ with a filtration $\mathbb{F} = (\mathcal{F}_t)_{t \in \mathbb{T}}$ satisfying $\mathcal{F}_0 = \{\Omega, \emptyset\}$, $\mathcal{F}_{T^*} = \mathcal{F}$.

Let r be an adapted positive process, the risk-free rate, and define B by

$$B_t := \prod_{j=0}^{t-1}(1 + r_j) \ , \ t \leq T^* .$$

A *zero-coupon* of maturity $T \leq T^*$ and nominal 1 is a financial product paying (to its buyer) 1 at T. We denote by $B_t(T)$ its price at $t \leq T$.

We assume that, for each $T \leq T^*$, there exist positive processes $u(T)$ and $d(T)$ such that

$$\mathbb{P}\left[\left\{B_{t+1}(T) = \frac{u_t(T)B_t(T)}{B_t(t+1)}\right\} \cup \left\{B_{t+1}(T) = \frac{d_t(T)B_t(T)}{B_t(t+1)}\right\}\right] = 1$$

$\forall\, 0 \leq t < T$,

$$u_t(T) > d_t(T) \quad \mathbb{P} - \text{a.s.} \;, \quad \forall\, 0 \leq t < T - 1 \;.$$

We shall denote by π the adapted process defined as

$$\pi_t(T) := \mathbb{P}\left[B_{t+1}(T) = u_t(T)B_t(T)/B_t(t+1) \mid \mathcal{F}_t\right] \;,$$

and we assume that it takes values in $]0, 1[$ for $t < T - 1$. In all this exercise, we shall assume that there are no arbitrage opportunities.

1. What is the value of $B_T(T)$? Show that $u_t(t+1) = d_t(t+1) = 1\ \mathbb{P} - \text{a.s.}$ for all $t < T^*$. From now on, we shall use the convention $B_t(T) := B_t/B_T$ if $t > T$.
2. A financial strategy $\phi = (\phi(1), \ldots, \phi(T^*))$ is an adapted process with values in $\mathbb{R}^{T^*}$. Namely, $\phi_t(T)$ stands for the amount invested between t and $t + 1$ in the zero-coupon maturing at T. We denote by $V^{v,\phi}$ the wealth process induced by an initial endowment $v \in \mathbb{R}$, at 0, and the strategy ϕ. Give the dynamics of $V^{v,\phi}$ under the self-financing condition.
3. Fix $t < T \leq T^*$.

   a. Show that $B_t(t+1) = B_t/B_{t+1} = (1 + r_t)^{-1}$.
   b. Recall the notion of arbitrage opportunities.
   c. Let $t < T - 1$. Show that $\mathbb{P}[d_t(T) > 1] = 1$ implies $B_{t+1}(T) > B_t(T)B_{t+1}/B_t\ \mathbb{P} - \text{a.s.}$ Deduce that the latter implies the existence of an arbitrage, and provide the corresponding strategy.
   d. Answer to the above question again but under the weaker condition $\mathbb{P}[d_t(T) \geq 1] > 0$.
   e. Show that one can neither have: $\mathbb{P}[u_t(T) \leq 1] > 0$ for $t < T - 1$.
   f. Deduce a necessary condition on u and d leading to the absence of arbitrage.
4. We now assume that $u_t(T) > 1 > d_t(T)\ \mathbb{P} - \text{a.s.}$ for all $t < T - 1 \leq T^*$.

   a. Show that, for all $T \leq T^*$, there exists a unique adapted process $q(T)$ with values in $]0, 1[$ such that

   $$q_t(T)u_t(T)\frac{1}{B_t(t+1)} + (1 - q_t(T))d_t(T)\frac{1}{B_t(t+1)}$$

   $$= \frac{B_{t+1}}{B_t} \quad \mathbb{P} - \text{a.s.} \;, \quad \forall\, t < T - 1$$

   and $q_t(T) = 1$ if $t + 1 = T$.

b. Let H be defined on $\mathbb{T}$ by

$$H_t(T^*) := \prod_{j=0}^{t-1} \left(\frac{q_j(T^*)}{p_j(T^*)} \mathbf{1}_{A_j(T^*)} + \frac{(1 - q_j(T^*))}{1 - p_j(T^*)} \mathbf{1}_{A_j^c(T^*)} \right)$$

in which $A_j(T) := \{B_{j+1}(T) = u_j(T)B_j(T)/B_j(j+1)\}$. Show that $H(T^*)$ is a $\mathbb{P}$-martingale on $\mathbb{T}$.

c. Let $\mathbb{Q}$ be the measure of density $H_{T^*}(T^*)$ with respect to $\mathbb{P}$. Show that it is equivalent to $\mathbb{P}$.

d. Show that $\mathbb{Q}$ is the unique equivalent probability measure under which the process $B^{-1} B(T^*)$ is a martingale.

e. Deduce that $B^{-1} B(T)$ is a $\mathbb{Q}$-martingale for all $T \leq T^*$ and that

$$q_t(T^*)u_t(T) + (1 - q_t(T^*))d_t(T) = 1 \quad \mathbb{P} - \text{a.s.}, \; \forall t < T \leq T^*,$$

if there is no arbitrage.

5. We assume from now on that $u_t(T) = U(T-t)$ and $d_t(T) = D(T-t)$, in which U and D are deterministic function, and we define the sequence of random variables $(\xi_t(\delta))$ by

$$\xi_{t+1}(\delta) := U(\delta)\mathbf{1}_{A_t(t+\delta)} + D(\delta)\mathbf{1}_{A_t^c(t+\delta)} \;, \; t < T^* \;, \; \delta \leq T^* - t \;,$$

so that

$$B_t(T) = \xi_t(T - t + 1) \frac{B_{t-1}(T)}{B_{t-1}(t)} \;.$$

a. Show that

$$\prod_{j=0}^{t-1} B_j(j + 1) = B_0(t) \prod_{j=1}^{t-1} \xi_j(t + 1 - j) \;, \; t \leq T^*.$$

b. Deduce that

$$B_t(T) = \frac{B_0(T)}{B_0(t)} \frac{\prod_{j=1}^{t} \xi_j(T - j + 1)}{\prod_{j=1}^{t-1} \xi_j(t - j + 1)} \;, \; t \leq T \leq T^* \;.$$

c. Express r_t as a function of ξ and B_0.

1.7 (Kreps-Yan theorem in finite dimension) Let us denote by x^j the j-th component of a vector x of $\mathbb{R}^d$. Let C be a closed and convex subset of $\mathbb{R}^d$ such that

(i) $\lambda x \in C$ if $\lambda \geq 0$ and $x \in C$.

(ii) $(-\infty, 0]^d \subset C.$
(iii) $C \cap [0, \infty)^d = \{0\}$

Let us denote by e_i the i-th canonical vector of $\mathbb{R}^d$ (i.e. its i-th component is one, the others are 0).

1. By using the Hahn-Banach separation theorem, show that, for all $i \leq d$, there exists $Y_i \in \mathbb{R}^d$ such that

$$\sup_{x \in C} \langle x, Y_i \rangle < \langle e_i, Y_i \rangle.$$

2. By using (i), show that $\sup_{x \in C} \langle x, Y_i \rangle = 0 < \langle e_i, Y_i \rangle$.
3. Deduce that $Y_i^i > 0$ and show that $Y_i \in [0, \infty)^d$ (use (ii)).
4. Deduce that there exists $Y \in (0, \infty)^d$ such that $\langle x, Y \rangle \leq 0$ for all $x \in C$.
5. Provide a financial interpretation of this result (what do C and Y represent?).

1.8 (Callable American options) Let us assume that Ω is finite, just to avoid integrability issues. We set the risk-free interest rate to 0 for simplicity and consider only one risky asset described by the adapted process $S = (S_t)_{t \leq T}$. We assume that S is a martingale under $\mathbb{P}$ and that the market is complete, namely any random variable X admits a representation of the form $X = \mathbb{E}[X] + \sum_{t=0}^{T-1} \phi_t(S_{t+1} - S_t)$ for some $\phi \in \mathcal{A}$, the set of real-valued adapted processes.

Let us define $\mathcal{T}$ as the collection of stopping times with values in $\{0, \ldots, T\}$. Given $\tau, \theta \in \mathcal{T}$, we set $G(\tau, \theta) := \ell_\tau \mathbf{1}_{\{\tau \leq \theta\}} + L_\theta \mathbf{1}_{\{\tau > \theta\}}$ in which ℓ, L are two non-negative adapted processes satisfying $\ell_t \leq L_t$ $\mathbb{P}-$a.s. for all $t \leq T$.

1. Given $\theta, \tau \in \mathcal{T}$, we define Y^θ and X^τ by

$$Y_T^\theta = G(T, \theta) \, , \, Y_t^\theta = \max\{G(t, \theta) \, , \, \mathbb{E}\left[Y_{t+1}^\theta \mid \mathcal{F}_t\right]\},$$
$$X_T^\tau = G(\tau, T) \, , \, X_t^\tau = \min\{G(\tau, t) \, , \, \mathbb{E}\left[X_{t+1}^\tau \mid \mathcal{F}_t\right]\},$$

for all $t \leq T - 1$.

a. Express Y^θ and X^τ as Snell-envelopes of some processes.
b. Show that there exists two martingales M^θ and N^τ as well as two predictable non-decreasing processes A^θ and B^τ satisfying $A_0^\theta = B_0^\tau = 0$ such that $Y^\theta = M^\theta - A^\theta$ and $X^\tau = N^\tau + B^\tau$.
c. Show that $X_\theta^\tau \leq Y_\tau^\theta$ $\mathbb{P}-$a.s. and deduce that $X_{t \wedge \vartheta}^\tau \leq Y_{t \wedge \vartheta}^\theta$ for all $t \leq T$, in which $\vartheta := \tau \wedge \theta$.

2. Let us consider the callable American option of payoff process G. Here, τ should be interpreted as the exercise date of the buyer, and θ as the date at which it is bought back by the issuer. If $\tau \leq \theta$, the buyer receives ℓ_τ at τ. If $\tau > \theta$, namely if the issuer calls back the option before it has been exercised, then the buyer receives L_θ at time θ. We let p be the price at which the option has been sold at time 0.

a. Show that the issuer can make an arbitrage if $p > y_0 := \inf_{\theta \in \mathcal{T}} Y_0^\theta$.

b. Show that the buyer can make an arbitrage if $p < x_0 := \sup_{\tau \in \mathcal{T}} X_0^\tau$.

3. We now define the process Z by

$$Z_T = \ell_T, \; Z_t = \min\{L_t, \max\{\ell_t, \mathbb{E}[Z_{t+1}|\mathcal{F}_t]\}\}, \; t \le T-1.$$

Let us set $\hat\tau := \inf\{t \ge 0 : Z_t = \ell_t\}$, $\hat\theta := \inf\{t \ge 0 : Z_t = L_t\} \wedge T$ and $\hat\vartheta := \hat\tau \wedge \hat\theta$.

a. Show that $(Z_{t \wedge \hat\theta})_{t \le T}$ is a supermartingale that is bounded from below by $(G(t, \hat\theta))_{t \le T}$ and that $(Z_{t \wedge \hat\tau})_{t \le T}$ is a submartingale that is bounded from above by $(G(\hat\tau, t))_{t \le T}$.

b. Deduce that $Y^{\hat\theta}_{t \wedge \hat\vartheta} = Z_{t \wedge \hat\vartheta} = X^{\hat\tau}_{t \wedge \hat\vartheta}$ $\mathbb{P}$ − a.s. for all $t \le T$.

4. We are now in position to determine the prices that are compatible with the absence of arbitrage opportunities.

a. Show that neither the issuer nor the buyer can make an arbitrage if the option is sold at the price Z_0 at time 0.

b. Show that there exist a martingale $\bar{M}$ and a predictable process $\bar{A}$, equal to 0 at 0, such that $Z = \bar{M} + \bar{A}$, $\bar{A}_{t \wedge \hat\tau} \ge 0$ $\mathbb{P}$ − a.s. and $\bar{A}_{t \wedge \hat\theta} \le 0$ $\mathbb{P}$ − a.s. for all $t \le T$.

c. Show that the issuer should call back the option at time $\hat\theta$ (if not exercised yet) and characterise the hedging strategy in terms of the representation of $\bar{M}$ as a stochastic integral with respect to S

d. When should the buyer exercise ? Provide a brief justification.

1.9 (Swing options) Let us assume that Ω is finite, just to avoid integrability issues. We set the risk-free interest rate to 0 for simplicity and consider only one risky asset described by the adapted process $S = (S_t)_{t \le T}$. We assume that S is a martingale under $\mathbb{P}$ and that the market is complete, namely any random variable X admits a representation of the form $X = \mathbb{E}[X] + \sum_{t=0}^{T-1} \phi_t(S_{t+1} - S_t)$ for some $\phi \in \mathcal{A}$, the set of real-valued adapted processes.

Let G be an adapted process taking non-negative values. A swing option of payoff G with at most N exercise dates ($0 \le N \le T$) is an American type option that can be exercised N times before T. At each exercise date τ, the buyer receives G_τ. It cannot be exercised more than once at a given date.

Let us define $\mathcal{T}^N$ as the collection of N-tuples of stopping times $(\tau_N, \ldots, \tau_1)$ with values in $\{0, \ldots, T\}$ and such that

$$\tau_{i+1} < \tau_i \; \mathbb{P} - \text{a.s. and } \tau_{i+1} \le T - i \text{ for all } i < N.$$

They will be used to model the different exercise dates: τ_1 will stand for the last time at which the buyer exercises, τ_2 is the time before the last one, and so on. As the payoff is non-negative, we can always impose that there should not remain more than i exercises after $T - i$.

1. Let Y^1 be the process defined by

$$Y_T^1 = G_T,$$

$$Y_t^1 = \max\left\{G_t, \mathbb{E}\left[Y_{t+1}^1 \mid \mathcal{F}_t\right]\right\} \quad \text{if } t < T.$$

   a. What is the financial interpretation of Y_t^1?
   b. What is the financial interpretation of $\tilde{\tau}^1 := \inf\{t \geq 0 \mid Y_t^1 = G_t\}$?

We now define the sequence $(Y^n)_{1 \leq n \leq N}$ as follows: Y^1 is defined as above and, for $1 \leq n \leq N$,

$$Y_t^{n+1} = Y_t^n \quad \text{if } T - n < t \leq T \tag{1.24}$$

and

$$Y_t^{n+1} = \max\left\{G_t + \mathbb{E}\left[Y_{t+1}^n \mid \mathcal{F}_t\right], \mathbb{E}\left[Y_{t+1}^{n+1} \mid \mathcal{F}_t\right]\right\} \quad \text{otherwise.} \tag{1.25}$$

2. We now study the supermartingale property of the Y^n.

   a. Show that Y^n is a non-negative process, for all $n \leq N$.
   b. Show that $(Y_{t \wedge (T-n+1)}^{n+1})_{t \leq T}$ is a supermartingale for all $0 \leq n \leq N$ (hint: write its evolution between t and $t + 1$ and isolate the case $t \leq T - n$).
   c. With the help of (1.24) and of the previous question, show that

$$Y_{T-n+1}^{n+1} = Y_{T-n+1}^n \geq \mathbb{E}\left[Y_{T-n+2}^n \mid \mathcal{F}_{T-n+1}\right] = \mathbb{E}\left[Y_{T-n+2}^{n+1} \mid \mathcal{F}_{T-n+1}\right],$$

     for $2 \leq n \leq N$.
   d. Show by induction that Y^n is a supermartingale on $[0, T]$ for all $1 \leq n \leq N$.

3. Given $v := (\tau_N, \ldots, \tau_1) \in \mathcal{T}^N$, we define

$$C_s^v := \sum_{i=N}^{1} G_{\tau_i} \mathbf{1}_{\tau_i \leq s} \quad s \leq T.$$

   a. Provide a financial interpretation of $V^{v,\phi} - C^v$, given $v \in \mathbb{R}$, $\phi \in \mathcal{A}$ and $v \in \mathcal{T}^N$.
   b. Show that there exists $\phi \in \mathcal{A}$ such that

$$V_{\tau_N}^{Y_0^N, \phi} - C_{\tau_N}^v \geq \mathbb{E}\left[Y_{\tau_N+1}^{N-1} \mid \mathcal{F}_{\tau_N}\right],$$

     whatever is $v := (\tau_N, \ldots, \tau_1) \in \mathcal{T}^N$ (hint: recall that $\tau_N \leq T - N + 1$ by definition).

c. Deduce that there exists $\bar{\phi} \in \mathcal{A}$ such that

$$V_{\tau_N+1}^{Y_0^N, \bar{\phi}} - C_{\tau_N+1}^{\nu} \geq Y_{\tau_N+1}^{N-1} - G_{\tau_{N-1}} \mathbf{1}_{\{\tau_{N-1}=\tau_N+1\}}$$

and

$$V_{\tau_{N-1}}^{Y_0^N, \bar{\phi}} - C_{\tau_{N-1}}^{\nu} \geq \mathbb{E}\left[Y_{\tau_{N-1}+1}^{N-2} \mid \mathcal{F}_{\tau_{N-1}}\right],$$

whatever is $\nu := (\tau_N, \ldots, \tau_1) \in \mathcal{T}^N$.

d. Use an induction to exhibit $\hat{\phi} \in \mathcal{A}$ such that $V_{\tau_{k+1}}^{Y_0^N, \hat{\phi}} - C_{\tau_{k+1}}^{\nu} \geq$ $\mathbb{E}\left[Y_{\tau_{k+1}+1}^k \mid \mathcal{F}_{\tau_{k+1}}\right]$ for all $k < N$, whatever is $\nu := (\tau_N, \ldots, \tau_1) \in \mathcal{T}^N$.

e. Deduce that

$$Y_0^N \geq p_N := \inf\{x \geq 0 \ : \ \exists \, \phi \in \mathcal{A} \text{ s.t. } V_s^{x,\phi} - C_s^{\nu} \geq 0 \ \forall \, s \leq T \, , \ \forall \, \nu \in \mathcal{T}^N\}.$$

4. Let $\hat{\nu} := (\hat{\tau}_N, \ldots, \hat{\tau}_1)$ be the N-tuple of stopping times defined for $2 \leq n \leq N$ by

$$\hat{\tau}_n := \inf\{t > \hat{\tau}_{n+1} \mid Y_t^n = G_t + \mathbb{E}\left[Y_{t+1}^{n-1} \mid \mathcal{F}_t\right] \mathbf{1}_{t<T}\},$$

where

$$\hat{\tau}_1 := \inf\{t > \hat{\tau}_2 \mid Y_t^1 = G_t\} \quad \text{and} \quad \hat{\tau}_{N+1} := -1 \, ,$$

with the convention $\inf \emptyset = +\infty$.

a. By using the fact that $G \geq 0$, show that $\hat{\tau}_n \leq T - n + 1$ for all $1 \leq n \leq N$.
b. Show that $(Y_{t \wedge \hat{\tau}_n}^n)_{t \leq T}$ is a martingale, for all $1 \leq n \leq N$.
c. Let $\phi \in \mathcal{A}$ and $\nu \in \mathbb{R}$ be such that $V_s^{\nu,\phi} \geq C_s^{\hat{\nu}}$ for all $s \leq T$.

   i. Show that

   $$V_{\hat{\tau}_1}^{\nu,\phi} \geq C_{\hat{\tau}_1}^{\hat{\nu}} = C_{\hat{\tau}_1 - 1}^{\hat{\nu}} + G_{\hat{\tau}_1} = C_{\hat{\tau}_2}^{\hat{\nu}} + Y_{\hat{\tau}_1}^1.$$

   ii. Deduce that

   $$V_{\hat{\tau}_2}^{\nu,\phi} \geq C_{\hat{\tau}_2}^{\hat{\nu}} + \mathbb{E}\left[Y_{\hat{\tau}_2+1}^1 \mid \mathcal{F}_{\hat{\tau}_2}\right] \geq C_{\hat{\tau}_3}^{\hat{\nu}} + Y_{\hat{\tau}_2}^2 \, .$$

   iii. By using an induction, deduce that $V_0^{x,\phi} \geq Y_0^N$.

5. Deduce from the above that $Y_0^N = p_N$.
6. Express the hedging strategy in terms of the representation of the martingale parts of the different Y^n.
7. What are the rational exercise times for the holder of the option? Briefly justify.

1.10 (Imperfect information) Let us assume that Ω is finite to avoid integrability issues. We consider a market made of a risk-free asset B and d risky assets $S = (S^1, \ldots, S^d)$, in which S is $\mathbb{F}$-adapted. The dynamics of B is given by $B_t = (1 + r_{t-1})B_{t-1}$ ($t \geq 1$) with $r = (r_t)_{t \geq 0}$ $\mathbb{F}$-adapted and non-negative and $B_0 = 1$. A portfolio strategy is a $\mathbb{G}$-adapted process ϕ with values in $\mathbb{R}^d$: ϕ_t^i is the number of shares of the risky asset S^i held on the time period $[t-1, t]$. Let us define $\mathcal{A}$ as the collection of such processes. In the above, the filtration $\mathbb{G} = (\mathcal{G}_t)_{t \geq 0}$ might be different of $\mathbb{F}$.

1. Write down the dynamics of the portfolio process $V^{v,\phi}$ associated to the strategy $\phi \in \mathcal{A}$ and the initial endowment $v \in \mathbb{R}$.
2. If $\phi \in \mathcal{A}$ and $\mathcal{G}_t \subset \mathcal{F}_t$ for all t, to which filtration is $V^{v,\phi}$ adapted?
3. To a probability measure $\mathbb{Q} \sim \mathbb{P}$, we associate the process $\bar{S}_t^{\mathbb{Q}} := \mathbb{E}^{\mathbb{Q}}[\tilde{S}_t \mid \mathcal{G}_t]$ in which $\tilde{S}$ is the discounted price process. Let us define $\mathcal{M}(\mathbb{G})$ as the set of measures $\mathbb{Q} \sim \mathbb{P}$ such that $\bar{S}^{\mathbb{Q}} := (\bar{S}_t^{\mathbb{Q}})_{t \geq 0}$ is a $(\mathbb{G}, \mathbb{Q})$-martingale. Show that for all $\phi \in \mathcal{A}$ and $\mathbb{Q} \in \mathcal{M}(\mathbb{G})$, $(\mathbb{E}^{\mathbb{Q}}[\tilde{V}_t^{v,\phi} \mid \mathcal{G}_t])_{t \geq 0}$ is a $(\mathbb{G}, \mathbb{Q})$-martingale. Here, $\tilde{V}^{v,\phi}$ is the discounted wealth process.
4. Deduce a sufficient condition for the absence of arbitrage opportunities in this model.

1.11 (A model with price impact #1) Let us consider a very simple probability space $\Omega = \{\omega_1, \omega_2\}$ endowed with $\mathbb{P}$ defined by $\mathbb{P}[\omega_i] = 1/2$ for $i = 1, 2$.

We consider a one period market in which the strategy of a large investor has an impact on the prices. More precisely, we assume that the risk-free rate is zero and that there is only one risky asset. Its reference price (mid bid-ask spread price for instance) at 0 is $S_0 > 0$ but the buying price of $\phi_0 = |\phi_0| \geq 0$ shares is $|\phi_0|(S_0 + \lambda|\phi_0|)$, in which $\lambda > 0$, and the gain made from a sell of $-\phi_0 = |\phi_0| \geq 0$ shares is $|\phi_0|(S_0 - \lambda|\phi_0|)$. As usual, $\phi_0 \geq 0$ means that this is a buy of $|\phi_0|$ shares at 0, while $\phi_0 \leq 0$ means a sell of $|\phi_0|$ shares at 0.

The reference price at time 1 depends on the impact of ϕ_0 at 0: $S_1^{\phi_0}(\omega_i) = \alpha_i(S_0 + \lambda\phi_0/2)$ for $i = 1, 2$, with $\alpha_1 > \alpha_2 > 0$.

At time 1, the cost of buying $\phi_1 = |\phi_1| \geq 0$ shares is $|\phi_1|(S_1^{\phi_0} + \lambda|\phi_1|)$ and the gain made from a sell of $-\phi_1 = |\phi_1| \geq 0$ shares is $|\phi_1|(S_1^{\phi_0} - \lambda|\phi_1|)$.

1. Consider the strategy of buying $q \in \mathbb{R}$ shares at 0 and $-q$ at 1. Show that the corresponding gain is:

$$V_1^q(\omega_i) = q(\alpha_i - 1)S_0 + q^2\lambda(\alpha_i - 4)/2 = q\{(\alpha_i - 1)S_0 + q\lambda(\alpha_i - 4)/2\},$$

for $i = 1, 2$.

In the following, we will say that there is no arbitrage, i.e. NA holds, if:

$$\nexists q \in \mathbb{R} \text{ s.t. } V_1^q(\omega_i) \geq 0, i = 1, 2, \text{ and } V_1^q(\omega_1) + V_1^q(\omega_2) > 0.$$

2. Show that we can find an arbitrage by choosing $|q|$ small if $\alpha_2 > 1$ or if $1 > \alpha_1$ (recall that $\alpha_1 > \alpha_2 > 0$ and that $S_0 > 0$).

3. Let us assume in this question that $\alpha_2 \leq 1 \leq \alpha_1 \leq 4$. Show that $V_1^q(\omega_2) \geq 0$ implies $q \leq 0$, and that $q < 0$ implies $V_1^q(\omega_1) < 0$. Deduce that $\alpha_2 \leq 1 \leq \alpha_1 \leq 4$ implies NA.

4. Deduce a necessary and sufficient condition on α_1 and α_2 which ensures NA in the case $\alpha_1 \leq 4$. Does it imply the existence of a measure $\mathbb{Q} \sim \mathbb{P}$ such that $\mathbb{E}^{\mathbb{Q}}[S_1^0] = S_0$? Comment.

5. Let us assume in this question that $\alpha_2 \leq 1 \leq \alpha_1$ and $\alpha_1 > 4$.

   a. Show that $V_1^q(\omega_i) \geq 0$ for $i = 1, 2$ implies that $q \leq 0$ and

$$m := \frac{2(\alpha_1 - 1)S_0}{\lambda(\alpha_1 - 4)} \leq |q| \leq \frac{2|\alpha_2 - 1|S_0}{\lambda|\alpha_2 - 4|} =: M.$$

   b. Show that one can construct an arbitrage if $m < M$.

   c. How can we interpret this in terms of price manipulation?

1.12 (Model with price impact #2) We consider a very simple probability space $\Omega = \{\omega_1, \omega_2\}$ endowed with the probability $\mathbb{P}$ defined by $\mathbb{P}[\omega_i] = 1/2$ for $i = 1, 2$.

We consider a one period financial model in which the strategy of a (large) trader has an impact on prices and generates a liquidity cost. More precisely, we assume that the risk free interest rate is zero and that the market is composed of one stock. The reference price (mid bid-ask spread price for instance) at time 0 before the initial trade is S_{0-}. The strategy is described by a process $\phi = (\phi_0, \phi_1)$, where $\phi_0 \in \mathbb{R}$ is the number of shares bought at time 0 and ϕ_1 (which is $\mathcal{F}_1$-measurable) is the number of shares bought at time one. When buying ϕ_0 shares (this is a sell if $\phi_0 < 0$), the trader moves the price from S_{0-} to $S_0^\phi = S_{0-} + \lambda\phi_0$, where $\lambda > 0$ is a given constant. The cost of buying is $\phi_0 S_{0-} + \frac{1}{2}\lambda\phi_0^2 = \phi_0\frac{1}{2}(S_{0-} + S_0^\phi)$ (again a sell corresponds to $\phi_0 < 0$). Before trading, the price at time 1 is $S_{1-}^\phi(\omega_i) = S_0^\phi + \alpha_i$, $i = 1, 2$, where $\alpha_1 > \alpha_2$. If the trader buys ϕ_1 shares at time 1, the price is moved again to $S_1^\phi = S_{1-}^\phi + \lambda\phi_1$ and the cost of trading is $\phi_1 S_{1-}^\phi + \frac{1}{2}\lambda\phi_1^2$.

1. We denote by C_1^ϕ the amount of cash in the portfolio at time 1 after the last transaction, when starting with an initial wealth equal to 0. Show that

$$C_1^\phi(\omega_i) = -\phi_1(S_{0-} + \lambda\phi_0 + \alpha_i) - \phi_0 S_{0-} - \frac{1}{2}\lambda\phi_0^2 - \frac{1}{2}\lambda\phi_1^2.$$

In the following, we say that there is no arbitrage if

NA: we cannot find ϕ such that $\phi_0 + \phi_1 = 0$, $C_1^\phi(\omega_i) \geq 0$ for $i = 1, 2$

and $C_1^\phi(\omega_1) + C_1^\phi(\omega_2) > 0$.

2. Comment this no-arbitrage condition.

3. Compute C_1^ϕ when $\phi_0 + \phi_1 = 0$ and show that NA is equivalent to $\alpha_2 < 0 < \alpha_1$. Comment.

In the following, we assume that NA holds.

4. What is the super-hedging price and the super-hedging strategy of a call of payoff $[S_1 - K]^+$ paid in cash at 1, in which $S_1 = S_1^\phi$ is the price observed at time 1 if we use the strategy ϕ?

5. We now consider a call option with delivery. Namely, the trader has to deliver 1 unit of stock at time 1 and receives K in cash, if the stock value at 1 (after all trades) is larger than K. We want to compute the minimal amount of cash needed at time 0 to cover this claim (assuming that the initial position in stock is 0).

   a. Show that the super-hedging price $\hat{p}$ is given by the minimal p such that there exists $\phi_0 \in \mathbb{R}$ and $(\phi_1^i)_{i=1,2} \in \mathbb{R}^2$ satisfying for $i = 1, 2$:

   $$\phi_0 + \phi_1^i \geq \mathbf{1}_{\{S_{0-}+\alpha_i+(\phi_0+\phi_1^i)\lambda \geq K\}},$$

   $$p - (\phi_1^i + \phi_0)S_{0-} - \phi_1^i \alpha_i - \tfrac{1}{2}\lambda(\phi_0 + \phi_1^i)^2 \geq -K\mathbf{1}_{\{S_{0-}+\alpha_i+(\phi_0+\phi_1^i)\lambda \geq K\}}.$$

   b. Set

   $$F^i(\gamma^i, \phi_0) := \gamma^i S_{0-} + (\gamma^i - \phi_0)\alpha_i + \frac{1}{2}\lambda(\gamma^i)^2 - K\mathbf{1}_{\{S_{0-}+\alpha_i+\gamma^i\lambda \geq K\}},$$

   and deduce that

   $$\hat{p} = \inf_{\phi_0 \in \mathbb{R}} \max_{i \in \{1,2\}} \inf\{F^i(\gamma^i, \phi_0) : \gamma^i \geq \mathbf{1}_{\{S_{0-}+\alpha_i+\gamma^i\lambda \geq K\}}\}$$

   (the last inf is the inf of $F^i(\gamma^i, \phi_0)$ taken over γ^i such that $\gamma^i \geq \mathbf{1}_{\{S_{0-}+\alpha_i+\gamma^i\lambda \geq K\}}$).

6. From now on we assume that $0 < S_{0-} + \alpha_2 < K < S_{0-} + \alpha_1$.

   a. Show that, in the formula for $\hat{p}$, we must restrict to $\gamma^1 \geq 0$ and that the optimum is achieved by $\hat{\gamma}^1 = 1$.
   b. Show that, in the formula for $\hat{p}$, we must restrict to $\gamma^2 \geq 0$ and that the optimum is achieved by $\hat{\gamma}^2$ equal to 0 or equal to $(K - S_{0-} - \alpha_2)/\lambda$.
   c. Let $\hat{\phi}_0$ be such that

   $$F^1(\hat{\gamma}^1, \hat{\phi}_0) = F^2(\hat{\gamma}^2, \hat{\phi}_0).$$

   Deduce from the above that

   $$\hat{p} = S_{0-} + (1 - \hat{\phi}_0)\alpha_1 + \frac{1}{2}\lambda - K.$$

   d. What is the super-hedging strategy? (in terms of $\hat{\phi}_0$, $\hat{\gamma}^1$ and $\hat{\gamma}^2$)
   e. What is the difference with the case where the payoff is paid in cash only (no delivery but cash settlement)? (you can restrict to the case $\hat{\gamma}^2 = 0$)

Corrections

1.1

1. If $1 + r \le de^b < ue^b$, one can perform an arbitrage by buying one unit of the risky asset at time 0, and clearing the position at time T (i.e. sell one unit of the risky asset at time 1). With our general notations, this corresponds to $\phi_t = 1$ for all $t < T$. Then, the wealth process starting from a zero initial endowment satisfies

$$V_T^\phi = S_T - S_0(1+r)^T$$
$$= S_0(e^{bT}d^T - (1+r)^T) + S_0 e^{bT} d^T ((u/d)^{N_T} - 1).$$

The latter is non-negative for any $\omega \in \Omega$ under the condition $1 + r \le de^b < ue^b$, and is (strictly) positive for all $\omega \in \Omega$ such that at least one component ω_t is equal to u. Since each element of Ω has a positive probability, this is an arbitrage.

Similarly, if $de^b < ue^b \le 1 + r$, one can perform an arbitrage by selling one unit of the risky asset at time 0 and by clearing the position at T.

2. Under the condition (1.20), the quantity q defined in (1.22) belongs to $(0, 1)$. It follows that $\mathbb{Q}$ defined in (1.21) is a probability measure that is equivalent to $\mathbb{P}$ (as it assigns a positive weight to each $\omega \in \Omega$ and $\mathbb{Q}[\Omega] = 1$). It remains to show that it turns $\tilde{S}$ into a martingale. To see this, let us first compute, for $0 \le t < T$,

$$\mathbb{E}^{\mathbb{Q}}[\tilde{S}_{t+1}|\mathcal{F}_t](\omega) = \frac{\tilde{S}_t(\omega)e^b}{1+r} \{u\mathbb{Q}[\omega_{t+1} = u|\mathcal{F}_t](\omega) + d\mathbb{Q}[\omega_{t+1} = d|\mathcal{F}_t](\omega)\}$$

$$= \frac{\tilde{S}_t(\omega)e^b}{1+r} \{uq + d(1-q)\}.$$

We then observe that (1.22) is equivalent to $e^b \{uq + d(1-q)\}/(1+r) = 1$, which implies that $\tilde{S}$ is a $\mathbb{Q}$-martingale. Conversely, the above computation shows that an equivalent probability measure $\mathbb{Q}'$ such that $\tilde{S}$ is a $\mathbb{Q}'$-martingale should satisfy $\mathbb{Q}'[\omega_{t+1} = u|\mathcal{F}_t](\omega) = q$ for all $t < T$, and therefore should be equal to $\mathbb{Q}$.

3. We use the result of Sect. 1.3.4. The hedging portfolio V satisfies the backward induction

$$V_t(\omega) = \frac{qV_{t+1}(\omega)\mathbf{1}_{\{\omega_{t+1}=u\}} + (1-q)V_{t+1}(\omega)\mathbf{1}_{\{\omega_{t+1}=d\}}}{1+r}, \quad t < T,$$

with the terminal condition $V_T(\omega) = G(\omega)$, for $\omega \in \Omega$. The corresponding hedging strategy ϕ is defined by

$$\phi_t(\omega) = \frac{V_{t+1}(\omega) - (1+r)V_t(\omega)}{S_{t+1}(\omega) - (1+r)S_t(\omega)} = \mathbf{1}_{\{\omega_{t+1}=u\}} \frac{(1-q)V_{t+1}(\omega)}{S_t(\omega)(ue^b - (1+r))}.$$

4. One can assume that $G = (G_t)_{t \in T}$ takes non-negative values (otherwise we replace G_t by $G_t \vee 0$ in the following). We use the results of Sect. 1.5.1. The hedging portfolio V satisfies the backward induction

$$V_t(\omega) = \max \left\{ G_t(\omega), \frac{qV_{t+1}(\omega)\mathbf{1}_{\{\omega_{t+1}=u\}} + (1-q)V_{t+1}(\omega)\mathbf{1}_{\{\omega_{t+1}=d\}}}{1+r} \right\}, \ t < T,$$

with the terminal condition $V_T(\omega) = G(\omega)$, for $\omega \in \Omega$. The corresponding hedging strategy ϕ is defined by

$$\phi_t(\omega) = \frac{V_{t+1}(\omega) - (1+r)V_t(\omega)}{S_{t+1}(\omega) - (1+r)S_t(\omega)} = \mathbf{1}_{\{\omega_{t+1}=u\}} \frac{(1-q)V_{t+1}(\omega)}{S_t(\omega)(ue^b - (1+r))}.$$

1.2 Questions 1, 2 and 3 are applications of the definitions. In particular, this model does not exhibit any possible arbitrage. As for the questions 4, 5 and 6, it suffices to apply the algorithms of Exercise 1.1 after having extended them to the case of two risky assets (you can obviously check whether the constructed hedging strategies are actually correct, by checking that they lead to the correct replication).

1.3

1. This is a no-arbitrage condition under a no short-selling constraint.
2. Argue as in Exercise 1.1 to show that, if this condition is violated, then one can build an arbitrage by first buying and then selling one unit of S.
3. This is an immediate consequence of the condition $d < R$.
4. It turns $\tilde{S}$ into a supermartingale (and not a martingale).
5. If $\tilde{S}$ is a supermatingale under the weights associated to π, so is any discounted admissible wealth process V. Then, $V_1 \geq 0$ implies $V_1 = 0$, ruling out arbitrages.
6. Take any model with $d < u \leq R$. A short-selling strategy leads to an arbitrage, also NA$_+$ holds.

1.4 Given a process ξ, we write $\Delta \xi_{t+1}$ for $\xi_{t+1} - \xi_t$.

1. By definition, this is V_t^1/S_t.
2. We have $V_{t+1}^0 = V_t^0 + (1-\lambda)\Delta L_{t+1} - (1+\lambda)\Delta M_{t+1}$ and $V_{t+1}^1 = (V_t^1/S_t)S_{t+1} - \Delta L_{t+1} + \Delta M_{t+1}$.
3. By the above $K := \{(v^0, v^1) : v^1 + a \geq 0 \text{ and } v^0 - a - \lambda|a| \geq 0 \text{ for some } a \in \mathbb{R}\} = \{(v^0, v^1) : v^0 + v^1 - \lambda|v^1| \geq 0\}$.
4. We have

$$\langle Z_{t+1}, V_{t+1}\rangle - \langle Z_t, V_t \rangle = \Delta Z_{t+1}^0 V_t^0 + \Delta(Z^1 S)_{t+1}(V_t^1/S_t) + \langle Z_{t+1}, B_{t+1}\rangle$$

where $B_{t+1} = ((1-\lambda)\Delta L_{t+1} - (1+\lambda)\Delta M_{t+1}, -\Delta L_{t+1} + \Delta M_{t+1}) \in -K_t$ by construction. Hence, $\langle Z_{t+1}, B_{t+1}\rangle \leq 0$ by definition of K^o. The remaining part is a martingale by definition of $\mathcal{Z}$.

5. If $V_T \in \mathbb{R}_+^d$ with $\mathbb{P}[V_T \neq 0] > 0$ then $\langle Z_T, V_T \rangle \geq 0$ and $\mathbb{P}[\langle Z_T, V_T \rangle \neq 0] > 0$. Since $\langle Z, V \rangle$ is a supermartingale, this implies that $\langle Z_0, V_0 \rangle > 0$. Since $V_0 = V_{0-} + B_0$ with $B_0 \in -K_0$, and since $Z_0 \in K_0^o$, this entails that $\langle Z_0, V_{0-} \rangle > 0$, a contradiction to the fact that $V_{0-} = 0$.

1.5

1. Direct from computations.
2. Argue as in Exercise 1.1. Because of the fixed cost, we only need $1 + r \in [d, u]$ in place of $1 + r \in (d, u)$.
3. This is obvious.
4. We deduce from the above that the condition $1 + r \in [d, u]$ should hold.
5. a. Direct from computations.
   b. Use the previous result and recall that $\mathbb{E}^{\mathbb{Q}}[V_1^{0,\theta}/(1 + r)] \leq 0$.
   c. If this is not true, the above would be violated by some θ.
   d. Take the weights q as in c. above.
   e. Write that $q, (1 - q) \in (0, 1)$.
6. a. Use the perfect hedging strategy in the model without fixed costs, and add the fixed costs.
   b. Use the perfect hedging strategy in the model without fixed costs, and add the fixed costs.
   c. It must be $[p^-, p^+]$. Argue as in the proof of Theorem 1.4, but observe that here p^+ and p^- just allows one to replicate perfectly G and $-G$.

1.6

1. $\tilde{V}_t^{v,\phi} = v + \sum_{s<t} \sum_{T \leq T^*} \phi_s(T)(\tilde{B}_{s+1}(T) - \tilde{B}_s(T))$.
2. Fix $t < T \leq T^*$.

   a. If not, an arbitrage can be constructed.
   b. Definition 1.1.
   c. Since $u \geq d$, $\mathbb{P}[d_t(T) > 1] = 1$ implies $B_{t+1}(T) > B_t(T)/B_t(t + 1) \mathbb{P}$ – a.s., and we have shown that $B_t(t + 1) = B_t/B_{t+1}$ in question 1. The arbitrage is then obtained by buying $B(T)$ at t and selling it at $t + 1$.
   d. Since d_t is known at t, we do the above strategy on $\{d_t(T) \geq 1\} \subset \{u_t(T) > 1\}$ only. The gain is made on $\{d_t(T) \geq 1\} \cap \{B_{t+1}(T) = u_t(T)B_t(T)/B_t(t + 1)\}$, recall that $\pi_t(T) > 0$.
   e. Otherwise, we do the same except that we sell and buy back instead of buying first.
   f. A necessary condition is $d_t(T) < 1 < u_t(T) \mathbb{P}$ – a.s. if $t < T - 1$.

3. a. This follows from $u_t(T) > 1 > d_t(T) \mathbb{P}$ – a.s. for all $t < T$, $u_{T-1}(T) = d_{T-1}(T) = 1$ and $B_t(t + 1) = B_t/B_{t+1}$.
   b. The definition of π implies

   $$\mathbb{E}[H_{t+1}(T^*) \,|\, \mathcal{F}_t]$$

$$= H_t(T^*)\mathbb{E}\Big[\Big(\frac{q_{t-1}(T^*)}{\pi_{t-1}(T^*)}\mathbf{1}_{A_{t-1}(T^*)} + \frac{(1-q_{t-1}(T^*))}{1-\pi_{t-1}(T^*)}\mathbf{1}_{A^c_{t-1}(T^*)}\Big)\ |\mathcal{F}_t\Big]$$

$$= H_t(T^*)(q_{t-1}(T^*) + (1-q_{t-1}(T^*))).$$

  c. We have $H_{T^*}(T^*) > 0$.

  d. This is the only one whose weights satisfy the condition of question 4a.

  e. The first assertion follows from the first fundamental theorem of asset pricing. Writing down what it means leads to the second assertion.

4. Use simple algebra and the fact that $B_t(t+1) = (1+r_t)^{-1}$.

1.7

1. This a consequence of $C \cap \{e_i\} = \emptyset$.
2. (i) implies that $\lambda\langle x, Y_i\rangle < \langle e_i, Y_i\rangle$ for all $\lambda > 0$ and $x \in C$. By sending $\lambda \to \infty$, we obtain $\langle x, Y_i\rangle \le 0$. Since $0 \in C$ by (iii), the required result holds.
3. $Y_i \in [0,\infty)^d$ because of the above and the fact that $-\mathbf{1}_{Y_i<0} \in C$ by (ii). Then, $Y_i^i = \langle e_i, Y_i\rangle > 0$ by the above.
4. Take $Y := \sum_{i=1}^d Y_i > 0$ as all the terms are non-negative and $Y_i^i > 0$ for each $i \le d$.
5. The arguments used here are similar to the ones used in the proof of Proposition 1.2. Considering a one step model, we can interpret C as the set of the discounted values of the claims that can be super-hedged from 0, and Y as the density (up to a normalisation) of a martingale measure.

1.8 Note that our assumptions implies that the risk-free interest rate is zero.

1. a. Y^θ is the Snell-envelope of $G(\cdot, \theta)$ and X^τ the Snell-envelope of $G(\tau, \cdot)$.
  b. Apply the Doob-Meyer decomposition to the supermartingales Y^θ and $-X^\tau$.
  c. $X^\tau_\theta \le G(\tau, \theta) \le Y^\theta_\tau$ by construction. One can then iterate backward by using the dynamic programming algorithm they satisfy.
2. a. Take $\theta \in \mathcal{T}$ such that $p > Y^\theta_0$ and construct an arbitrage by following the proof of (ii) in Theorem 1.9 but by buying back the option at θ.
  b. Argue as above but exercise at τ such that $p < X^\tau_\theta$.
3. a. Use the fact that $Z_t = \min\{L_t, \ \mathbb{E}[Z_{t+1}|\mathcal{F}_t]\}$ on $\{t \le \hat\tau\}$, while $Z_t = \max\{\ell_t, \ \mathbb{E}[Z_{t+1}|\mathcal{F}_t]\}$ on $\{t \le \hat\theta\}$.
  b. The above implies that $Z_{\cdot\wedge\hat\vartheta}$ is a martingale with $Z_{\hat\vartheta} = \ell_\vartheta \mathbf{1}_{\{\vartheta=\hat\tau\}} + L_\vartheta \mathbf{1}_{\{\vartheta<\hat\tau\}}$.
4. a. Let V be a wealth process starting from Z_0 and stopped at the call-back time θ, such that $V_{\hat\tau\wedge\theta} \ge G(\hat\tau, \hat\tau \wedge \theta)$ $\mathbb{P}$-a.s. Then, $Z_0 \ge \mathbb{E}[V_{\hat\tau\wedge\theta}] \ge \mathbb{E}[G(\hat\tau, \hat\tau \wedge \theta)] \ge \mathbb{E}[Z_{\hat\tau\wedge\theta}] \ge Z_0$ by definition of $\hat\theta$ and since $Z_{\hat\tau\wedge\cdot}$ is a submartingale. Hence, $V_{\hat\tau\wedge\theta} = G(\hat\tau, \hat\tau \wedge \theta)$ $\mathbb{P}$-a.s. The other way around for the buyer.
  b. Use the Doob-Meyer decomposition and 3.a.
  c. Argue as in Sect. 1.5.5.
  d. At $\hat\tau$.

1.9

1. Y_t^1 is the price at time t of the American option of payoff G, $\tilde{\tau}^1$ is the exercise time.
2. We first use that $Y^1 \geq G \geq 0$ and $Y^{n+1} \geq Y^n$. The remaining is by construction.
3. a. C_s^ν is the cumulated amount at s that has been paid to the buyer if he uses the exercise times ν.
   b. Note that $C_{\tau_N}^\nu = G_{\tau_N}$, hence we have to show that $V_{\tau_N}^{Y_0^N, \phi} \geq G_{\tau_N} + \mathbb{E}\left[Y_{\tau_N+1}^{N-1} \mid \mathcal{F}_{\tau_N}\right]$. Since Y^N is the Snell envelope of $G + \mathbb{E}\left[Y_{\cdot+1}^{N-1} \mid \mathcal{F}.\right]$ on $[0, T-N+1]$ and $\tau_N \leq T-N+1$ by construction, the result follows from the Doob-Meyer decomposition of Y^N and the martingale representation theorem.
   c. By the above and the martingale representation theorem, there exists $\bar{\phi}_{\tau_N} \in \mathbb{L}^0(\mathcal{F}_{\tau_n})$ such that, $\mathbb{E}\left[Y_{\tau_N+1}^{N-1} \mid \mathcal{F}_{\tau_N}\right] + \bar{\phi}_{\tau_N}(S_{\tau_N+1} - S_{\tau_N}) \geq Y_{\tau_N+1}^{N-1}$. This is the same as the first required result. As for the second, Y^{N-1} is the Snell envelope of $G + \mathbb{E}\left[Y_{\cdot+1}^{N-2} \mid \mathcal{F}.\right]$ on $[0, T-N+2]$ and we can then find $\hat{\phi}$ such that $Y_{\tau_N+1}^{N-1} + \sum_{\tau_N+1 \leq s < \tau_{N-1}} \hat{\phi}_s(S_{s+1} - S_s) \geq G_{\tau_{N-1}} + \mathbb{E}\left[Y_{\tau_{N-1}+1}^{N-2} \mid \mathcal{F}_{\tau_{N-1}}\right]$. Then, we combine with the above and use the definition of C^ν.
   d. We have started the induction in 3.c. Go on similarly between $N-2$ and $N-3$, and so on.
   e. The above shows that $V_s^{Y_0^N, \hat{\phi}} - C_s^\nu \geq 0$, $\forall s \leq T$, $\forall \nu \in \mathcal{T}^N$.
4. a. Since $Y_{T-n+1}^n = Y_{T-n+1}^{n+1}$, the fact that $G \geq 0$ and the definition of Y^{n+1} imply that $Y_{T-n}^{n+1} = G_{T-n} + \mathbb{E}\left[Y_{T-n+1}^{n-1} \mid \mathcal{F}_{T-n}\right]$.
   b. Argue as in the proof of Proposition 1.10.
   c. Let $\phi \in \mathcal{A}$ and $\nu \in \mathbb{R}$ be such that $V_s^{\nu, \phi} \geq C_s^{\hat{\nu}}$ for all $s \leq T$.

      i. Use the definition of C^ν and $G_{\hat{\tau}_1} = Y_{\hat{\tau}_1}^1$.
      ii. Take expectation given $\mathcal{F}_{\hat{\tau}_2}$ on both sides and use the martingale property of $Y_{\cdot \wedge \hat{\tau}_1}^1$ to obtain $V_{\hat{\tau}_2}^{\nu, \phi} \geq C_{\hat{\nu}_2}^{\hat{\nu}} + \mathbb{E}\left[Y_{\hat{\tau}_2+1}^1 \mid \mathcal{F}_{\hat{\tau}_2}\right]$. Then, use the definition of $\hat{\tau}_2$.
      iii. Repeat the above argument forward on $n = 2, 3, \cdots, N$.
5. Combine 4.c.iii with 3.e.
6. Express the hedging strategy in terms of the representation of the martingale parts of the different Y^n.
7. The same analysis as in Sect. 1.5.5 can be done by induction: Y^{N-1} is the price of the American option of payoff $G + \mathbb{E}\left[Y_{\cdot+1}^{N-2} \mid \mathcal{F}.\right]$ and maturity $T-N+2$. The buyer should exercise at the times $\hat{\nu}$.

1.10

1. As in Sect. 1.1.
2. To $\mathbb{F}$ only.
3. $\mathbb{E}^\mathbb{Q}[\mathbb{E}^\mathbb{Q}[\tilde{V}_{t+1}^{\nu, \phi} \mid \mathcal{G}_{t+1}] \mid \mathcal{G}_t] = \mathbb{E}^\mathbb{Q}[\tilde{V}_t^{\nu, \phi} \mid \mathcal{G}_t] + \langle \phi_t, \mathbb{E}^\mathbb{Q}[\tilde{S}_{t+1} - \tilde{S}_t \mid \mathcal{G}_t]\rangle = \mathbb{E}^\mathbb{Q}[\tilde{V}_t^{\nu, \phi} \mid \mathcal{G}_t]$.

4. If such a $\mathbb{Q}$ exists, then $\tilde{V}_0^{v,\phi} = \mathbb{E}^{\mathbb{Q}}[V_T^{v,\phi}]$, by the above. This rules out any arbitrage opportunity. Hence a sufficient condition[7] is $\mathcal{M}(\mathbb{G}) \neq \emptyset$.

1.11

1. Just do the accounting and re-arrange terms.
2. We want to find q such that $V_1^q \geq 0$ and $V_1^q(\omega_i) > 0$ for at least one i. Use the previous question and do the analysis on q for ω_1 and ω_2.
3. Just do the computations using the assumption. If $q = 0$ then $V_1^q = 0$, hence the above shows that we cannot make an arbitrage with $q \leq 0$. But $V_1^q(\omega_2) < 0$ if $q > 0$. A sufficient condition for NA is then $\alpha_2 \leq 1 \leq \alpha_1 \leq 4$.
4. The necessary and sufficient condition $\alpha_2 \leq 1 \leq \alpha_1$ by the previous questions. Yes, but it applies only to the strategy $q = 0$.
5. a. This is obtained by immediate computations.
   b. Pick up any $q < 0$ such that $m < |q| < M$ and use the definitions of m and M to show that it leads to an arbitrage.
   c. By selling an appropriate amount, we push the price down sufficiently to make a gain when buying back the position.

1.12

1. Just do the accounting and re-arrange terms.
2. It means that we can start with 0, end up with a 0 position in S at 1 and have a probability of making a strictly positive gain without taking any risk.
3. $C_1^\phi(\omega_i) = \phi_0 \alpha_i$ when $\phi_0 + \phi_1 = 0$. Then, $\alpha_2 < 0 < \alpha_1$ is clearly equivalent to the fact that $C_1^\phi \geq 0$ implies $C_1^\phi = 0$ for all ϕ such that $\phi_0 + \phi_1 = 0$.
4. It is $\inf\{p : \exists \phi \text{ s.t. } p + C_1^\phi \geq [S_1^\phi - K]^+ \text{ and } \phi_0 + \phi_1 = 0\}$. We have $S_1^\phi = S_1^0$. Hence, combined with the previous question, this implies that the super-hedging price is $\inf\{p : \exists \phi_0 \text{ s.t. } p + \phi_0(S_1^0 - S_0^0) \geq [S_1^0 - K]^+\}$. It is the hedging price in a one period tree. This p and $\phi_0 = -\phi_1$ are obtained by solving $p + \phi_0(S_1^0 - S_0^0) = [S_1^0 - K]^+$ for ω_1 and ω_2. We do not see any effect of the price impact...
5. a. The super-hedging price is

$$\inf\{p : \exists \phi \text{ s.t. } p + C_1^\phi \geq -K\mathbf{1}_{\{S_1 \geq K\}} \text{ and } \phi_0 + \phi_1 = \mathbf{1}_{\{S_1 \geq K\}}\}.$$

   Then, expand all the quantities.
   b. This ϕ_1 can be chosen at time 1, we can choose the minimal $\phi_1 + \phi_0$ at 1 depending on ω_i, this is γ^i. Then, the max over i comes from the fact that the inequalities should hold for each ω_i. This produces a hedging price which depends on ϕ_0, on which we can optimise again.
6. a. $\gamma^1 \geq \mathbf{1}_{\{S_{0-}+\alpha_1+\gamma^1\lambda \geq K\}}$ so $\gamma^1 \geq 0$, and $\gamma^1 \geq \mathbf{1}_{\{S_{0-}+\alpha_1+\gamma^1\lambda \geq K\}} = 1$ since $S_{0-} + \alpha_1 \geq K$. Hence, $\gamma^1 \geq 1$. On the other hand, F^1 in non-decreasing in γ_1.

---

[7]It is indeed equivalent. This can be shown by similar arguments as in the case $\mathbb{G} = \mathbb{F}$ studied above.

b. Use similar arguments as above.

c. Just do the computations.

d. Buy $\hat{\phi}_0$ units at 0 and buy $\hat{\gamma}^1 \mathbf{1}_{\{\omega=\omega_1\}} + \hat{\gamma}^2 \mathbf{1}_{\{\omega=\omega_2\}}$ at 1.

e. Now the price impact plays a real role. Prices can be compared by computing $\hat{\phi}_0$.

Chapter 2
Continuous Time Models

In this chapter, we extend the results obtained in discrete time markets to a continuous time setting. We work with Itô semimartingale models in which the risky assets are modeled as a diffusion driven by a Brownian motion. Note however that most of the results presented below remain true in much more general setting, see e.g. [23] and [24]. The most technical results will be stated without proofs.

2.1 Financial Asset and Portfolio Strategies

From now on, we work on a complete probability space $(\Omega, \mathcal{F}, \mathbb{P})$ supporting a n-dimensional standard Brownian motion $W = (W^1, \ldots, W^n)$. We denote by $\mathbb{F} = (\mathcal{F}_t)_{t \in \mathbb{T}}$ the augmented raw filtration generated by W. In this chapter, $\mathbb{T} = [0, T]$ in which $T > 0$ is a fixed time horizon. We also assume that $\mathcal{F}_T = \mathcal{F}$, i.e. all the randomness is generated by W.

2.1.1 Financial Assets

As in discrete time, we assume that there exists a *risk-free interest rate*. It is modelled as a $\mathbb{F}$-*predictable*[1] process $r = (r_t)_{t \in \mathbb{T}}$. One dollar invested at time s at this rate produces at time $t > s$ an amount equal to $e^{\int_s^t r_u du}$.

---

[1]This means that $(t, \omega) \mapsto r_t(\omega)$ is measurable with respect to the tribe generated by left-continuous processes $\xi = (\xi_t)_{t \leq T}$ such that ξ_t is $\mathcal{F}_t$-measurable for each $t \leq T$, i.e. such that ξ is $\mathbb{F}$-*adapted*. This implies in particular that the process r is $\mathbb{F}$-*progressively measurable* in the sense that $(t, \omega) \in [0, s] \times \Omega \mapsto r_t(\omega)$ is $\mathcal{B}_{[0,s]} \otimes \mathcal{F}_s$-measurable for each s, in which $\mathcal{B}_{[0,s]}$ is the Borel tribe of $[0, s]$.

© Springer International Publishing Switzerland 2016
B. Bouchard, J.-F. Chassagneux, *Fundamentals and Advanced Techniques in Derivatives Hedging*, Universitext, DOI 10.1007/978-3-319-38990-5_2

This leads to the following definition of the *discounting process* β:

$$\beta_t := e^{-\int_0^t r_u du} \quad, \; t \in \mathbb{T}.$$

For sake of simplicity, we shall assume that

$$\sup_{t \in \mathbb{T}} |r_t| < \infty \quad \mathbb{P} - a.s.$$

The market also consists of d *risky assets*, whose price process $S = (S^1, \ldots, S^d)$ is given by an Itô semimartingale of the form

$$S_t = S_0 + \int_0^t b_s ds + \int_0^t \sigma_s dW_s \quad, \; t \in \mathbb{T}. \tag{2.1}$$

Here, $S_0 \in \mathbb{R}_+^d$ is given, (b, σ) is a predictable process with values in $\mathbb{R}^d \times \mathbb{M}^{d,n}$. Such a process is well defined under the additional integrability condition

$$\int_0^T \left(\|b_t\| + \|\sigma_t\|^2 \right) dt < \infty \quad \mathbb{P} - a.s. \tag{2.2}$$

Note that the above representation is unique (see [42]).

Proposition 2.1 *Let (b^i, σ^i), $i = 1, 2$, be two predictable processes with values in $\mathbb{R}^d \times \mathbb{M}^{d,n}$ such that*

$$\int_0^T \left(\|b_t^i\| + \|\sigma_t^i\|^2 \right) dt < \infty \quad \mathbb{P} - a.s., \; i = 1, 2.$$

If $\mathbb{P} - a.s.$ for all $t \in \mathbb{T}$

$$\int_0^t b_s^1 ds + \int_0^t \sigma_s^1 dW_s = \int_0^t b_s^2 ds + \int_0^t \sigma_s^2 dW_s$$

then

$$(b^1, \sigma^1) = (b^2, \sigma^2) \quad dt \times d\mathbb{P}-a.e.$$

In the following, we use the notation

$$\tilde{S} = \beta S$$

for the discounted price process. As a simple consequence of Itô's lemma,

$$\tilde{S}_t = S_0 + \int_0^t (\tilde{b}_s - r_s \tilde{S}_s) ds + \int_0^t \tilde{\sigma}_s dW_s \quad, \; t \in \mathbb{T}, \tag{2.3}$$

in which $(\tilde{b}, \tilde{\sigma}) = \beta(b, \sigma)$.

2.1.2 Portfolio Strategies

A portfolio strategy is a $\mathbb{F}$-predictable process (α, ϕ) with values in $\mathbb{R} \times \mathbb{R}^d$. For each $i \leq d$ and $t \in \mathbb{T}$, ϕ_t^i represents the number of shares of the asset S^i in the portfolio while α_t is the amount of money invested at the risk-free rate.

The instantaneous return of the amount α_t is $\alpha_t r_t dt$, while the amount $\phi_t^i S_t^i$ invested in the risky asset S^i induces a gain equal to $\phi_t^i dS_t^i$.

By similarity with discrete time models, we shall say the strategy is *self-financing* if the instantaneous variation of the wealth process V depends only on the returns generated by the investment in the risky assets and at the risk free rate, namely if[2]

$$dV_t = r_t \alpha_t dt + \phi_t' dS_t.$$

Since the value of the portfolio is

$$V_t = \alpha_t + \langle \phi_t, S_t \rangle,$$

this implies that

$$\alpha_t = V_t - \langle \phi_t, S_t \rangle,$$

and therefore

$$dV_t = (r_t V_t + \langle \phi_t, b_t - r_t S_t \rangle) \, dt + \phi_t' \sigma_t dW_t. \tag{2.4}$$

Moreover, it follows from Itô's lemma that the discounted wealth process $\tilde{V} = \beta V$ evolves according to

$$d\tilde{V}_t = \langle \phi_t, \tilde{b}_t - r_t \tilde{S}_t \rangle dt + \phi_t' \tilde{\sigma}_t dW_t = \phi_t' d\tilde{S}_t. \tag{2.5}$$

As in discrete time the amount α invested in the risk free asset is fully determined by (V, ϕ), and a strategy can be simply identified to ϕ. From now on, we denote by $V^{v,\phi}$ (resp. $\tilde{V}^{v,\phi}$) the value of the (resp. discounted) portfolio process induced by a strategy ϕ and valued v at 0.

In order to be able to give a sense to (2.4) and (2.5), we restrict to the predictable processes ϕ such that

$$\int_0^T \left(\| \langle \phi_t, b_t \rangle \| + \| \langle \phi_t, S_t \rangle \| + \| \phi_t' \sigma_t \|^2 \right) dt < \infty \ \mathbb{P} - \text{a.s.},$$

---

[2]More rigorously, this can be seen as the limit of the wealth dynamics in a discrete time model when the duration between two trading times goes to 0. This follows from the construction of the stochastic integral. Note that any continuous time model, possibly with frictions or other non standard feature, should come from the limit of a discrete time setting to be considered as an approximation of the reality, in which trading cannot be continuous.

recall that r is $\mathbb{P}$ − a.s. bounded. Moreover, in order to avoid the so-called *doubling strategies*, see the introduction in [23] and Exercise 2.18 below, that can lead to arbitrages if one can borrow money without restriction to cover intermediate losses, we impose a *finite credit line condition* of the form

$$\tilde{V}_t^{0,\phi} \geq -c_\phi^0 - \sum_{i=1}^d c_\phi^i \tilde{S}_t^i \quad \mathbb{P} - \text{a.s. for all } t \in \mathbb{T} \tag{2.6}$$

in which $c_\phi^0, \ldots, c_\phi^d$ are constants that can depend on ϕ. It means that potential intermediate losses are controlled by a certain bound (possibly depending on S) that should be known at time 0.

A strategy ϕ is said to be an *admissible strategy* if the above conditions are satisfied. The collection of admissible strategies is denoted $\mathcal{A}$.

Remark 2.2 The condition (2.6) implies that

$$\mathbb{P}\left[\forall\, q \text{ rational number in } \mathbb{T} \,:\, \tilde{V}_q^{0,\phi} \geq -c_\phi^0 - \sum_{i=1}^d c_\phi^i \tilde{S}_q^i\right] = 1 \,.$$

Since $\tilde{V}^{0,\phi}$ and $\tilde{S}$ have $\mathbb{P}$ − a.s. continuous paths,[3] this condition is equivalent to

$$\tilde{V}_t^{0,\phi} \geq -c_\phi^0 - \sum_{i=1}^d c_\phi^i \tilde{S}_t^i \quad \text{for all } t \in \mathbb{T} \; \mathbb{P} - \text{a.s.}$$

2.2   Absence of Arbitrage and Martingale Measures

The Definition 1.1 of the absence of arbitrage opportunities of Chap. 1 can be extended to our continuous time setting in a natural way.

Definition 2.3 (NA) There is no *arbitrage opportunity* if

(NA) $: \forall\, \phi \in \mathcal{A}, V_T^{0,\phi} \geq 0 \,\mathbb{P} - \text{a.s.} \;\Rightarrow\; V_T^{0,\phi} = 0 \,\mathbb{P} - \text{a.s.}$

2.2.1   Necessary Condition

Let us first discuss a necessary condition that characterises the absence of arbitrage opportunities by the existence of a *risk premium*.

---

[3]See [51, Corollary of Theorem 30, Chapter IV].

Theorem 2.4 *If* (NA) *holds, then there exists a* $\mathbb{F}$-*predictable process* λ *such that*

$$\sigma\lambda = b - rS \quad dt \times d\mathbb{P}\text{-a.e.}$$

This process is called: the risk premium.

The proof of the above relies on the fact that the return of a *risk free portfolio*, i.e. such that $\phi'\sigma \equiv 0$, is necessarily given by r under the condition (NA).

Proposition 2.5 *If* (NA) *holds, then one cannot find a process* $\phi \in \mathcal{A}$ *such that*

$$\phi'\sigma = 0 \quad dt \times d\mathbb{P}\text{-a.e.} \quad and \quad \mathbb{P}\left[\int_0^T \mathbf{1}_{\{\phi_t'(b_t - r_t S_t) > 0\}} dt > 0\right] > 0.$$

Proof Assume to the contrary that there exists some $\phi \in \mathcal{A}$ such that

$$\phi'\sigma = 0 \quad dt \times d\mathbb{P}\text{-a.e.} \quad and \quad \mathbb{P}\left[\int_0^T \mathbf{1}_{\{\phi_t'(b_t - r_t S_t) > 0\}} dt > 0\right] > 0.$$

Let us then define $\bar{\phi} = \phi \mathbf{1}_{\{\phi'(b - rS) > 0\}}$. In view of (2.5),

$$\tilde{V}_t^{0,\bar{\phi}} = \int_0^t \phi_s'(\bar{b}_s - r_s \tilde{S}_s) \mathbf{1}_{\{\phi_s'(b_s - r_s S_s) > 0\}} ds \geq 0$$

and $\mathbb{P}\left[\tilde{V}_T^{0,\phi} > 0\right] > 0$. Since $\bar{\phi} \in \mathcal{A}$, this contradicts (NA). $\square$

To complete the proof of Theorem 2.4, we now appeal to *Farkas' Lemma*.

Lemma 2.1 (Farkas' Lemma) *Fix* $(a, A) \in \mathbb{R}^d \times \mathbb{M}^{d,n}$. *If there is no* $x \in \mathbb{R}^d$ *such that* $x'A = 0$ *and* $x'a > 0$, *then there exists* $y \in \mathbb{R}^n$ *such that* $a = Ay$. *Moreover,* y *depends on* (a, A) *in a Borel-measurable way.*

Proof Let $\xi_1, \ldots, \xi_n$ denotes the columns of A. The assumption of the lemma means that $\text{Vect}\{\xi_1, \ldots, \xi_n\}^\perp \subset \text{Vect}\{a\}^\perp$ so that $a \in \text{Vect}\{\xi_1, \ldots, \xi_n\}$. We can thus find $y \in \mathbb{R}^n$ such that $a = Ay$. After possibly considering an orthogonal basis of $\text{Vect}\{\xi_1, \ldots, \xi_n\}$ by a Gram and Schmidt's algorithm, we can assume that the ξ_is are orthogonal. Then, $y^i = \langle a, \xi_i\rangle / \|\xi_i\|^2$ if $\xi_i \neq 0$, $y^i = 0$ otherwise. $\square$

Proof of Theorem 2.4 By Proposition 2.5, the following holds $dt \times d\mathbb{P}$-a.e.: $x'\sigma_t = 0$ implies $x'(b_t - r_t S_t) = 0$. Hence, it follows from Lemma 2.1 that, $dt \times d\mathbb{P}$-a.e., there exists a λ_t such that $(b_t - r_t S_t) = \sigma_t \lambda_t$. Since λ is a measurable function of the predictable process $(b - rS, \sigma)$, it is predictable. $\square$

Remark 2.6

1. If $n = d$ and σ_t is invertible then $\lambda_t = \sigma_t^{-1}(b_t - r_t S_t)$.
2. If $\sigma_t'\sigma_t$ is invertible then $\lambda_t = (\sigma_t'\sigma_t)^{-1}\sigma_t'(b_t - r_t S_t)$.

2.2.2 Sufficient Condition

We now provide a sufficient condition, the existence of a sufficiently integrable *risk premium* λ, recall the definition in Theorem 2.4.

Unlike in the discrete time setting, it is not always possible to find an element of the set $\mathcal{M}(\tilde{S})$ of equivalent probability measures $\mathbb{Q} \sim \mathbb{P}$ such that $\tilde{S}$ is a $\mathbb{Q}$-martingale.[4] Otherwise stated, there does not always exist an *equivalent martingale measure*. In general, one needs to consider a bigger set: the collection $\mathcal{M}_{\mathrm{loc}}(\tilde{S})$ of probability measures $\mathbb{Q} \sim \mathbb{P}$ such that $\tilde{S}$ is a $\mathbb{Q}$-*local martingale*, i.e. such that there exists a sequence of $\mathbb{F}$-stopping times[5] $(\tau_k)_{k \geq 1}$ such that $\tau_k \to \infty$ $\mathbb{Q}$-a.s and $(\tilde{S}_{t \wedge \tau_k})_{t \in \mathbb{T}}$ is a $\mathbb{Q}$-martingale, for all $k \geq 1$.

This is the so-called set of *equivalent local martingale measures*.

Theorem 2.7 *If there exists a risk premium λ such that the solution H to*

$$H_t = 1 - \int_0^t H_s \lambda_s' dW_s, \ t \in [0, T],$$

is a $\mathbb{P}$-martingale, then the equivalent measure $\mathbb{Q} \sim \mathbb{P}$ of density H_T with respect to $\mathbb{P}$ belongs to $\mathcal{M}_{\mathrm{loc}}(\tilde{S})$.

If $\tilde{V}^{v,\phi}$ is bounded from below by a $\mathbb{Q}$-martingale, then $\tilde{V}^{v,\phi}$ is a $\mathbb{Q}$-supermartingale, $\forall \ (v, \phi) \in \mathbb{R} \times \mathcal{A}$. It is in particular the case if $\mathbb{Q} \in \mathcal{M}(\tilde{S})$, and then (NA) is satisfied.

Remark 2.8 By applying Itô's Lemma, one can verify that the process H introduced in Theorem 2.7 is given by

$$H_t = e^{-\frac{1}{2} \int_0^t \|\lambda_s\|^2 ds - \int_0^t \lambda_s' dW_s}.$$

Saying that H is a martingale is then equivalent to $\mathbb{E}[H_T] = 1$. A sufficient condition for this is the *Novikov's condition*

$$\mathbb{E}\left[e^{\frac{1}{2} \int_0^t \|\lambda_s\|^2 ds} \right] < \infty,$$

see [42, Proposition 5.12]. It is obviously satisfied if λ is bounded.

The proof of Theorem 2.7 relies on Girsanov's theorem and a few technical lemmata.

---

[4]We recall that X is a $\mathbb{P}$-*supermartingale* if for all $s \leq t \leq T : X_t^- \in \mathbb{L}^1(\mathbb{P})$ and $\mathbb{E}[X_t \mid \mathcal{F}_s] \leq X_s$ $\mathbb{P}$ − a.s. The process X is a $\mathbb{P}$-*martingale* if X and $-X$ are both supermartingales.

[5]We recall that a $\mathbb{F}$-*stopping time* τ is a non-negative random variable such that $\{\tau \leq t\} \in \mathcal{F}_t$ for all $t \in \mathbb{T}$.

Theorem 2.9 (Girsanov's theorem) *Let ξ be an adapted process that is $\mathbb{P} - a.s.$ square integrable and such that the solution H to*

$$H_t = 1 - \int_0^t H_s \xi_s' dW_s$$

is a $\mathbb{P}$-martingale, then the process $W^{\bar{\mathbb{Q}}}$ defined by

$$W^{\bar{\mathbb{Q}}} = W + \int_0^\cdot \xi_s ds$$

is a Brownian motion under the measure $\bar{\mathbb{Q}}$ with density H_T with respect to $\mathbb{P}$.

Proof See [41, Theorem 5.1]. $\square$

We now provide two technical results.

Lemma 2.2 *If ξ is $\mathbb{P} - a.s.$ square integrable and $\bar{W}$ is a n-dimensional Brownian motion under the measure $\bar{\mathbb{Q}} \sim \mathbb{P}$, then the process X defined by*

$$X_t = \int_0^t \xi_s' d\bar{W}_s \quad , t \in \mathbb{T} \,,$$

is a $\bar{\mathbb{Q}}$-local martingale.

Proof See [42, Proposition 2.24]. $\square$

Theorem 2.10 (Optional sampling) *Let $\bar{\mathbb{Q}} \sim \mathbb{P}$ and X be a $\bar{\mathbb{Q}}$- super-martingale. If τ_1 and τ_2 are two bounded stopping times such that $\tau_1 \leq \tau_2$ $\bar{\mathbb{Q}}$-a.s., then*

$$\mathbb{E}^{\bar{\mathbb{Q}}}[X_{\tau_2} \mid \mathcal{F}_{\tau_1}] \leq X_{\tau_1} \,.$$

In particular, if τ is a bounded stopping time, then the stopped process $(X_{t \wedge \tau})_{t \in \mathbb{T}}$ is a $\bar{\mathbb{Q}}$-supermartingale.

Proof All martingales are continuous since the filtration is generated by a Brownian motion, see [51, Chapter IV, Corollary 1]. Our first assertion is thus a consequence of [51, Chapter 1, Theorem 16]. The second assertion then follows immediately. $\square$

Remark 2.11 It is clear that the above can be applied to martingales as well, by definition of martingales in terms of supermartingales.

Lemma 2.3 *Let $\bar{\mathbb{Q}} \sim \mathbb{P}$, X be a $\bar{\mathbb{Q}}$-local super-martingale and M be a $\bar{\mathbb{Q}}$-martingale such that $X_t \geq M_t$ $\forall\, t \in \mathbb{T}$ $\mathbb{P}$-a.s. Then, X is a $\bar{\mathbb{Q}}$-supermartingale.*

Proof Let $(\tau_k)_{k \geq 1}$ be a sequence of stopping times such that $\tau_k \to \infty$ $\bar{\mathbb{Q}}$-a.s and $(X_{t \wedge \tau_k})_{t \in \mathbb{T}}$ is a $\bar{\mathbb{Q}}$-martingale, for all $k \geq 1$. Fix $t \geq s$. For all $k \geq 1$, we deduce from

Theorem 2.10 that

$$\mathbb{E}^{\bar{\mathbb{Q}}}[X_{t \wedge \tau_k} - M_{t \wedge \tau_k} \mid \mathcal{F}_s] \le X_{s \wedge \tau_k} - M_{s \wedge \tau_k} .$$

Since $X - M \ge 0$ et $(X, M)._{\cdot \wedge \tau_k} \to (X, M)$ $\bar{\mathbb{Q}}$-a.s., Fatou's Lemma implies

$$\mathbb{E}^{\bar{\mathbb{Q}}}[X_t - M_t \mid \mathcal{F}_s] \le X_s - M_s .$$

Since M is a $\bar{\mathbb{Q}}$-martingale, this provides the required result. $\square$

Proof of Theorem 2.7 Theorem 2.9 implies that $W^{\mathbb{Q}}$ defined by

$$W^{\mathbb{Q}} = W + \int_0^{\cdot} \lambda_s ds$$

is a Brownian motion under the equivalent measure $\mathbb{Q}$ with density H_T. By (2.3),

$$\tilde{S}_t = S_0 + \int_0^t \tilde{\sigma}_s dW_s^{\mathbb{Q}} \quad , t \in \mathbb{T} ,$$

and the condition (2.2) implies that $\tilde{S}$ is a $\mathbb{Q}$-local martingale. In view of (2.5), the dynamics of the discounted wealth process is

$$\tilde{V}_t^{0,\phi} = \int_0^t \phi_s' \tilde{\sigma}_s dW_s^{\mathbb{Q}} .$$

Lemmas 2.2 and 2.3 then imply that the discounted wealth process is a $\mathbb{Q}$-super-martingale whenever it is bounded from below by a $\mathbb{Q}$-martingale. Given the admissibility condition (2.6) and Remark 2.2, it is the case when $\mathbb{Q} \in \mathcal{M}(\tilde{S})$. This implies that $\mathbb{E}^{\mathbb{Q}}\left[\tilde{V}_T^{0,\phi} \right] \le 0$ and therefore that $\tilde{V}_T^{0,\phi} = 0 \; \mathbb{P}-$a.s. if $\tilde{V}_T^{0,\phi} \ge 0 \; \mathbb{P}-$a.s. The condition (NA) is thus satisfied. $\square$

2.2.3 Necessary and Sufficient Condition

In general, the condition (NA) is not enough to obtain the existence of an element of $\mathcal{M}(\tilde{S})$ and one needs to consider a slightly stronger condition, based on a weaker notion of arbitrage. The most natural and popular one is the *no free-lunch with vanishing risk* condition that was first proposed by [23].

In the following, we denote by $\mathcal{A}_b$ the set of strategies $\phi \in \mathcal{A}$ such that $\tilde{V}^{0,\phi}$ (essentially) bounded from below by a constant c_ϕ, which can depend on ϕ.

Definition 2.12 (NFLVR) We say that the condition *(NFLVR)* is satisfied if there does not exist a sequence $(\phi_n)_{n \geq 1}$ in $\mathcal{A}_b$ such that

1. $(V_T^{0,\phi_n})^- \to 0$ for the $\mathbb{L}^\infty$-norm;
2. $V_T^{0,\phi_n} \to f \; \mathbb{P} - $ a.s. in which $f \in \mathbb{L}^0(\mathbb{R}_+) \setminus \{0\}$.

This condition means that it is not possible to construct asymptotically an arbitrage, by considering a sequence of portfolios with risk vanishing to 0 uniformly. Intuitively, we cannot be as close as we want to an arbitrage.

The following version of the *first fundamental theorem of asset pricing* is due to [23].

Theorem 2.13 (First fundamental theorem of asset pricing) (NFLVR) $\Leftrightarrow$ $\mathcal{M}_{loc}(\tilde{S}) \neq \emptyset$.

Remark 2.14 It is clear that (NFLVR) implies the absence of arbitrage if we restrict to strategies $\phi \in \mathcal{A}$ such that $\tilde{V}^{0,\phi}$ is (essentially) bounded from below by a constant c_ϕ, i.e. $\phi \in \mathcal{A}_b$. In particular, if $\phi \in \mathcal{A}_b$ then $\tilde{V}^{0,\phi}$ is a super-martingale under $\mathbb{Q}$ for all $\mathbb{Q} \in \mathcal{M}_{loc}(\tilde{S})$.

The following provides an easy condition for an element of $\mathcal{M}_{loc}(\tilde{S})$ to belong to $\mathcal{M}(\tilde{S})$.

Proposition 2.15 *Let* $\mathbb{Q} \in \mathcal{M}_{loc}(\tilde{S})$ *be such that* $\mathbb{E}^\mathbb{Q}\left[\sup_{t \in \mathbb{T}} \|\tilde{S}_t\|\right] < \infty$, *then* $\mathbb{Q} \in \mathcal{M}(\tilde{S})$.

Proof Since each component of $\tilde{S}$ (resp. $-\tilde{S}$) is (essentially) bounded from below by the martingale $(-\mathbb{E}^\mathbb{Q}\left[\sup_{s \in \mathbb{T}} \|\tilde{S}_s\| \mid \mathcal{F}_t\right])_{t \in \mathbb{T}}$, it follows from Lemma 2.3 that each component of $\tilde{S}$ (resp. $-\tilde{S}$) is a $\mathbb{Q}$-supermartingale. $\square$

2.3 Pricing by Super-hedging

In this section, we only consider strategies of the set $\mathcal{A}_b$ as defined in Remark 2.14 above. As in discrete time models, there exist two natural notions of price for European derivatives.

Definition 2.16 (Viable/super-hedging price)

(i) A price p is a *viable price* for the European option of payoff $G \in \mathbb{L}^0$ if one cannot make an arbitrage by buying or selling this option, i.e. if

$$\nexists \; \phi \in \mathcal{A}_b \text{ and } \epsilon \in \{-1, 1\} \text{ s.t. } V_T^{\epsilon p, \phi} - \epsilon \, G \in \mathbb{L}^0(\mathbb{R}_+) \setminus \{0\} \, . \quad (2.7)$$

(ii) The *super-hedging price* of G is the smallest initial capital required to super-hedge G, i.e.

$$p(G) := \inf\{p \in \mathbb{R} \; : \; \exists \, \phi \in \mathcal{A}_b \text{ s.t. } V_T^{p,\phi} - G \in \mathbb{L}^0(\mathbb{R}_+)\} \, . \quad (2.8)$$

Unlike in discrete time, it is quite difficult to work with option payoffs in L_S, in a continuous time setting. In the following, we shall therefore confine ourselves to consider payoffs $G \in \mathbb{L}^0$ such that $\tilde{G}^-$ belong to $\mathbb{L}^\infty$.

Theorem 2.17 (Dual formulation, [28]) *Assume that $\mathcal{M}_{\text{loc}}(\tilde{S}) \neq \emptyset$ then, for all $G \in \mathbb{L}^0$ such that $\tilde{G}^- \in \mathbb{L}^\infty$,*

$$p(G) = \sup_{\mathbb{Q} \in \mathcal{M}_{\text{loc}}(\tilde{S})} \mathbb{E}^{\mathbb{Q}}[\beta_T G].$$

Moreover, there exists $\phi \in \mathcal{A}_b$ such that $V_T^{p(G),\phi} \geq G\, \mathbb{P} - a.s.$ whenever $p(G) < \infty$.

This theorem allows us to provide a precise description of the collection of viable prices.

Corollary 2.18 *Assume that $\mathcal{M}_{\text{loc}}(\tilde{S}) \neq \emptyset$ then, for all $G \in \mathbb{L}^\infty$, the set of viable price is the relative interior of the interval $[-p(-G), p(G)]$. Moreover, there exists $\phi \in \mathcal{A}_b$ such that $V_T^{p(G),\phi} = G\, \mathbb{P} - a.s.$ if and only if $-p(-G) = p(G)$.*

Proof This follows from Theorem 2.17 by the same arguments as the one used in the proof of Theorem 1.4. □

The characterisation of the super-hedging price of an American option obtained in Theorem 1.8 can also be extended to the continuous time setting. The proof is very technical and will only be provided for complete markets in Sect. 2.4.4 below. In order to simplify the formulation, we restrict here to payoffs with continuous paths.

Theorem 2.19 (Dual formulation) *Assume that $\mathcal{M}_{\text{loc}}(\tilde{S}) \neq \emptyset$. Let $G = (G_t)_{t \in \mathbb{T}}$ be a continuous adapted process, such that $\sup_{t \leq T} |\tilde{G}| \in \mathbb{L}^\infty$. Then,*

$$p^{US}(G) := \inf\left\{ p \in \mathbb{R} : \exists \phi \in \mathcal{A} \ s.t. \ V_t^{p,\phi} \geq G_t \ \forall t \in \mathbb{T}, \ \mathbb{P} - a.s. \right\}$$

$$= \sup_{\tau \in \mathcal{T}_0, \mathbb{Q} \in \mathcal{M}_{\text{loc}}(\tilde{S})} \mathbb{E}^{\mathbb{Q}}[\tilde{G}_\tau].$$

We also refer to Sect. 2.4.4 for a discussion on the optimal exercise strategy.

2.4 Complete Markets

Roughly speaking, a market is said to be complete when all the sources of risk can be offset by trading the liquid assets.

2.4.1 Characterisation

Definition 2.20 (Complete/incomplete markets) A market is said to be complete if for all $G \in \mathbb{L}^0$ such that $|G| \in \mathbb{L}^\infty$, there exists a couple $(p, \phi) \in \mathbb{R} \times \mathcal{A}_b$ such that $V_T^{p,\phi} = G \; \mathbb{P} - $ a.s. In this case, one says that G is *attainable* or *replicable*.

As in discrete time, see Chap. 1, the completeness is characterised by the uniqueness of the (local) martingale measure.

Theorem 2.21 (Second fundamental theorem of asset pricing) *Assume that* $\mathcal{M}_{\text{loc}}(\tilde{S}) \neq \emptyset$. *Then the market is complete if and only if* $\mathcal{M}_{\text{loc}}(\tilde{S})$ *is a singleton.*

Proof This is an immediate consequence of Theorem 2.17 and Corollary 2.18. $\square$

The previous theorem shows that, if $\mathbb{Q}$ is the unique local martingale measure and if $\tilde{S}$ is a $\mathbb{Q}$-martingale, then any $\mathbb{Q}$-martingale admits a representation in terms of $\tilde{S}$. This follows from the so-called *martingale representation theorem*.

Theorem 2.22 (Martingales representation theorem) *If M is a $\mathbb{P}$-martingale, then there exists a predictable process ξ which is $\mathbb{P} - $ a.s.-square integrable and such that*

$$M = M_0 - \int_0^\cdot \xi_s' dW_s .$$

If $H \in \mathbb{L}^1(\mathbb{P})$ satisfies $H > 0 \; \mathbb{P}-$a.s., then there exists a predictable process ξ which is $\mathbb{P} - $ a.s.-square integrable and such that

$$H = \mathbb{E}[H] \, e^{-\frac{1}{2} \int_0^\cdot \|\xi_s\|^2 ds - \int_0^\cdot \xi_s' dW_s} .$$

Proof See [51, Chapter IV, Corollary 3 and 4, p.156]. $\square$

Corollary 2.23 *Assume that* $\mathcal{M}_{\text{loc}}(\tilde{S}) = \mathcal{M}(\tilde{S}) = \{\mathbb{Q}\}$. *Then,*

(i) *there exists a* risk premium *associated to $\mathbb{Q}$ which satisfies the conditions of Theorem 2.7;*
(ii) *every $\mathbb{Q}$-martingale M satisfying $|M_T| \in L_S$ admits a representation of the form $M = \tilde{V}^{M_0,\phi}$ for some $\phi \in \mathcal{A}$;*
(iii) *$\tilde{V}^{p,\phi}$ is a $\mathbb{Q}$-martingale for all $\phi \in \mathcal{A}$ and $p \in \mathbb{R}$ such that $|\tilde{V}_T^{p,\phi}| \in L_S$.*

Proof

(i) Let H_T be the density of $\mathbb{Q}$ with respect to $\mathbb{P}$. Let ξ be the predictable process in the representation of H_T of Theorem 2.22. Then, $H := (H_t)_{t \in \mathbb{T}}$ defined by $H_t = \mathbb{E}[H_T|\mathcal{F}_t]$ is a martingale (a positive local martingale such that $\mathbb{E}[H_T] = H_0 = 1$) satisfying

$$H_t = 1 - \int_0^t H_s \xi_s' dW_s .$$

Saying that $\tilde{S}$ is a $\mathbb{Q}$-martingale is equivalent to the fact that $H\tilde{S}$ is a $\mathbb{P}$-martingale. On the other hand, Itô's Lemma and (2.3) imply that

$$(H\tilde{S})_t = S_0 + \int_0^t H_s(\tilde{b}_s - r_s\tilde{S}_s - \tilde{\sigma}_s\xi_s)ds + \int_0^t (H_s\tilde{\sigma}_s - H_s\tilde{S}_s\xi_s')dW_s \ .$$

We then deduce from Proposition 2.1 and Theorem 2.22 that ξ is a *risk premium* associated to $\mathbb{Q}$ which satisfies the conditions of Theorem 2.7.

(ii) By Theorems 2.21, and Corollary 2.18, there exists $\phi \in \mathcal{A}$ such that $M_T = \tilde{V}_T^{M_0,\phi}$. Since Theorem 2.7 implies that $\tilde{V}^{M_0,\phi}$ is a $\mathbb{Q}$-supermartingale, we deduce that $M_t = \mathbb{E}^{\mathbb{Q}}[M_T|\mathcal{F}_t] = \mathbb{E}^{\mathbb{Q}}\left[\tilde{V}_T^{M_0,\phi}|\mathcal{F}_t\right] \leq \tilde{V}_t^{M_0,\phi} \ \mathbb{P} - $ a.s. On the other hand, $\mathbb{E}^{\mathbb{Q}}\left[M_t - \tilde{V}_t^{M_0,\phi}\right] \geq M_0 - M_0 = 0$, so that $M_t = \tilde{V}_t^{M_0,\phi} \ \mathbb{P} - $ a.s. for all $t \leq T$. It is therefore $\mathbb{P} - $ a.s. true for all rational numbers of $\mathbb{T}$. Since $V^{M_0,\phi}$ and M have continuous path, recall Theorem 2.22, the equality holds for all $t \in \mathbb{T} \ \mathbb{P} - $ a.s.

(iii) The last assertion is proved similarly.

$\square$

2.4.2 The Case of an Invertible Volatility

We now provide a sufficient condition in terms of r, b and σ, which can be easily checked in practice. The material presented below also allows us to extend the result of Corollary 2.23 to a wider class of martingales.

We assume that there exists a *risk premium* λ such that the solution H to

$$H_t = 1 - \int_0^t H_s\lambda_s'dW_s$$

is a $\mathbb{P}$-martingale.

Let us denote by $\mathbb{Q}$ the equivalent measure whose density is H_T. In view of Theorem 2.9, the process

$$W^{\mathbb{Q}} = W + \int_0^{\cdot} \lambda_s ds$$

is a Brownian motion under $\mathbb{Q} \in \mathcal{M}_{\text{loc}}(\tilde{S})$, see also Theorem 2.7.

Recall that the wealth dynamics can be written in terms of $W^{\mathbb{Q}}$ in the form

$$\tilde{V}_t^{0,\phi} = \int_0^t \phi_s'\tilde{\sigma}_s dW_s^{\mathbb{Q}} \ . \tag{2.9}$$

From now on, we shall work under the following conditions:

S0. $\mathbb{Q} \in \mathcal{M}(\tilde{S})$;

S1. $\tilde{\sigma}_t$ is invertible for all $t \in \mathbb{T} \, \mathbb{P} - \text{a.s.}$;

S2. Its inverse $(\tilde{\sigma}_t)^{-1}$ is uniformly bounded for all $t \in \mathbb{T} \, \mathbb{P} - \text{a.s.}$

Remark 2.24 If λ satisfies the *Novikov's condition*, see Remark 2.8, then H is a $\mathbb{P}$-martingale. It is the case if λ is bounded. If moreover, $\mathbb{E}^{\mathbb{Q}}\left[\int_0^T \|\tilde{\sigma}_t\|^2 dt\right] < \infty$ then S0 holds. It is the case if $H_T \in \mathbb{L}^2$ and $\mathbb{E}\left[\int_0^T \|\tilde{\sigma}_t\|^4 dt\right] < \infty$.

Proposition 2.25 *Under the conditions S0, S1 et S2, the market is complete. Any G such that $\tilde{G}^- \in L_S$ can be written as the terminal value of a portfolio starting from $\mathbb{E}^{\mathbb{Q}}[\tilde{G}]$ and following a strategy in $\mathcal{A}$. Moreover, if M is a $\mathbb{Q}$-martingale such that*

$$M_t \geq -c_M^0 - \sum_{i=1}^d c_M^i \tilde{S}_t^i \quad \mathbb{P} - \text{a.s. for all } t \in \mathbb{T},$$

for some real numbers $c_M^0, \cdots, c_M^d$, then there exists $\phi \in \mathcal{A}$ such that $\tilde{V}^{M_0, \phi} = M \, \mathbb{P} - \text{a.s.}$

Proof Let G be such that $|\tilde{G}| \in L_S$. Since $\tilde{S}$ is a martingale under $\mathbb{Q}$, $\tilde{G} \in \mathbb{L}^1(\mathbb{Q})$. The process $M := (\mathbb{E}^{\mathbb{Q}}[\tilde{G}|\mathcal{F}_t])_{t \in \mathbb{T}}$ is thus a $\mathbb{Q}$-martingale which moreover satisfies

$$M_t \geq -c_G^0 - \sum_{i=1}^d c_G^i \tilde{S}_t^i \quad \mathbb{P} - \text{a.s. for all } t \in \mathbb{T},$$

for some real numbers $c_G^0, \cdots, c_G^d$. On the other hand, Theorem 2.22 implies that M can be written in the form

$$M = \mathbb{E}^{\mathbb{Q}}[\tilde{G}] + \int_0^{\cdot} \xi_s' dW_s^{\mathbb{Q}},$$

for some predictable process ξ. One then deduces from (2.9) that the process $\phi := (\xi'(\tilde{\sigma})^{-1})'$ verifies

$$\tilde{V}^{\mathbb{E}^{\mathbb{Q}}[\tilde{G}], \phi} = \mathbb{E}^{\mathbb{Q}}[\tilde{G}] + \int_0^{\cdot} \phi_s' \tilde{\sigma}_s dW_s^{\mathbb{Q}}$$

$$= \mathbb{E}^{\mathbb{Q}}[\tilde{G}] + \int_0^{\cdot} \xi_s' dW_s^{\mathbb{Q}}$$

$$= M$$

together with the admissibility condition for $\mathcal{A}$. Finally, since ξ is $\mathbb{P} - \text{a.s.-square}$ integrable and $(\tilde{\sigma})^{-1}$ is bounded $\mathbb{P} - \text{a.s.}$, one deduces that ϕ is $\mathbb{P} - \text{a.s.-square}$ integrable as well. $\qquad \square$

2.4.3   Hedging and Malliavin Calculus

We have seen in Proposition 2.25 that, under the conditions of Sect. 2.4.2, any European option G satisfying $\tilde{G}^- \in L_S$ can be replicated: there exists $\phi \in \mathcal{A}$ such that $V_T^{\mathbb{E}^{\mathbb{Q}}[\tilde{G}],\phi} = G$.

In practice, one needs to know the hedging strategy ϕ. We shall explain in this section how it can be obtained by using the Malliavin calculus, more precisely the Clark-Ocone formula.

In Chap. 4 below, we will provide another way to compute it, based on the *delta* of the hedging price.

2.4.3.1   Introduction to Malliavin Calculus

We start with a brief introduction to Malliavin calculus. It can be seen as a differential calculus on the path of the Brownian motion. This tool has been popularised in finance by the paper [32], see also [31].

One needs first to define the notion of Malliavin derivative for a class of simple random variables. We assume for the moment that $W^{\mathbb{Q}}$ is a n-dimensional Brownian motion under the risk neutral probability measure $\mathbb{Q}$ of the previous section.

Definition 2.26 A random variable F in $\mathbb{L}^2(\mathbb{Q})$ is *simple* if there exists an integer k_F, a sequence $0 \leq s_1^F \ldots < s_{k_F}^F \leq T$ and a function ϕ^F from $(\mathbb{R}^n)^{k_F}$ into $\mathbb{R}$, continuous and C^1, such that, for all $t \in [0, T]$,

$$F = \phi^F\left(W_{s_1^F}^{\mathbb{Q}}, \ldots, W_{s_{k_F}^F}^{\mathbb{Q}}\right) \text{ and } \sum_{j=1}^{k_F} \partial_{x^j}\phi^F\left(W^{\mathbb{Q}}\right) \mathbf{1}_{[t,T]}(s_j^F) \in \mathbb{L}^2 .$$

The collection of such random variables is denoted $\mathcal{S}$.

Definition 2.27 (Malliavin derivative) Fix $F \in \mathcal{S}$. Then, for all $t \in [0, T]$, the Malliavin derivative at time t with respect to the i-th component of $W^{\mathbb{Q}}$ is defined as

$$D_t^i F := \lim_{\varepsilon \to 0} \frac{\phi^F\left(W^{\mathbb{Q}} + \varepsilon e_i \mathbf{1}_{[t,T]}\right) - \phi^F\left(W^{\mathbb{Q}}\right)}{\varepsilon}$$

$$= \sum_{j=1}^{k_F} \partial_{x^j}\phi^F\left(W^{\mathbb{Q}}\right) \mathbf{1}_{[t,T]}(s_j^F) \quad \mathbb{P}\text{-a.s.}$$

in which $e_i^j = 1_{i=j}$. We set $DF = (D^i F)_{i \leq d}$, viewed as a row vector. The random variable $D_t F$ is called the Malliavin derivative of F at t.

The above definition should be understood as follows: The path of the Brownian motion $W^{\mathbb{Q}}$ is slightly shifted by εe_i en t, it is replaced at t by the path $W^{\mathbb{Q}} +$

$\varepsilon e_i \mathbf{1}_{[t,T]}$. The Malliavin derivative provides the impact of such a shift on F for an infinitesimal ε.

Example 2.28 Fix $s, t \in [0, T]$, and, for sake of simplicity, we assume that $n = 1$.

1/ $s \in \mathbb{R}$ does not depend on $W^{\mathbb{Q}}$ and therefore $D_t s = 0$.

2/ If $F = W_s^{\mathbb{Q}}$,

$$D_t F = \lim_{\varepsilon \to 0} \frac{W_s^{\mathbb{Q}} + \varepsilon \mathbf{1}_{[t,T]}(s) - W_s^{\mathbb{Q}}}{\varepsilon} = \mathbf{1}_{[t,T]}(s) \, .$$

3/ In the Black and Scholes' model,

$$D_t x e^{(r-\sigma^2/2)s + \sigma W_s^{\mathbb{Q}}} = \sigma x e^{(r-\sigma^2/2)s + \sigma W_s^{\mathbb{Q}}} \mathbf{1}_{[t,T]}(s) \, .$$

The last example shows that $D_t F$ is in general not $\mathcal{F}_t$-measurable. This is not surprising since the derivative evaluates the impact on F of a shock on the path of $W^{\mathbb{Q}}$.

We now extend the notion of Malliavin derivative to a wider class of random variables. For this purpose, let us introduce the norm

$$\|F\|_{\mathbb{D}_{1,2}} := \left(\mathbb{E}^{\mathbb{Q}}[|F|^2] + \mathbb{E}^{\mathbb{Q}}[\int_0^T \|D_t F\|^2 dt] \right)^{\frac{1}{2}}$$

and denote by $\mathbb{D}_{1,2}$ the closure of the set of simple random variables $\mathcal{S}$ for $\|\cdot\|_{\mathbb{D}_{1,2}}$. If $F \in \mathbb{D}_{1,2}$, then one can find a sequence $(F_n)_{n \geq 1} \subset \mathcal{S}$ such that $F_n \to F$ in $\mathbb{L}^2(\mathbb{Q})$ and DF_n converges in $\mathbb{L}^2([0,T] \times \Omega, dt \times d\mathbb{Q})$ to a process called DF. One can also show that the limit DF does not depend on the approximating sequence $(F_n)_{n \geq 1}$, so that the Malliavin derivative DF is uniquely defined.

The following properties are easy consequences of the previous definition.

Proposition 2.29 *Fix F et $G \in \mathbb{D}_{1,2}$ and $\psi \in C_b^1$. Then,*

(i) *$FG \in \mathbb{D}_{1,2}$ and $D(FG) = (DF)G + F(DG)$ whenever $F, G \in \mathbb{L}^\infty$;*
(ii) *$\psi(F) \in \mathbb{D}_{1,2}$ and $D\psi(F) = \partial \psi(F)DF$.*

Remark 2.30 The assertion (ii) can be extended to the case where ψ is only Lipschitz and C_b^1 on its support $\mathrm{supp}(\psi)$. In this case, one can check that $D\psi(F) = \partial \psi(F)DF \mathbf{1}_{\{F \in \mathrm{supp}(\psi)\}}$. It is in particular the case for the payoff function of a call $x \mapsto [x - K]^+$, in which $K > 0$: $D[F - K]^+ = DF \mathbf{1}_{\{F > K\}}$. See for example [48, Propositions 1.2.3 and 1.3.7].

Remark 2.31 One can show that, if α is a deterministic process with continuous path, then $\int_0^T \alpha_s dW_s^{\mathbb{Q}} \in \mathbb{D}_{1,2}$ and $D_t \int_0^T \alpha_s dW_s^{\mathbb{Q}} = \alpha_t$.

2.4.3.2 Clark-Ocone Formula

The strength of the notion of Malliavin derivative is that it allows to characterise the hedging strategy.

Theorem 2.32 (Clark-Ocone formula) *Fix $\tilde{G} \in \mathbb{D}_{1,2}$, then*

$$\tilde{G} = \mathbb{E}^{\mathbb{Q}}[\tilde{G}] + \int_0^T \mathbb{E}^{\mathbb{Q}}[D_t\tilde{G}|\mathcal{F}_t]dW_t^{\mathbb{Q}} .$$

Proof For sake of simplicity, let us confine ourselves to the case $n = 1$ and to a simple random variable $F \in \mathcal{S}$. We also assume that $\phi^F \in C_b^2$. The general case is obtained by an approximation argument. Given $s_{i-1}^F < t \leq s_i^F$, $\mathbb{E}^{\mathbb{Q}}[F|\mathcal{F}_t]$ is a function $\psi(t, W_t^{\mathbb{Q}}, (W_{s_j^F}^{\mathbb{Q}})_{j\leq i-1})$. Let ψ' denotes the derivative with respect to the second argument. Then,

$$\psi'(t, w, z) =$$
$$\lim_{\varepsilon \to 0} \varepsilon^{-1}\mathbb{E}^{\mathbb{Q}}\left[\phi^F\left(z, \tilde{\omega}^i + w + \varepsilon, \ldots, \tilde{\omega}^{k_F} + w + \varepsilon\right) - \phi^F\left(z, \tilde{\omega}^i + w, \ldots, \tilde{\omega}^{k_F} + w\right)\right],$$

where $\tilde{\omega} := (\tilde{\omega}^i, \ldots, \tilde{\omega}^{k_F})$ is a random variable with values in $\mathbb{R}^{k_F-i+1}$ which is distributed under $\mathbb{Q}$ according to the Gaussian distribution $\mathcal{N}(0, \Sigma)$ with

$$\Sigma^{lk} = \min\{s_l^F - t , s_k^F - t\} .$$

Hence, it follows from the dominated convergence theorem that

$$\psi'(t, w, z) = \mathbb{E}^{\mathbb{Q}}\left[\sum_{j=i}^{k_F} \partial_{x^j}\phi^F\left(z, \tilde{\omega}^i + w, \ldots, \tilde{\omega}^{k_F} + w\right)\right],$$

so that for $t \in (s_{i-1}^F, s_i^F]$

$$\psi'\left(t, W_t, (W_{s_j^F})_{j\leq i-1}\right) = \mathbb{E}^{\mathbb{Q}}\left[\sum_{j=i}^{k_F} \partial_{x^j}\phi^F\left(W^{\mathbb{Q}}\right)|\mathcal{F}_t\right] = \mathbb{E}^{\mathbb{Q}}[D_t\phi^F\left(W^{\mathbb{Q}}\right)|\mathcal{F}_t] .$$

Since $\phi^F \in C_b^2$, one can similarly check that ψ is $C_b^{1,2}$ with respect to its two first arguments. By applying Itô's formula to the martingale $\psi(t, W_t, (W_{s_j^F})_{j\leq i-1})$ on

$(s_{i-1}^F, s_i^F]$, we then obtain

$$\mathbb{E}^{\mathbb{Q}}\left[F|\mathcal{F}_{s_i^F}\right] = \mathbb{E}^{\mathbb{Q}}\left[F|\mathcal{F}_{s_{i-1}^F}\right] + \int_{s_{i-1}^F}^{s_i^F} \psi'\left(t, W_t^{\mathbb{Q}}, (W_{s_j^F}^{\mathbb{Q}})_{j \le i-1}\right) dW_t^{\mathbb{Q}}$$

$$= \mathbb{E}^{\mathbb{Q}}\left[F|\mathcal{F}_{s_{i-1}^F}\right] + \int_{s_{i-1}^F}^{s_i^F} \mathbb{E}^{\mathbb{Q}}\left[D_t\phi^F\left(W^{\mathbb{Q}}\right)|\mathcal{F}_t\right] dW_t^{\mathbb{Q}}.$$

This relation is true for all $i \in \{1, \ldots, k_F\}$, from which we deduce by summing up the above over i that

$$F = \mathbb{E}^{\mathbb{Q}}\left[F|\mathcal{F}_{s_k^F}\right] = \mathbb{E}^{\mathbb{Q}}[F|\mathcal{F}_0] + \int_0^{s_{k_F}^F} \mathbb{E}^{\mathbb{Q}}\left[D_t\phi^F\left(W^{\mathbb{Q}}\right)|\mathcal{F}_t\right] dW_t^{\mathbb{Q}}$$

$$= \mathbb{E}^{\mathbb{Q}}[F] + \int_0^T \mathbb{E}^{\mathbb{Q}}\left[D_t\phi^F\left(W^{\mathbb{Q}}\right)|\mathcal{F}_t\right] dW_t^{\mathbb{Q}}.$$

$\square$

Example 2.33 Let us consider a one-dimensional setting in which

$$S_T = S_0 e^{\int_0^T (r_s - \frac{1}{2}\sigma_s^2)ds + \int_0^T \sigma_s dW_s^{\mathbb{Q}}},$$

where $S_0 > 0$ is a constant, and r, σ are two deterministic continuous functions. Fix $K > 0$. By Remarks 2.30 and 2.31,

$$D_t\beta_T[S_T - K]^+ = \beta_T D_t S_T \mathbf{1}_{\{S_T > K\}} = \beta_T S_T \sigma_t \mathbf{1}_{\{S_T > K\}}.$$

Then, it follows from Clark-Ocone formula that the hedging strategy satisfies

$$\beta_T[S_T - K]^+ = \mathbb{E}^{\mathbb{Q}}\left[\beta_T[S_T - K]^+\right] + \int_0^T \beta_T \mathbb{E}^{\mathbb{Q}}\left[S_T \mathbf{1}_{\{S_T > K\}}|\mathcal{F}_t\right] \sigma_t dW_t^{\mathbb{Q}}$$

in which $\mathbb{E}^{\mathbb{Q}}\left[S_T \mathbf{1}_{\{S_T > K\}}|\mathcal{F}_t\right]$ can be computed explicitly in term of the cumulative function of the standard normal distribution.

2.4.4 American Options: Hedging and Exercise Strategy

In this section, we provide the dual formulation for the super-hedging price of an *American option*. It is the counterpart of the result obtained for discrete time models in Chap. 1.

We shall work under the following conditions

S0'. $\{\mathbb{Q}\} = \mathcal{M}(\tilde{S}) = \mathcal{M}_{\mathrm{loc}}(\tilde{S})$,

S1. $\tilde{\sigma}_t$ is invertible for all $t \in \mathbb{T}$ $\mathbb{P}$ – a.s.,

S2. Its inverse $(\tilde{\sigma}_t)^{-1}$ is uniformly bounded for all $t \in \mathbb{T}$ $\mathbb{P}$ – a.s.,

so that there exists a *risk premium* associated to $\mathbb{Q}$ which satisfies the condition of Theorem 2.7, see also Corollary 2.23.

The American option is modeled as an adapted process $G = (G_t)_{t \in \mathbb{T}}$ which we assume to have continuous path, for sake of simplicity. We also impose that

$$\mathbb{E}^{\mathbb{Q}}\left[\sup_{t \in \mathbb{T}} |\tilde{G}_t|\right] < \infty \; , \; |\tilde{G}| \le c_G^0 + \sum_{i=1}^{d} c_G^i \tilde{S}^i \;\; \mathbb{P} - \text{a.s.} \tag{2.10}$$

where $\tilde{G} = \beta G$ and $c_G^0, \dots, c_G^d$ are given real numbers.

The notion of viable price is defined as in Sect. 1.5.

Definition 2.34 (Viable price) The price p is a *viable price* for the American option G if

(i) $\nexists \, \phi \in \mathcal{A}$ s.t. $V_\tau^{x,\phi} - G_\tau \in \mathbb{L}^0(\mathbb{R}_+) \setminus \{0\}$ for all $\tau \in \mathcal{T}_0$;

(ii) $\nexists \, (\phi, \tau) \in \mathcal{A} \times \mathcal{T}_0$ s.t. $V_\tau^{-x,\phi} + G_\tau \in \mathbb{L}^0(\mathbb{R}_+) \setminus \{0\}$;

in which $\mathcal{T}_t$ denotes the set of stopping times with values in $[t, T]$, $t \in \mathbb{T}$.

As in discrete time models, there is only one viable price when the market is complete, and it coincides with the super-hedging price. It admits a dual formulation in terms of an *optimal stopping* problem.

Theorem 2.35 *Let S0', S1 and S2 hold. Then, the unique viable price is*

$$p^{US}(G) := \inf\left\{ p \in \mathbb{R} \; : \; \exists \, \phi \in \mathcal{A} \text{ s.t. } V_t^{p,\phi} \ge G_t \; \forall \, t \in \mathbb{T} \, , \; \mathbb{P} - \text{a.s.} \right\}$$

$$= \sup_{\tau \in \mathcal{T}_0} \mathbb{E}^{\mathbb{Q}}\left[\tilde{G}_\tau\right] =: Y_0 \; .$$

The proof is slightly more technical than the one of Chap. 1 but follows similar ideas. We split it in several intermediate results. Some of them will not be proved.

Proposition 2.36 *If S0', S1 and S2 holds, then $p^{US}(G) \ge Y_0$.*

Proof If $p \in \mathbb{R}$ and $\phi \in \mathcal{A}$ are such that $V^{p,\phi} \ge G$ $\mathbb{P}$ – a.s., then one deduces from S0', (2.10), Theorems 2.7 and 2.10 that $p \ge \mathbb{E}^{\mathbb{Q}}\left[\tilde{V}_\tau^{p,\phi}\right] \ge \mathbb{E}^{\mathbb{Q}}\left[\tilde{G}_\tau\right]$ for all $\tau \in \mathcal{T}_0$. This readily implies that $p^{US}(G) \ge Y_0$. $\qquad\qquad\square$

The converse inequality is more difficult to obtain. It relies on a *dynamic programming principle* satisfied by the family $(\bar{Y}_\vartheta, \; \vartheta \in \mathcal{T}_0)$ defined

$$\bar{Y}_\vartheta := \operatorname*{ess\,sup}_{\tau \in \mathcal{T}_\vartheta} \mathbb{E}^{\mathbb{Q}}\left[\tilde{G}_\tau | \mathcal{F}_\vartheta\right], \quad \vartheta \in \mathcal{T}_0,$$

where $\mathcal{T}_\vartheta$ is the set of stopping times in $\mathcal{T}_0$ that are $\mathbb{P}-$ a.s. bigger than ϑ. This is the continuous time counterpart of the backward algorithm (1.17).

Proposition 2.37 (Dynamic programming principle) *For all $\vartheta \in \mathcal{T}_0$ and $\theta \in \mathcal{T}_\vartheta$,*

$$\bar{Y}_\vartheta = \underset{\tau \in \mathcal{T}_\vartheta}{\text{ess sup}}\ \mathbb{E}^{\mathbb{Q}}\left[\bar{Y}_\theta \mathbf{1}_{\tau > \theta} + \tilde{G}_\tau \mathbf{1}_{\tau \le \theta} | \mathcal{F}_\vartheta\right] \mathbb{P}-a.s.$$

Moreover, $\bar{Y}_\vartheta \ge \tilde{G}_\vartheta\ \mathbb{P}-a.s.$

Proof Since $\vartheta \in \mathcal{T}_\vartheta$ it is clear that $\bar{Y}_\vartheta \ge \mathbb{E}^{\mathbb{Q}}\left[\tilde{G}_\vartheta | \mathcal{F}_\vartheta\right] = \tilde{G}_\vartheta$ and that

$$\mathbb{E}^{\mathbb{Q}}\left[\tilde{G}_\tau | \mathcal{F}_\theta\right]\mathbf{1}_{\tau > \theta} = \mathbb{E}^{\mathbb{Q}}\left[\tilde{G}_{\tau \vee \theta} | \mathcal{F}_\theta\right]\mathbf{1}_{\tau > \theta} \le \bar{Y}_\theta \mathbf{1}_{\tau > \theta}$$

so that

$$\bar{Y}_\vartheta \le \underset{\tau \in \mathcal{T}_\vartheta}{\text{ess sup}}\ \mathbb{E}^{\mathbb{Q}}\left[\bar{Y}_\theta \mathbf{1}_{\tau > \theta} + \tilde{G}_\tau \mathbf{1}_{\tau \le \theta} | \mathcal{F}_\vartheta\right].$$

We now prove the converse inequality. We first note that $F := \{J_\mu := \mathbb{E}^{\mathbb{Q}}\left[\tilde{G}_\mu | \mathcal{F}_\theta\right] : \mu \in \mathcal{T}_\theta\}$ is *directed upward* in the sense of Definition 1.8: if $\mu_1, \mu_2 \in \mathcal{T}_\theta$, then there exists $\mu_3 \in \mathcal{T}_\theta$ such that $J_{\mu_3} \ge \max\{J_{\mu_1}, J_{\mu_2}\}\ \mathbb{P}-$ a.s. It suffices to choose $\mu_3 = \mu_1 \mathbf{1}_A + \mu_2 \mathbf{1}_{A^c}$ with $A := \{J_{\mu_1} \ge J_{\mu_2}\} \in \mathcal{F}_\theta$. Hence, there exists a sequence $(\mu_n)_n$ in $\mathcal{T}_\theta$ such that

$$\lim_{n \to \infty} \uparrow J_{\mu_n} = \underset{\mu \in \mathcal{T}_\theta}{\text{ess sup}}\ J_\mu\ \mathbb{P}-a.s.,$$

see Proposition 1.9. By monotone convergence, it follows that

$$\mathbb{E}^{\mathbb{Q}}\left[\bar{Y}_\theta \mathbf{1}_{\tau > \theta} + \tilde{G}_\tau \mathbf{1}_{\tau \le \theta} | \mathcal{F}_\vartheta\right]$$
$$= \mathbb{E}^{\mathbb{Q}}\left[\lim_{n \to \infty} \uparrow \mathbb{E}^{\mathbb{Q}}\left[\tilde{G}_{\mu_n} | \mathcal{F}_\theta\right]\mathbf{1}_{\tau > \theta} + \tilde{G}_\tau \mathbf{1}_{\tau \le \theta} | \mathcal{F}_\vartheta\right]$$
$$= \lim_{n \to \infty} \uparrow \mathbb{E}^{\mathbb{Q}}\left[\mathbb{E}^{\mathbb{Q}}\left[\tilde{G}_{\mu_n} | \mathcal{F}_\theta\right]\mathbf{1}_{\tau > \theta} + \tilde{G}_\tau \mathbf{1}_{\tau \le \theta} | \mathcal{F}_\vartheta\right]$$
$$= \lim_{n \to \infty} \uparrow \mathbb{E}^{\mathbb{Q}}\left[\tilde{G}_{\mu_n} \mathbf{1}_{\tau > \theta} + \tilde{G}_\tau \mathbf{1}_{\tau \le \theta} | \mathcal{F}_\vartheta\right].$$

Since $\mu_n \mathbf{1}_{\tau > \theta} + \tau \mathbf{1}_{\tau \le \theta} \in \mathcal{T}_\vartheta$, this concludes the proof. $\square$

It remains to show that $p(G) \le Y_0$. One needs the following result which is proved in [43].

Proposition 2.38 *There exists a $\mathbb{Q}$-supermartingale Y with right-continuous path which aggregates $\bar{Y}$ in the sense that $Y_\tau = \bar{Y}_\tau\ \mathbb{P}-a.s.$ for all $\tau \in \mathcal{T}_0$.*

We also need the extension of the discrete time Doob-Meyer's decomposition of Remark 1.11.

Theorem 2.39 (Doob-Meyer's decomposition) *Let X be a $\mathbb{Q}$-super-martingale with right-continuous path such that the family $\{X_\tau, \ \tau \in \mathcal{T}_0\}$ is uniformly integrable. Then, there exists a $\mathbb{Q}$-martingale M and a non-decreasing process A such that $X = M - A$. One can take $A_0 = 0$.*

Corollary 2.40 *Under S0', S1 et S2, $p^{US}(G) \leq Y_0$.*

Proof This is an immediate consequence of Propositions 2.25, 2.37 and Theorem 2.39. Indeed, the condition (2.10) and Theorem 2.10 imply $Y \geq -c_G^0 - \sum_{i=1}^d c_G^i \tilde{S}^i \ \mathbb{P}$ − a.s. Hence, it suffices to represent by a portfolio process $\tilde{V}^{M_0,\phi}$ the martingale $M \geq Y$ which appears in the Doob-Meyer decomposition of Y and to use the fact that $Y = M - A \geq \tilde{G}$ which implies $M \geq \tilde{G}$ since $A \geq 0$. This shows that $p^{US}(G) \leq M_0 = Y_0$.   □

The fact that the viable price is unique follows from the above and similar arguments as those used for discrete time models.

Remark 2.41 One can show that

$$\hat{\tau} := \inf\left\{t \in \mathbb{T} \ : \ Y_t = \tilde{G}_t\right\}$$

is optimal in the sense that

$$p^{US}(G) = Y_0 = \mathbb{E}^{\mathbb{Q}}\left[\tilde{G}_{\hat{\tau}}\right] ,$$

see [43]. This implies in particular that $(Y_{t\wedge\hat{\tau}})_{t\in\mathbb{T}}$ is a $\mathbb{Q}$-martingale.

As in discrete time models, one can show that the *rational exercise strategy* for the buyer consists in exercising the option at time $\hat{\tau}$.

2.5 Portfolio Constraints

As in Sect. 1.6, we now introduce *portfolio constraints*. The vector of risky asset holdings is restricted to take values in a set K, which is assumed to be a convex and closed subset of $\mathbb{R}^d$ such that $0 \in K$.

The corresponding set of admissible strategies is denoted by $\mathcal{A}_K^b$, it is the subset of processes $\phi \in \mathcal{A}_b$ such that

$$\phi \in K \ \ dt \times d\mathbb{P} - \text{a.e.}$$

We assume that the market is complete, in particular that the conditions S0', S1 and S2 of Sect. 2.4.4 hold.

The super-hedging problem with constraints consists in studying

$$p_K(G) := \inf\left\{p \in \mathbb{R} \ : \ \exists \ \phi \in \mathcal{A}_K^b \ \text{s.t.} \ V_T^{p,\phi} \geq G \ \mathbb{P} - \text{a.s.}\right\} . \tag{2.11}$$

2.5.1 Dual Formulation of the Super-hedging Price

In this section, we will provide a *dual formulation*. This requires some basic notions of convex analysis. We first recall that the *support function* of the convex set K is defined by

$$z \in \mathbb{R}^d \mapsto \delta_K(z) := \sup_{p \in K} p'z$$

and that

$$\hat{K} := \{z \in \mathbb{R}^d : \delta_K(z) < \infty\}$$

is its domain.

Let us now denote by $\hat{\mathcal{K}}_b$ the collection of bounded progressively measurable processes with values in $\hat{K}$. To $\nu \in \hat{\mathcal{K}}_b$, we associate

$$Z^\nu := \int_0^\cdot \delta_K(\nu_s)ds \,, \lambda^\nu := \lambda - \tilde{\sigma}^{-1}\nu \text{ where } \lambda := \sigma^{-1}(b - rS),$$

$$\frac{d\mathbb{Q}^\nu}{d\mathbb{P}} := H_T^\nu \text{ with } H^\nu = e^{-\int_0^\cdot \lambda_s^\nu dW_s - \frac{1}{2}\int_0^\cdot \|\lambda_s^\nu\|^2 ds}.$$

In the presence of constraints, the dual formulation is formulated in terms of the above quantities.

Theorem 2.42 *Let G be a $\mathcal{F}_T$-measurable random variable such that $\tilde{G}^- \in \mathbb{L}^\infty$. Then,*

$$p_K(G) = \sup_{\nu \in \hat{\mathcal{K}}_b} \mathbb{E}^{\mathbb{Q}^\nu}[\tilde{G} - Z_T^\nu] = \sup_{\nu \in \hat{\mathcal{K}}_b} \mathbb{E}\left[H_T^\nu(\tilde{G} - Z_T^\nu)\right].$$

If $p_K(G) < \infty$ then there exists $\phi \in \mathcal{A}_K^b$ such that $V_T^{p_K(G),\phi} \geq G$.

This theorem corresponds to Corollary 1.3. The additional term Z^ν comes from the fact that the constraints may not be conic as in Chap. 1, in which the family $\{H^\nu, \nu \in \hat{\mathcal{K}}_b\}$ is a subset of $\mathcal{H}_b^K(\tilde{S})$ which is sufficient for the characterisation of the absence of arbitrage.

Remark 2.43 As in Definition 2.16, one can define the notion of viable price by replacing $\mathcal{A}_b$ by $\mathcal{A}_K^b$. By similar arguments as in Corollary 2.18, the interval of viable prices is

$$] - p_K(-G), p_K(G)[\cup \{p_K(G)\}\mathbf{1}_{\{p_K(G) = \mathbb{E}^{\mathbb{Q}_0}[\tilde{G}]\}} \cup \{-p_K(-G)\}\mathbf{1}_{\{p_K(-G) = \mathbb{E}^{\mathbb{Q}_0}[-\tilde{G}]\}}.$$

See Exercise 2.19 below.

2.5.2 An Auxiliary Family of Unconstrained Problems

In order to prove Theorem 2.42, we shall use the standard technique of constraints relaxation, which explains the introduction of the family $\{\mathbb{Q}^\nu, \nu \in \hat{\mathcal{K}}_b\}$.

We first provide a supermartingale type property satisfied by any admissible wealth process. Its proof uses

$$W^{\mathbb{Q}^\nu} := W + \int_0^\cdot \lambda_s^\nu ds = W^{\mathbb{Q}} - \int_0^\cdot \tilde{\sigma}_s^{-1} \nu_s ds,$$

defined for $\nu \in \hat{\mathcal{K}}_b$. Note that Girsanov's theorem shows that $W^{\mathbb{Q}^\nu}$ is a $\mathbb{Q}^\nu$-Brownian motion.

We shall also appeal to the following relation between K and the support function δ_K.

Lemma 2.4 (Constraints characterisation)

$$p \in K \Leftrightarrow \delta_K(z) - p'z \geq 0, \ \forall z \in \hat{K} \Leftrightarrow \inf_{\|z\|=1} \delta_K(z) - p'z \geq 0.$$

Proposition 2.44 *For all* $\nu \in \hat{\mathcal{K}}_b$, $\phi \in \mathcal{A}_K^b$ *and* $v \in \mathbb{R}$, *the process* $\tilde{V}^{v,\phi} - Z^\nu$ *is a* $\mathbb{Q}^\nu$-*surmartingale.*

Proof Let us observe that

$$d(\tilde{V}_t^{v,\phi} - Z_t^\nu) = (\phi_t' \nu_t - \delta_K(\nu_t))dt + \phi_t' \tilde{\sigma}_t dW_t^{\mathbb{Q}^\nu}.$$

By Lemma 2.4, $(\phi_t' \nu_t - \delta_K(\nu_t)) \leq 0$. This provides the $\mathbb{Q}^\nu$-local supermartingale property. By using the fact that ν is bounded and the admissibility condition on ϕ, one verifies that $\tilde{V}_t^{v,\phi} - Z_t^\nu \geq -c_\phi - c_\nu T$, for some constants c_ϕ and c_ν. This shows that the $\mathbb{Q}^\nu$-local supermartingale $\tilde{V}^{v,\phi} - Z^\nu$ is bounded from below and is therefore a supermartingale, see Lemma 2.3. □

Proposition 2.45 *Let* G *be an* $\mathcal{F}_T$-*measurable random variable such that* $\tilde{G}^- \in \mathbb{L}^\infty$. *Then*

$$p_K(G) \geq \sup_{\nu \in \tilde{\mathcal{K}}_b} \mathbb{E}^{\mathbb{Q}^\nu}[\tilde{G} - Z_T^\nu] = \sup_{\nu \in \tilde{\mathcal{K}}_b} \mathbb{E}[H_T^\nu(\tilde{G} - Z_T^\nu)].$$

Proof If $p_K(G) = \infty$ then the inequality holds. Otherwise, let v, ϕ be such that $\tilde{V}_T^{v,\phi} \geq \tilde{G}$. By Proposition 2.44, $\tilde{V}^{v,\phi} - Z^\nu$ is a $\mathbb{Q}^\nu$-supermartingale for all $\nu \in \hat{\mathcal{K}}_b$. Hence,

$$v \geq \mathbb{E}^{\mathbb{Q}^\nu}[\tilde{V}_T^{v,\phi} - Z_T^\nu] \geq \mathbb{E}^{\mathbb{Q}^\nu}[\tilde{G} - Z_T^\nu].$$

The proof is concluded by taking the supremum over ν and then the infimum over $\nu > p_K(G)$. □

2.5.3 Study of the Dual Problem

In this section, we show that the reverse inequality also holds. We now assume that $\sup_{\nu \in \hat{\mathcal{K}}_b} \mathbb{E}^{\mathbb{Q}^\nu}[\tilde{G} - Z_T^\nu] < \infty$ (otherwise the inequality in Theorem 2.42 is trivial, see Proposition 2.45).

We first define a dynamical version of the dual problem: for $\tau \in \mathcal{T}_0$ we set

$$
\tilde{Y}_\tau := \operatorname*{ess\,sup}_{\nu \in \hat{\mathcal{K}}_b} \mathbb{E}^{\mathbb{Q}^\nu}\left[\tilde{G} - (Z_T^\nu - Z_\tau^\nu)|\mathcal{F}_\tau\right]
$$

$$
= \operatorname*{ess\,sup}_{\nu \in \hat{\mathcal{K}}_b} \mathbb{E}\left[\frac{H_T^\nu}{H_\tau^\nu}(\tilde{G} - (Z_T^\nu - Z_\tau^\nu))|\mathcal{F}_\tau\right]. \tag{2.12}
$$

As in the study of American options, see Sect. 2.4.4, the key lies in the *dynamic programming principle*.

Lemma 2.5 (Dynamic programming principle) *Fix* $\theta \in \mathcal{T}_\tau$, *then*

$$
\tilde{Y}_\tau = \operatorname*{ess\,sup}_{\nu \in \hat{\mathcal{K}}_b} \mathbb{E}^{\mathbb{Q}^\nu}\left[\tilde{Y}_\theta - (Z_\theta^\nu - Z_\tau^\nu)|\mathcal{F}_\tau\right]
$$

$$
= \operatorname*{ess\,sup}_{\nu \in \hat{\mathcal{K}}_b} \mathbb{E}\left[\frac{H_\theta^\nu}{H_\tau^\nu}(\tilde{Y}_\theta - (Z_\theta^\nu - Z_\tau^\nu))|\mathcal{F}_\tau\right].
$$

Proof

1. We first proceed by successive conditioning:

$$
\tilde{Y}_\tau = \operatorname*{ess\,sup}_{\nu \in \hat{\mathcal{K}}_b} \mathbb{E}^{\mathbb{Q}^\nu}\left[\tilde{G} - (Z_T^\nu - Z_\theta^\nu) - (Z_\theta^\nu - Z_\tau^\nu)|\mathcal{F}_\tau\right]
$$

$$
= \operatorname*{ess\,sup}_{\nu \in \hat{\mathcal{K}}_b} \mathbb{E}^{\mathbb{Q}^\nu}\left[\mathbb{E}^{\mathbb{Q}^\nu}[\tilde{G} - (Z_T^\nu - Z_\theta^\nu)|\mathcal{F}_\theta] - (Z_\theta^\nu - Z_\tau^\nu)|\mathcal{F}_\tau\right]
$$

$$
\leq \operatorname*{ess\,sup}_{\nu \in \hat{\mathcal{K}}_b} \mathbb{E}^{\mathbb{Q}^\nu}\left[\tilde{Y}_\theta - (Z_\theta^\nu - Z_\tau^\nu)|\mathcal{F}_\tau\right].
$$

2. It remains to prove the reverse inequality. Let us now define, for $\nu \in \hat{\mathcal{K}}_b$,

$$
J_\theta^\nu := \mathbb{E}\left[\frac{H_T^\nu}{H_\theta^\nu}\left(\tilde{G} - (Z_T^\nu - Z_\theta^\nu)\right)\Big|\mathcal{F}_\theta\right].
$$

Observe that $\{J^\nu_\theta, \nu \in \hat{\mathcal{K}}_b\}$ is directed upward, see Definition 1.8. Proposition 1.9 then implies that there exists $(\nu^k)_{k \geq 1} \subset \hat{\mathcal{K}}_b$ such that

$$\tilde{Y}_\theta := \operatorname*{ess\,sup}_{\nu \in \hat{\mathcal{K}}_b} J^\nu_\theta = \lim_{k \to \infty} \uparrow J^{\nu^k}_\theta.$$

Given $\mu \in \hat{\mathcal{K}}_b$ and a control ν^k, we now form a new control μ^k equal to μ on $[\tau, \theta)$ and to ν^k on $[\theta, T]$. By definition,

$$\tilde{Y}_\tau \geq J^{\mu^k}_\tau \geq \mathbb{E}\left[\frac{H^\mu_\theta}{H^\mu_\tau} J^{\nu^k}_\theta | \mathcal{F}_\tau\right] - \mathbb{E}\left[\frac{H^\mu_\theta}{H^\mu_\tau}(Z^\mu_\theta - Z^\mu_\tau) | \mathcal{F}_\tau\right].$$

It then follows from the monotone convergence theorem applied to $(J^{\nu^k}_\theta - J^{\nu^1}_\theta)_k$, that

$$\tilde{Y}_\tau \geq \mathbb{E}\left[\frac{H^\mu_\theta}{H^\mu_\tau}(\tilde{Y}_\theta - (Z^\mu_\theta - Z^\mu_\tau)) | \mathcal{F}_\tau\right].$$

The proof is concluded by taking the esssup over $\mu \in \hat{\mathcal{K}}_b$.   □

The next result follows directly from the last Proposition.

Corollary 2.46 *The $\mathbb{Q}^\nu$-supermartingale property holds for the family $\{\tilde{Y}_\tau - Z^\nu_\tau, \ \tau \in \mathcal{T}_0\}$, for all $\nu \in \mathcal{K}_b$.*

We can now use an aggregation result [42].

Theorem 2.47 *There exists a càdlàg[6] process Y that aggregates the family $\{\tilde{Y}_\tau, \ \tau \in \mathcal{T}_0\}$, i.e. such that $Y_\tau = \tilde{Y}_\tau \ \mathbb{P} - a.s.$ for all $\tau \in \mathcal{T}_0$. Moreover, $Y - Z^\nu$ is a $\mathbb{Q}^\nu$-super-martingale for all $\nu \in \hat{\mathcal{K}}_b$.*

The proof of Theorem 2.42 is concluded by applying the next Proposition to the process Y, $\tilde{Y}_0$ is therefore the super-hedging price.

Proposition 2.48 *Let X be an adapted càdlàg process that is bounded from below and such that $X - Z^\nu$ is a $\mathbb{Q}^\nu$-supermartingale for all $\nu \in \hat{\mathcal{K}}_b$. Then there exists a càdlàg predictable and non-decreasing process C and $\phi \in \mathcal{A}^b_K$ such that $X = \tilde{V}^{X_0,\phi} - C$. Moreover, $C_0 = 0$.*

Proof

1. We first apply the Doob-Meyer decomposition, see Theorem 2.39, to $X = X - Z^0$ that is a $\mathbb{Q}^0$-supermartingale, to obtain

$$X_t = X_0 + \int_0^t \psi_s dW^{\mathbb{Q}^0}_s - C^0_t,$$

---

[6]This is the French acronym for *right-continuous with left-limits*.

where ψ is $\mathbb{P}$ − a.s. square integrable and $\int_0^\cdot \psi_s dW_s^{\mathbb{Q}^0}$ is a $\mathbb{Q}^0$-martingale. By setting $\phi' = \psi \tilde{\sigma}^{-1}$, we obtain $X = \tilde{V}^{X_0, \phi} - C^0$.

2. We now apply the Doob-Meyer decomposition to $X - Z^\nu$ that is also a $\mathbb{Q}^\nu$-supermartingale:

$$X_t - Z_t^\nu = X_0 + M_t^\nu - C_t^\nu \,.$$

But it follows from step 1 that

$$X_t - Z_t^\nu = X_0 + \int_0^t \psi_s dW_s^{\mathbb{Q}^0} - C_t^0 - Z_t^\nu$$

$$= X_0 + \int_0^t \psi_s dW_s^{\mathbb{Q}^\nu} - C_t^0 + \int_0^t \left(\phi_s' \nu_s - \delta_K(\nu_s) \right) ds.$$

By identifying the Brownian diffusion and the bounded variation parts of these two decompositions, we deduce that

$$\int_0^t \left(\phi_s' \nu_s - \delta_K(\nu_s) \right) ds - C_t^0 = -C_t^\nu$$

which leads to

$$C_T^0 \geq \int_0^t \left(\phi_s' \nu_s - \delta_K(\nu_s) \right) ds \,,$$

since C^0 and C^ν are both non-decreasing and start from 0. Upon replacing ν by $k \times \nu$ and then sending k to infinity, we obtain

$$\int_0^t \left(\phi_s' \nu_s - \delta_K(\nu_s) \right) ds \leq 0,$$

for all $\nu \in \hat{\mathcal{K}}_b$.

To show that $\phi \in K$, we now define $\bar{\nu} := \operatorname{argmin}_{\|\zeta\|=1} (\delta_K(\zeta) - \phi'\zeta)$ and use the previous inequality with

$$\nu := \mathbf{1}_{\{\delta_K(\bar{\nu}) - \phi'\bar{\nu} < 0\}}.$$

We deduce that $\min_{\|\zeta\|=1} (\delta_K(\zeta) - \phi'\zeta) \geq 0 \, dt \times d\mathbb{P}$, which, by Lemma 2.4, implies that $\phi \in K \, dt \times d\mathbb{P}$. Moreover, X is bounded from below, $C^0 \geq 0$ and therefore $\tilde{V}^{X_0, \phi} = X + C^0$, see step 1. above, is also bounded from below. Hence, $\phi \in \mathcal{A}_K^b$.
□

2.6 Problems

2.1 (Zero-coupon pricing) A *zero-coupon* of maturity T is a contract that pays 1 at T in exchange of $B_t(T)$ at $t \leq T$. We assume that $\mathcal{M}(\tilde{S}) \neq \emptyset$.

1. What is the link between β, $\mathcal{M}(\tilde{S})$ and the interval of viable prices at 0 for the zero-coupon of maturity T?
2. We assume that the risk-free interest rate r takes values between two constants $\underline{r} < \bar{r}$. Show that $B_t(T)$ belongs to the set $[e^{-(T-t)\bar{r}}, e^{-(T-t)\underline{r}}]$.
3. What is $B_t(T)$ if r is constant?

2.2 (Swap) We assume that $d = 1$. A swap is a product according to which two entities, say X and Y, exchange financial flows (called legs). X pays a fix flow ρ (fix leg) and Y pays in exchange a variable flow V (variable leg). At the origination of the contract the two legs have the same value. One needs to calculate the value of ρ, called *swap rate*, ensuring this. We assume that the payment dates are $0 < t_1 < \ldots < t_k = T$ and that the flow of the variable leg at t_i is L_{t_i}. Compute the swap rate ρ when the market is complete and the risk neutral measure is $\mathbb{Q}$.

2.3 (No early exercise for American calls) We consider a single risky asset S with price S_t at t. We denote by C_t the price at t of the European call maturing at T and of strike K, and by C_t^a the corresponding price for the American call. We denote by $B_t(T)$ the price at t of the zero-coupon paying 1 at T. The risk-free interest rate is assumed to be strictly positive. We finally assume that the market is arbitrage-free and that $\mathbb{P}[S_T > K \mid \mathcal{F}_t] > 0$ and $\mathbb{P}[S_T < K \mid \mathcal{F}_t] > 0$ $\mathbb{P} -$ a.s. for all $t < T$.

1. Show that $C_t^a \geq C_t$.
2. Show that $C_t > [S_t - KB_t(T)]^+$ for $t < T$.
3. Show that it is always better to sell the American call rather than to exercise it.
4. Deduce that $C_t^a = C_t$.

2.4 (Multivariate Black and Scholes model) The multivariate *Black and Scholes model* corresponds to the model of Sect. 2.1 in the case where r is a positive constant and (2.1) reads

$$S_t = S_0 + \int_0^t \text{diag}\,[S_s]\,\mu ds + \int_0^t \text{diag}\,[S_s]\,\sigma dW_s \ , \ t \in \mathbb{T}\,, \tag{2.13}$$

for two constants $\mu \in \mathbb{R}^d$ and $\sigma \in \mathbb{M}^{d,n}$ (recall that W is n-dimensional, while the dimension of S is d).

1. By applying Itô's Lemma, check that each component S^i of S is given by

$$S_t^i = S_0^i e^{(\mu^i - \frac{1}{2}\|\sigma^i\|^2)t + \langle \sigma^i, W_t \rangle} \ , \ t \in \mathbb{T}, \tag{2.14}$$

where σ^i denotes the i-th line of the matrix σ.
2. Does the condition (2.2) hold?

3. Deduce that

$$\tilde{S}^i_t = S^i_0 e^{(\mu^i - r - \frac{1}{2}\|\sigma^i\|^2)t + \langle \sigma^i, W_t \rangle}.$$ (2.15)

4. Show that (NA) implies

$$\exists\, z \in \mathbb{R}^d \text{ s.t. } z'\sigma = 0 \Rightarrow \langle z, \mu - r\mathbf{1}_d \rangle = 0\,,$$

where $\mathbf{1}_d = (1, \dots, 1) \in \mathbb{R}^d$.
5. Why can we reduce to the case $d \le n$?
6. Show that if

$$d \le n \text{ and } \{z'\sigma = 0 \Rightarrow \langle z, \mu - r\mathbf{1}_d \rangle = 0\}\ \forall\, z \in \mathbb{R}^d$$ (2.16)

then (NA) holds.
7. Construct the set of martingale measures under (2.16).
8. We assume that (2.16) holds and that $\sigma'\sigma$ is invertible. Show that the market is complete if and only if $d = n$. How can we interpret this result?
9. We assume that $d = n$ and that σ is invertible. Let $G \in \mathbb{L}^0(\mathbb{P})$, bounded from below, be the payoff of a European option. Show that

$$p(G) = \mathbb{E}^{\mathbb{Q}}\left[e^{-rT}G\right]$$ (2.17)

for a measure $\mathbb{Q}$. Express the density of $\mathbb{Q}$ with respect to $\mathbb{P}$.
10. Let us consider the preceding question in the case $G = g(S_T)$ where g is a non-negative function with linear growth that maps $\mathbb{R}^d$ to $\mathbb{R}$. Show that

$$p(G) = \mathbb{E}^{\mathbb{Q}}\left[e^{-rT}g(S_T)\right]$$
$$= \mathbb{E}\left[e^{-rT}g\left(\left(S_0 e^{(r-\|\sigma^i\|^2)T + \langle \sigma^i, N_T \rangle}\right)_{i=1,\dots,d}\right)\right],$$ (2.18)

where σ^i is the i-th line of σ and N_T denotes a random variable distributed according to the Gaussian law $\mathcal{N}(0, TI_d)$ under $\mathbb{P}$, in which I_d is the identity matrix.
11. From now on, we restrict to the univariate setting, i.e. $d = n = 1$. Let f denote the density of the Gaussian distribution with mean 0 and standard deviation 1. Let Φ be the corresponding cumulative function. We set

$$p^{\text{call}} = e^{-rT} \int_{-\infty}^{\infty} [S_0 e^{(r-\frac{1}{2}\sigma^2)T + \sigma\sqrt{T}x} - K]^+ f(x)dx,$$

where $K > 0$ is a constant.

a. What is the meaning of p^{call}?

b. Show that

$$p^{\text{call}} = \int_{-d_2}^{\infty} \left(S_0 e^{-\frac{1}{2}\sigma^2 T + \sigma\sqrt{T}x} - e^{-rT}K \right) f(x)\,dx$$

where

$$d_2 := \frac{\ln(S_0/K) + (r - \frac{1}{2}\sigma^2)T}{\sigma\sqrt{T}}.$$

c. Deduce that

$$p^{\text{call}} = S_0 \Phi(d_1) - e^{-rT}K\Phi(d_2)$$

where

$$d_1 := d_2 + \sigma\sqrt{T}.$$

12. Compute the right-hand side term in (2.18) explicitly in terms of Φ in the case where g is of the form

$$g(x) = \sum_{i=1}^{I} (a_i + b_i x)\mathbf{1}_{x \in]k_i, k_{i+1}]},$$

with $a_i, b_i \in \mathbb{R}$, $(k_i)_{i \le I}$ an increasing sequence of real numbers, $I \in \mathbb{N}$.
13. In the case $\sigma > 0$, compute the hedging price of the options with the following payoffs:

a. Digital option: $g(x) = \mathbf{1}_{x \ge K}$, $K > 0$.
b. Butterfly: $g(x) = (x - K + a)\mathbf{1}_{K-a \le x < K} + (a + K - x)\mathbf{1}_{K < x \le K+a}$, $K > a > 0$.
c. Straddle: $g(x) = (K - x)\mathbf{1}_{x \le K+a} + (x - K - 2a)\mathbf{1}_{K+a < x}$, $K, a > 0$.

2.5 (Time dependent coefficients) We consider a financial market in which the risk-free interest rate is set to zero, $r = 0$, and the dynamics of the only risky asset are

$$S_t = S_0 + \int_0^t S_s \mu_s\,ds + \int_0^t S_s \sigma_s\,dW_s, \quad t \le T.$$

The maps $t \in [0, T] \mapsto \mu_t \in \mathbb{R}$ and $t \in [0, T] \mapsto \sigma_t \ge 0$ are deterministic and continuous.

1. Absence of arbitrage opportunity:

a. Recall the definition of the no-arbitrage condition (NA), in this market.
b. Show that (NA) implies that $B := \{t \in [0, T] : \sigma_t = 0 \text{ and } \mu_t \ne 0\}$ has a zero Lebesgue measure.

c. Consider the measures defined by $\nu_\sigma(O) := \int_0^T \mathbf{1}_{t\in O} \ \sigma_t dt$ and $\nu_\mu(O) := \int_0^T \mathbf{1}_{t\in O} \ \mu_t dt$, for O a Borel set of $[0,T]$. Deduce from the above that ν_μ is dominated by μ_σ and deduce that there exists a measurable map λ such that $\mu = \sigma\lambda$, if (NA) holds.

d. Show that if there exists a bounded measurable map λ such that $\mu = \sigma\lambda$, then (NA) holds.

2. Completeness: We assume for this question that there exists at least one bounded measurable map λ such that $\mu = \sigma\lambda$.

a. Let $\mathbb{Q} \sim \mathbb{P}$ be a probability measure such that S is a $\mathbb{Q}$-martingale. Write down the dynamics of S in terms of a $\mathbb{Q}$-Brownian motion $W^\mathbb{Q}$.

b. Does the law of S under $\mathbb{Q}$ depend of $\mathbb{Q}$?

c. Can we hedge perfectly (replicate) the options of payoffs of the type $G := g(S_{t_1}, \ldots, S_{t_\kappa})$ (g being a continuous and bounded function, $t_1, \ldots, t_\kappa$ dates in $[0,T]$)?

2.6 (Barrier option in the Black and Scholes model) Our aim is to compute the hedging price p of an option whose payoff is $[S_T - K]^+ \mathbf{1}_{\{\tau > T\}}$ where $\tau := \inf\{t \geq 0 : S_t \geq A\}$ with $A > S_0 \vee K > 0$, $S_0, K > 0$ are constants. We consider the Black and Scholes model of Exercise 2.4 with $d = n = 1$.

1. Write down the density of the risk-neutral measure $\mathbb{Q}$ in the Black and Scholes model. We shall denote by $W^\mathbb{Q}$ the corresponding Brownian motion.

2. Find a measure $\bar{\mathbb{Q}} \sim \mathbb{Q}$ such that $\bar{W}$ defined by $\bar{W}_t := W_t^\mathbb{Q} + t(r - \sigma^2/2)/\sigma$ is a Brownian motion under $\bar{\mathbb{Q}}$.

3. Write down S and $H := d\bar{\mathbb{Q}}/d\mathbb{Q}$ in terms of $\bar{W}$. Show that

$$p = e^{-rT}\mathbb{E}^{\bar{\mathbb{Q}}}\left[H^{-1}[S_T - K]^+ \mathbf{1}_{\{\tau > T\}}\right].$$

4. Show that $\{\tau > T , S_T \geq K\} = \{\max_{t\in[0,T]} \bar{W}_t < \bar{A} , \bar{W}_T \geq \bar{K}\}$ where $\bar{A}, \bar{K}$ are real numbers to be computed.

5. We now compute the joint law of $\max_{t\in[0,T]} \bar{W}_t$ and $\bar{W}_T$ under $\bar{\mathbb{Q}}$. In the following, we fix $a, b \in \mathbb{R}$.

a. Let $\theta_a := \inf\{t \geq 0 : \bar{W}_t = a\}$ denote the first hitting time of $a > 0$ by $\bar{W}$, the first time at which W reaches the level a. Show that

$$\bar{\mathbb{Q}}\left[\max_{t\in[0,T]} \bar{W}_t \geq a , \bar{W}_T \leq b\right]$$

$$= \bar{\mathbb{Q}}\left[\theta_a \leq T , \bar{W}_T \leq b\right]$$

$$= \bar{\mathbb{Q}}\left[\theta_a \leq T , \bar{W}_T - \bar{W}_{\theta_a} \leq b - a\right].$$

b. By using the fact that θ_a is $\mathcal{F}_{\theta_a}$—measurable and that $\bar{W}_{T\vee\theta_a} - \bar{W}_{\theta_a}$ is independent of $\mathcal{F}_{\theta_a}$ and has the same law as $\bar{W}_{\theta_a} - \bar{W}_{T\vee\theta_a}$, show that

$$\bar{\mathbb{Q}}\left[\max_{t\in[0,T]} \bar{W}_t \geq a,\ \bar{W}_T \leq b\right] = \bar{\mathbb{Q}}\left[\theta_a \leq T,\ \bar{W}_T \geq 2a - b\right].$$

c. Deduce that, if $a > 0$ and $b \leq a$, then

$$\bar{\mathbb{Q}}\left[\max_{t\in[0,T]} \bar{W}_t \geq a,\ \bar{W}_T \leq b\right] = \bar{\mathbb{Q}}\left[\bar{W}_T \geq 2a - b\right].$$

d. Deduce the density f of $(\max_{t\in[0,T]} \bar{W}_t, \bar{W}_T)$ when $b \leq a$ and $a > 0$.
e. What can we say if $a \leq 0$?
f. What can we say if $b > a$?

6. Show that the price of the barrier option is given by

$$e^{-rT}\mathbb{E}^{\bar{\mathbb{Q}}}\left[\psi(\max_{t\in[0,T]} \bar{W}_t, \bar{W}_T)\right],$$

for some map ψ to be determined.
7. Do the explicit computations using the cumulative distribution function Φ of the standard normal distribution.
8. Compute the delta $\Delta(t, S_t)$ of the option at t, i.e. the derivative of the price at t with respect to the value of S_t.
9. Show that it can happen that $|\Delta(t, S_t)| \to \infty$ when $t \to T$ if $S_t \to A$ according to a certain speed.
10. Conclude on the difficulty to hedge such an option in practice.

2.7 (Option on a spread) We consider the Black and Scholes model of Exercise 2.4 with $d = n = 2$ and consider the option of payoff $g(S_T) = [S_T^1 - S_T^2]^+$. This is a *call on spread*. Set

$$\bar{H} := \tilde{S}_T^2/S_0^2.$$

1. Show that $\mathbb{E}^{\mathbb{Q}}\left[\bar{H}\right] = 1$.
2. Let $\bar{\mathbb{Q}}$ be the measure equivalent to $\mathbb{Q}$ with density $\bar{H}$ with respect to $\mathbb{Q}$. Show that the price of the option is

$$p = S_0^2\,\mathbb{E}^{\bar{\mathbb{Q}}}\left[\left[S_T^1/S_T^2 - 1\right]^+\right].$$

3. By using Girsanov's theorem, find a $\bar{\mathbb{Q}}$-Brownian motion and write down S_T^1/S_T^2 in terms of it.
4. Deduce a closed form formula for the price in terms of the cumulative distribution function Φ of the standard normal distribution.

2.8 (Forward contract on exchange rates) We consider a model in which two different markets are considered at the same time (domestic and foreign). The risk-free interest rate on the domestic market is r. The risk-free interest rate on the foreign market is r^f. Both are constant. We assume that one unit of the foreign currency is worth S units of the domestic currency, in which S follows (2.13) for $d = n = 1$ and $\sigma > 0$. We assume that there is no arbitrage. Let us consider a contract that pays at T one unit of the foreign currency.

1. What is worth this contract in the domestic currency 0 if the premium is paid at time 0? If the premium is paid at time T?
2. What relation should hold between r, r_f and S_0 if no arbitrage can be made?

2.9 (Call on forward contracts) Within the one dimensional Black and Scholes model, we denote by $\{F_t, t \in [0,T]\}$ the price of the forward contract on S of maturity $T' > 0$, i.e. that pays $S_{T'} - m$ at T', m being its premium. Given that its value is zero at $t = 0$, what is the value of m? Use the Black and Scholes formula to compute the price of a European call written on F with maturity $T \in [0,T']$ and strike $K > 0$ (i.e. defined by the payment $(F_T - K)^+$ at T).

2.10 (Asian option: geometric mean specification) [7] Let us consider the one dimensional Black and Scholes model with interest rate $r \geq 0$, and risky asset evolving according to

$$\frac{dS_t}{S_t} = b\,dt + \sigma dW_t \;,\; S_0 > 0,$$

where $b \in \mathbb{R}$, $\sigma > 0$ are given. An Asian option on geometric average is defined by the payoff paid at $T > 0$:

$$G := \left(\bar{S}_T - K\right)^+ \text{ where } \bar{S}_T = \exp\left(\frac{1}{T}\int_0^T \log(S_t)dt\right).$$

1. Let $\mathbb{Q}$ denote the risk neutral measure and $W^{\mathbb{Q}}$ be the corresponding Brownian motion. Write down the density of $\mathbb{Q}$ with respect to $\mathbb{P}$, and $W^{\mathbb{Q}}$ in terms of W.
2. Show that $\int_0^T W_t^{\mathbb{Q}}dt = \int_0^T (T-t)dW_t^{\mathbb{Q}}$.
3. Show that

$$\bar{S}_T = \bar{S}_0 e^{\bar{r}T - \frac{1}{2}\int_0^T \bar{\sigma}(t)^2 dt + \int_0^T \bar{\sigma}(t)dW_t^{\mathbb{Q}}},$$

---

[7]This kind of payoff does not exist in the market. However, it can be used as a control variate to price Asian options on the arithmetic average.

where

$$\bar{S}_0 = S_0 e^{-\sigma^2 T/12}, \ \bar{r} = \frac{r}{2}, \ \bar{\sigma}(t) = \sigma \left(1 - \frac{t}{T}\right).$$

4. Deduce from the above the law under $\mathbb{Q}$ of $\bar{S}_T$.
5. Deduce the price of the Asian option in terms of the Black and Scholes formula.
6. Explain how to construct an hedging portfolio.

2.11 (Chooser option)   Let us consider a risky asset S. We denote by $P(t, \theta, K)$ (resp. $C(t, \theta, K)$) the price at t of the European put (resp. call) of strike K and maturity θ, $0 < t \le \theta, K > 0$. We consider a *chooser option* of maturity $t_0 > 0$ on option of maturity $\theta > t_0$ and strike K, i.e. the buyer can choose at t_0 to receive a call or a put of strike K and maturity θ, written on S.

1. Show that this amounts to receiving at t_0 the payoff

$$\max \left\{P(t_0, \theta, K) , \ C(t_0, \theta, K)\right\} .$$

2. Show that, if no arbitrage is possible, then the value at t_0 of the chooser option is given by

$$P(t_0, \theta, K) + [S_{t_0} - K B_{t_0}(\theta)]^+ ,$$

where $B_s(\tau)$ is the price at s of the zero-coupon of maturity τ.
3. We assume (only for this question!) that the risk-free interest rate is deterministic. Show that in this case, and if no arbitrage can be made, the price at 0 of the chooser option is

$$P(0, \theta, K) + C(0, t_0, K B_{t_0}(\theta)) .$$

4. From now on, we assume that the dynamics of S and of the risk-free interest rate r are given by

$$S_t = S_0 + \int_0^t S_s r_s ds + \int_0^t S_s \sigma dW_s^1 , \ r_t = r_0 + bt + \rho W_t^1 + \nu W_t^2,$$

where $\sigma, \rho, \nu > 0$, $b \in \mathbb{R}$, and W^1, W^2 are two independent Brownian motions under the risk neutral measure $\mathbb{Q}$, that we assume to exist and be unique. We set $\vartheta := \sqrt{\rho^2 + \nu^2}$. By using the integration by parts formula, show that

$$\int_0^\theta r_s ds = \theta r_0 + b\frac{\theta^2}{2} + \int_0^\theta \rho(\theta - t) dW_t^1 + \int_0^\theta \nu(\theta - t) dW_t^2 .$$

Deduce that, for all $t \leq \tau$,

$$B_t(\tau) = e^{-r_t(\tau - t) + f(t, \tau)},$$

where f is a smooth deterministic function to be determined.

5. Deduce from the above the dynamics of $(B_t(\theta))_{t \leq \theta}$ and of $(S_t / B_t(\theta))_{t \leq \theta}$.
6. Show that the value at 0 of the payoff $[S_{t_0} - K]^+$ paid at t_0 is given by

$$p_0 := \mathbb{E}^{\mathbb{Q}} \left[e^{-\int_0^{t_0} r_s ds} [S_{t_0} - K B_{t_0}(\theta)]^+ \right].$$

7. Let $\mathbb{Q}_\theta$ denote the measure of density

$$\frac{d\mathbb{Q}_\theta}{d\mathbb{Q}} := e^{-\int_0^{t_0} r_s ds} B_{t_0}(\theta) / \mathbb{E}^{\mathbb{Q}} \left[e^{-\int_0^{t_0} r_s ds} B_{t_0}(\theta) \right].$$

Show that

$$p_0 = B_0(\theta) \mathbb{E}^{\mathbb{Q}_\theta} \left[[S_{t_0} / B_{t_0}(\theta) - K]^+ \right].$$

8. Show that

$$\frac{d\mathbb{Q}_\theta}{d\mathbb{Q}} = e^{-\frac{1}{2} \int_0^\theta (\rho^2 + v^2)(\theta - t)^2 dt - \int_0^\theta \rho(\theta - t) dW_t^1 - \int_0^\theta v(\theta - t) dW_t^2}.$$

9. Given $\kappa \geq 0$, $m \in \mathbb{R}$ and $v > 0$, we set

$$BS(m, v, \kappa) := \mathbb{E} \left[[e^Y - \kappa]^+ \right],$$

where $Y \sim \mathcal{N}(m, v)$, the Gaussian distribution with mean m and variance v. Compute p_0 in terms of $B_0(\theta)$, the function BS and the parameters of the model.

2.12 (Alternative formulation for the price of an American option) Let us consider a one dimensional setting, $d = n = 1$. The risk-free rate is set to zero, $r = 0$, and the risky asset evolves according to

$$S_t = S_0 + \int_0^t \sigma_s dW_s.$$

The process σ is assumed to be bounded predictable, and to verify $\inf_{0 \leq t \leq T} \sigma_t > 0$ $\mathbb{P}$-a.s. In the following, we let $\mathcal{A}$ be the set of real-valued predictable processes α such that $\int_0^T \alpha_s^2 ds < \infty$ $\mathbb{P}$-a.s.

1. Market completeness:

   a. Let $\mathbb{Q} \sim \mathbb{P}$. Show that there exists $\lambda \in \mathcal{A}$ such that $\mathbb{Q} = \mathbb{Q}^\lambda$, with $\mathbb{Q}^\lambda$ given by

   $$\frac{d\mathbb{Q}^\lambda}{d\mathbb{P}} = \exp\left(-\frac{1}{2}\int_0^T \lambda_s^2 ds + \int_0^T \lambda_s dW_s\right).$$

   b. Show that, for all $\lambda \in \mathcal{A}$ such that $\mathbb{Q}^\lambda$ is an equivalent probability measure, there exists a Brownian motion W^λ such that

   $$S_t = S_0 + \int_0^t \sigma_s \lambda_s ds + \int_0^t \sigma_s dW_s^\lambda.$$

   c. Deduce that $\mathbb{P}$ is the unique risk neutral measure.

   d. Deduce that any bounded $\mathbb{P}$-martingale M admits a predictable process $\phi^M \in \mathcal{A}$ such that

   $$M_t = M_0 + \int_0^t \phi_s^M dS_s, \quad 0 \le t \le T.$$

2. Alternative formulation for the American option price: We consider an American option that pays g_t if exercised at t. Here, $g = (g_t)_{t \le T}$ is bounded and predictable. We restrict to financial strategies $\phi \in \mathcal{A}$ for which there exists a constant $c_\phi \in \mathbb{R}$ such that

   $$V_t^{0,\phi} \ge -c_\phi \ \mathbb{P}-\text{a.s.} \ t \le T,$$

   and let $\mathcal{A}_b$ denote the corresponding set.

   Let p_0 be the super-hedging price of the American option. One has

   $$p_0 \le \sup_{\tau \in \mathcal{T}_0} \mathbb{E}\left[g_\tau\right].$$

   a. Why can we not work with strategies in $\mathcal{A}$? What is the role of the constraint imposed in the definition of $\mathcal{A}_b$?

   b. Using the very definition of the super-hedging price, explain why $p_0 \ge \sup_{\tau \in \mathcal{T}_0} \mathbb{E}\left[g_\tau\right]$.

   c. Let $\mathcal{M}_b(0)$ be the collection of all bounded martingales M starting from 0, i.e. $M_0 = 0$. Show that

   $$p_0 \le \sup_{\tau \in \mathcal{T}_0} \mathbb{E}\left[g_\tau - M_\tau\right],$$

and deduce that

$$p_0 \leq \mathbb{E}\left[\sup_{0 \leq t \leq T} (g_t - M_t)\right].$$

d. We admit that there exists a càdlàg adapted process Y^* such that

$$Y_t^* = \operatorname*{ess\,sup}_{\tau \in T_t} \mathbb{E}\left[g_\tau | \mathcal{F}_t\right], \quad 0 \leq t \leq T.$$

Briefly justify why Y^* is a bounded supermartingale.

e. Deduce that there exists a martingale M^* and a predictable non decreasing process A^* such that $Y^* = p_0 + M^* - A^*$ and $M_0^* = A_0^* = 0$ (in the following, we admit that M^* and A^* are bounded).

f. Justify the following identities

$$\sup_{0 \leq t \leq T} (Y_t^* - M_t^*) = \sup_{0 \leq t \leq T} (p_0 - A_t^*) = p_0.$$

g. Deduce from the above that

$$p_0 = \inf_{M \in \mathcal{M}_b(0)} \mathbb{E}\left[\sup_{0 \leq t \leq T} (g_t - M_t)\right] = \mathbb{E}\left[\sup_{0 \leq t \leq T} (g_t - M_t^*)\right].$$

h. Assuming that we can simulate the path of M^*, propose a numerical method to compute the price of the American option. What information does bring the above formulation in the case where M^* is not known explicitly?

2.13 (Call on zero-coupon in the Vasiček's model) Let us consider a market in which there exists a unique risk-neutral measure $\mathbb{Q}$. Under this measure the dynamics of the risk-free interest is

$$dr_t = (a - br_t)\, dt + \sigma dW_t^{\mathbb{Q}}, \tag{2.19}$$

where $\left(W_t^{\mathbb{Q}}\right)_{t \geq 0}$ is a Brownian motion, and where a, b, σ and r_0 are positive constants.

1. Show that the process X defined by $X_t = e^{bt} r_t$ satisfies

$$dX_t = e^{bt} a\, dt + e^{bt} \sigma dW_t^{\mathbb{Q}}, \quad X_0 = r_0.$$

2. Deduce that, for all $h > 0$,

$$r_{t+h} = e^{-bh} r_t + a\frac{1 - e^{-bh}}{b} + \int_t^{t+h} e^{-b(t+h-s)} \sigma dW_s^{\mathbb{Q}}.$$

3. Show that the random variable r_{t+h} has a Gaussian distribution conditionally to $\mathcal{F}_t$ under $\mathbb{Q}$, for $h > 0$. Compute its mean $m(t, t+h)$ and variance $v(t, t+h)$ in terms of r_t, a, b, σ and h.

  We now consider a zero-coupon of maturity T.

4. Explain why its price at t is

$$B_t(T) := \mathbb{E}^{\mathbb{Q}}\left[\exp\left(-\int_t^T r_s ds\right) | \mathcal{F}_t\right]. \tag{2.20}$$

5. By using the integration by parts formula, show that

$$\int_t^T \int_t^s e^{-b(s-u)} dW_u^{\mathbb{Q}} \, ds = \int_t^T \int_u^T e^{-b(s-u)} ds \, dW_u^{\mathbb{Q}} \tag{2.21}$$

  and deduce from this the law of $\int_t^T r_s ds$ conditionally to r_t.

6. Deduce that

$$B_t(T) = e^{-f_1(t,T)r_t + f_2(t,T)}$$

  where f_1 and f_2 are two functions that are C^2 with respect to t.

7. Let us consider the European call of strike K and maturity $\tau \in (0, T)$, i.e. of payoff $[B_\tau(T) - K]^+$ at τ. Show that its price at 0 is

$$C(0) = \mathbb{E}^{\mathbb{Q}}\left[\exp\left(-\int_0^\tau r_t dt\right)[B_\tau(T) - K]^+\right] = B_0(T)\mathbb{E}^{\mathbb{Q}_\tau}\left[[B_\tau(T) - K]^+\right]$$

  with[8]

$$\frac{d\mathbb{Q}_\tau}{d\mathbb{Q}} := \frac{\exp\left(-\int_0^\tau r_t dt\right)}{\mathbb{E}^{\mathbb{Q}}\left[\exp\left(-\int_0^\tau r_t dt\right)\right]}. \tag{2.22}$$

8. Recalling (2.21), find a process λ such that

$$\frac{d\mathbb{Q}_\tau}{d\mathbb{Q}} = \exp\left(-\frac{1}{2}\int_0^\tau \lambda_t^2 dt + \int_0^\tau \lambda_t dW_t^{\mathbb{Q}}\right).$$

9. Let W^τ be the $\mathbb{Q}_\tau$-Brownian motion associated to $W^{\mathbb{Q}}$ and λ by Girsanov's theorem. Provide the dynamics of the interest rate in terms of W^τ.

10. Provide the dynamics of $B(T)$ in terms of W^τ, and of f_1 and f_2.

11. Can we compute explicitly $C(0)$? If yes, explain briefly how.

---

[8]$\mathbb{Q}_\tau$ is called *forward risk-neutral measure* of maturity τ.

2.14 (Two factors Vasiček's model) We consider the two factors Vasiček's model in which the risk-free interest rate $(r_t)_{t \geq 0}$ evolves according to

$$r_t = r_0 + \int_0^t (a_s - b r_s) ds + \int_0^t \sigma dW_s^{\mathbb{Q}},$$

where the process $(a_t)_{t \geq 0}$ has the dynamics

$$a_t = a_0 + \int_0^t (\alpha - \beta a_s) ds + \int_0^t \gamma dZ_s^{\mathbb{Q}},$$

with $Z^{\mathbb{Q}} := \rho W^{\mathbb{Q}} + \sqrt{1 - \rho^2} \bar{W}^{\mathbb{Q}}$. Here, $W^{\mathbb{Q}}$ and $\bar{W}^{\mathbb{Q}}$ are two independent Brownian motions under the risk neutral measure $\mathbb{Q}$ (assumed to exist and be unique), and, $b, \beta, \alpha, \sigma, \gamma, r_0, a_0 > 0$ and $\rho \in [0, 1]$ are constants.

1. Provide the dynamics of $\check{r}$ defined by $\check{r}_t = e^{bt} r_t$, $t \geq 0$, in terms of $(\check{a}, \check{\sigma})$ defined by $(\check{a}_t, \check{\sigma}_t) = e^{bt}(a_t, \sigma)$, $t \geq 0$. Provide then the dynamics of $\hat{a}$ defined by $\hat{a}_t = e^{\beta t} a_t$, $t \geq 0$, in terms of $(\hat{\alpha}, \hat{\gamma})$ defined by $(\hat{\alpha}_t, \hat{\gamma}_t) = e^{\beta t}(\alpha, \gamma)$, $t \geq 0$.

2. By using the stochastic version of Fubini's lemma, deduce that there exist deterministic functions f_1, f_2, f_3 such that

$$r_s = e^{-b(s-t)} r_t + f_1(t,s) a_t + f_2(t,s) + e^{-bs} \left(\int_t^s \check{\sigma}_u dW_u + \int_t^s f_3(u,s) \hat{\gamma}_u dZ_u^{\mathbb{Q}} \right)$$

   for all $s \geq t$. (No need to compute them explicitly.)

3. Deduce that there exist deterministic functions m_r, m_a, m_0 such that

$$B_t(T) := \mathbb{E}^{\mathbb{Q}}[e^{-\int_t^T r_s ds} | \mathcal{F}_t] = e^{-m_r(t,T) r_t - m_a(t,T) a_t - m_0(t,T)}$$

   for all $T \geq t$. (No need to compute them explicitly.)

4. Deduce that $B(T)$ has the dynamics

$$\frac{dB_t(T)}{B_t(T)} = r_t dt - (m_r(t,T)\sigma + m_a(t,T)\rho\gamma) dW_t^{\mathbb{Q}} - m_a(t,T) \sqrt{1 - \rho^2} \gamma d\bar{W}_t^{\mathbb{Q}}.$$

5. Explain which change of measure allows one to pass from $\mathbb{Q}$ to the forward risk-neutral measure[9] of maturity $t \geq 0$.

6. Explain how one can deduce from the above a Black and Scholes type formula for the price at 0 of a call on zero-coupon delivered at t and of strike $K > 0$.

7. Can we hedge $[B_t(T) - K]^+$ from time 0 by only trading the zero-coupon maturing at T?

---

[9]It is defined in (2.22) above.

8. Show that there exist $p \in \mathbb{R}$ and a $\mathbb{R}^2$-valued square integrable predictable process $(\zeta, \bar{\zeta})$ such that

$$e^{-\int_0^t r_s ds}[B_t(T) - K]^+ = p + \int_0^t \zeta_s dW_s^{\mathbb{Q}} + \int_0^t \bar{\zeta}_s d\bar{W}_s^{\mathbb{Q}}.$$

9. How can we hedge the payoff $[B_t(T) - K]^+$ paid at t by trading two zero-coupons of different maturities? Provide in terms of $(\zeta, \bar{\zeta})$ the amount that should be invested at each time t in each of them.

2.15 (G2++ model and options on obligation) We consider the two factors G2++ model in which the dynamics of the interest rate $(r_t)_{t \geq 0}$ under the risk-neutral measure $\mathbb{Q}$ are

$$r_t = x_t + y_t + \varphi_t,$$

where φ is continuous deterministic process, x and y are solutions of

$$x_t = -\int_0^t ax_s ds + \sigma W_t^{\mathbb{Q}}, \ y_t = -\int_0^t by_s ds + \gamma Z_t^{\mathbb{Q}}, \ t \geq 0,$$

with $a, b, \sigma, \gamma > 0$, and $Z^{\mathbb{Q}} := \rho W^{\mathbb{Q}} + (1 - \rho^2)^{\frac{1}{2}} \bar{W}^{\mathbb{Q}}$ where $\rho \in [-1, 1]$. Here, $W^{\mathbb{Q}}$ and $\bar{W}^{\mathbb{Q}}$ are two independent Brownian motions under $\mathbb{Q}$.

1. Show that

$$r_t = x_s e^{-a(t-s)} + y_s e^{-b(t-s)} + \sigma \int_s^t e^{-a(t-u)} dW_u^{\mathbb{Q}} + \gamma \int_s^t e^{-b(t-u)} dZ_u^{\mathbb{Q}} + \varphi_t.$$

2. Deduce the law of r_t under $\mathbb{Q}$ given $\mathcal{F}_s$, $t \geq s$.
3. Let us fix $T \geq t$. Show that, given $\mathcal{F}_t$, $I(t, T) := \int_t^T (x_s + y_s) ds$ follows a Gaussian distribution under $\mathbb{Q}$, with mean

$$m(t, T; a)x_t + m(t, T; b)y_t,$$

where $m(t, T; c) := c^{-1}(1 - e^{-c(T-t)})$ for $c > 0$, and of variance

$$V(t, T) = \int_t^T \left[(\sigma m(u, T; a) + \rho \gamma m(u, T; b))^2 + (1 - \rho^2)\gamma^2 m^2(u, T; b) \right] du.$$

4. Show that the price of the zero-coupon of maturity T at $t \leq T$ is given by

$$B_t(T) := e^{-\int_t^T \varphi_s ds - m(t, T; a)x_t - m(t, T; b)y_t + \frac{1}{2}V(t, T)}.$$

5. Provide the dynamics of $(B_t(T))_{t \leq T}$ in terms of W^Q and Z^Q, $m(\cdot, T; a)$ and $m(\cdot, T; b)$.

6. Show that, for $t \leq T$,

$$e^{-\int_t^T \varphi_s ds} = \frac{B_0(T) e^{-\frac{1}{2} V(0,T)}}{B_0(t) e^{-\frac{1}{2} V(0,t)}}.$$

How can we use this to find φ given a, b, σ, γ and the prices observed on the market? How can we then fix the values of a, b, σ and γ based on the observation of liquid option prices?

7. Explain how to obtain a Black and Scholes type formula for the prices of calls and puts on zero-coupon.

8. We denote by $b(t, x, y, \tau)$ the price at t given $x_t = x$ and $y_t = y$ of the zero-coupon of maturity $\tau \in [t, T]$.

   a. Show that $x \mapsto b(t, x, y, \tau)$ is continuous, strictly decreasing, and maps $\mathbb{R}$ onto $(0, \infty)$, for each given $t < \tau$ and y.

   b. Let $o(t, x, y)$ be the price at t given $x_t = x$ and $y_t = y$ of the bond that pays a coupon $c_i > 0$ at each date t_i, $1 \leq i \leq N$, where $t < t_1 < \ldots < t_N \leq T$. Write down $o(t, x, y)$ in terms of $(b(t, x, y, t_i))_{1 \leq i \leq N}$ and show that there exists a function $\hat{x} : \mathbb{R} \mapsto \mathbb{R}$ such that $o(t, \hat{x}(y), y) = K$, where $K > 0$.

   c. Show that $K = \sum_{i=1}^{N} c_i K_i(y)$ where $K_i(y) := b(t, \hat{x}(y), y, t_i)$.

   d. Show that, for each $1 \leq i \leq N$, $o(t, x, y) \geq K$ if and only if $b(t, x, y, t_i) \geq K_i(y)$.

   e. Deduce that

$$\mathbb{E}^Q[e^{-\int_0^t r_s ds}[o(t, x_t, y_t) - K]^+] = \sum_{i=1}^{N} c_i \mathbb{E}^Q[e^{-\int_0^t r_s ds}[b(t, x_t, y_t) - K_i(y_t)]^+].$$

   f. How can one use this formula to compute in an efficient way the price of a call on a bond when $\rho = 0$?

9. Explain how to hedge a European option on a bond by trading in zero-coupons.

2.16 (Libor market model) The aim of this exercise is to construct a model for the forward Libor market which is consistent with market data. We assume in the following that the risk-free interest rate is a bounded continuous adapted process r and that $\mathbb{P}$ is already the risk neutral measure (for ease of notations). In the following, W is a d-dimensional Brownian motion.

1. Given $t \leq \tau$, we denote by $B_t(\tau)$ the value at t of the zero-coupon of maturity τ (paying 1). We assume that, for all $\tau \geq 0$, there exist two continuous square integrable and adapted processes μ^τ and Γ^τ with values in $\mathbb{R}$ and $\mathbb{R}^d$ such that

$$B_t(\tau) = B_0(\tau) + \int_0^t \mu_s^\tau ds + \int_0^t B_s(\tau) \Gamma_s^\tau dW_s, \quad t \leq \tau.$$

    a. Show that $\mu^\tau = rB(\tau)1_{[0,\tau]}$.
    b. What should be the value of Γ^τ after τ?

2. Fix $\tau \geq 0$. Set $\beta_\tau := e^{-\int_0^\tau r_s ds}$ and let $\mathbb{P}^\tau$ be defined by $d\mathbb{P}^\tau/d\mathbb{P} = H^\tau := \beta_\tau/B_0(\tau)$.

    a. Show that $\mathbb{P}^\tau$ is a $\mathbb{P}$-equivalent probability measure.
    b. By using the fact that $B_\tau(\tau) = 1$ and the dynamics of $B(\tau)$ above, show that

$$H^\tau = e^{-\frac{1}{2}\int_0^\tau |\Gamma_s^\tau|^2 ds + \int_0^\tau \Gamma_s^\tau dW_s}.$$

    c. What can we say on $W^\tau := W - \int_0^\cdot (\Gamma_s^\tau)' ds$?

3. We fix $\delta > 0$. Given $\tau \geq 0$, we now define the forward Libor rate process L^τ by

$$1 + \delta L^\tau := B(\tau)/B(\tau + \delta).$$

    a. Given $t \leq \tau$, provide a financial interpretation of L_t^τ.
    b. Show that L^τ is a martingale under $\mathbb{P}^{\tau+\delta}$ on $[0, \tau]$.
    c. Deduce that

$$dL_t^\tau = (1 + \delta L_t^\tau)(\Gamma_t^\tau - \Gamma_t^{\tau+\delta})dW_t^{\tau+\delta}, \quad t \leq \tau.$$

4. We now consider a family of deterministic maps $\lambda^\tau : \mathbb{R}_+ \to \mathbb{R}, \tau \geq 0$. One would like to write down a model in which

$$dL_t^\tau = L_t^\tau \lambda_t^\tau dW_t^{\tau+\delta}, \quad t \leq \tau. \qquad (2.23)$$

    a. Show that L_t^τ has a log-normal distribution, for each $t \leq \tau$. Why this specification can be useful in terms of option pricing?
    b. What should be the link between $(\Gamma^\tau)_{\tau\geq 0}$ and $(\lambda^\tau)_{\tau\geq 0}$?
    c. If $W^{\tau+\delta}$ and λ^τ are given, how can we deduce W^τ so that the previous relation is satisfied? (Write down W^τ in terms of $W^{\tau+\delta}$, λ^τ and L^τ.)
    d. Briefly explain how to construct a model in which (2.23) is satisfied for all $\tau \geq 0$ given $(\lambda^\tau)_{\tau\geq 0}$.

2.17 (Swap market model) The aim of this exercise is to construct a log normal model for the *forward swap rate*. We fix the payment dates $\tau_i = \tau_0 + i\delta, i = 1, \ldots, \kappa$ with $\kappa \geq 1$, an integer, and $\tau_0 > 0$. We let $(r_t)_{t\geq 0}$ be the risk-free interest rate, which we assume to be adapted, strictly positive and satisfying $\mathbb{E}\left[\int_0^\infty |r_s|^p ds\right] < \infty$ for all $p \geq 1$. Let $\beta_t := e^{-\int_0^t r_s ds}$. Given $i \geq 0$ and $0 \leq t \leq \tau_i$, we denote by $B_t(\tau_i) \in \mathbb{L}^\infty(\mathcal{F}_t)$ the price at t of the zero-coupon of maturity τ_i (paying 1), and by $L_{\tau_i} \in \mathbb{L}^p(\mathcal{F}_{\tau_i})$, for all $p \geq 1$, the Libor rate[10] determined at τ_i for the time period

---

[10]See Exercise 2.16.

$[\tau_i, \tau_{i+1}]$. Given $j \geq 0$ and $t \leq \tau_j$, we let $\rho_t^j \in \mathbb{L}^0(\mathcal{F}_t)$ be the value[11] of the swap rate fixed at t so that the value of the swap on Libor paying at τ_{k+1} the amount $\delta(L_{\tau_k} - \rho_t^j)$ for each $k \in \{j, \ldots, \kappa - 1\}$ is zero. We admit that ρ^j takes positive values, for all $j \leq \kappa - 1$.

We also assume that the market is complete and that the processes $\beta B(\tau_j)$ are $\mathbb{P}$-martingales on $[0, \tau_j]$, for all $j \leq \kappa$, i.e. $\mathbb{P}$ is already the risk-neutral measure. In the following, W is n-dimensional Brownian motion.

1. *Computation of the forward swap rate*: Let us fix $t \leq \tau_j$ with $j < \kappa$.

   a. Explain why ρ^j must satisfy

   $$0 = \sum_{i=j}^{\kappa-1} \mathbb{E}\left[\beta_{\tau_{i+1}}(L_{\tau_i} - \rho_t^j)|\mathcal{F}_t\right]/\beta_t.$$

   b. Explain the following identities, for $i \geq j$,

   $$\mathbb{E}\left[\beta_{\tau_{i+1}}\rho_t^j|\mathcal{F}_t\right] = \beta_t B_t(\tau_{i+1})\rho_t^j,$$

   $$\mathbb{E}\left[\beta_{\tau_{i+1}}L_{\tau_i}|\mathcal{F}_t\right] = \mathbb{E}\left[\beta_{\tau_i}B_{\tau_i}(\tau_{i+1})(\frac{1}{\delta B_{\tau_i}(\tau_{i+1})} - \frac{1}{\delta})|\mathcal{F}_t\right] = \frac{\beta_t}{\delta}(B_t(\tau_i) - B_t(\tau_{i+1})).$$

   c. Deduce that

   $$\rho_t^j = \left(B_t(\tau_j) - B_t(\tau_\kappa)\right)/(\delta\hat{N}_t^j) \text{ where } \hat{N}_t^j := \sum_{i=j}^{\kappa-1} B_t(\tau_{i+1}).$$

2. *Martingale property and dynamics*: Given $j \leq \kappa - 1$, we now define $\bar{\mathbb{P}}^j \sim \mathbb{P}$ by

   $$\frac{d\bar{\mathbb{P}}^j}{d\mathbb{P}} = N^j/\hat{N}_0^j \text{ where } N^j := \sum_{i=j}^{\kappa-1} \beta_{\tau_{i+1}}.$$

   a. Show that ρ^j is a $\bar{\mathbb{P}}^j$-martingale on $[0, \tau_j]$.
   b. By using the martingale representation theorem, show that there exists a predictable process α^j, $\mathbb{P}$ − a.s. square integrable, such that

   $$\frac{d\bar{\mathbb{P}}^j}{d\mathbb{P}} = e^{-\frac{1}{2}\int_0^{\tau_\kappa} \|\alpha_s^j\|^2 ds + \int_0^{\tau_\kappa} \alpha_s^j dW_s}.$$

   c. What can we say on $\bar{W}^j$ defined by $\bar{W}_t^j = W_t - \int_0^t (\alpha_s^j)' ds$ for $t \leq \tau_\kappa$?

---

[11]The process ρ^j is the forward swap rate process associated to the swap starting at τ_j.

d. By using the martingale representation theorem again, show that there exists
a predictable process λ^j such that

$$d\rho_t^j = \rho_t^j \lambda_t^j d\bar{W}_t^j, \; t \le \tau_j. \tag{2.24}$$

3. *Link between* $\bar{\mathbb{P}}^j$ *and* $\bar{\mathbb{P}}^{j-1}$, for $j \in \{1,\dots,\kappa-1\}$ *given:* We set $Z^j := B(\tau_j)/\hat{N}^j$
and $Z^{j,-1} := B(\tau_j)/\hat{N}^{j-1}$.

a. Show that Z^j (resp. $Z^{j,-1}$) is a $\bar{\mathbb{P}}^j$-martingale (resp. $\bar{\mathbb{P}}^{j-1}$-martingale) on $[0,\tau_j]$
and deduce that there exists an adapted process σ^j such that $dZ_t^j = \sigma_t^j d\bar{W}_t^j$, on
$[0,\tau_j]$.
b. Show that $Z^{j,-1} = Z^j/(1+Z^j)$.
c. Deduce that, on $[0,\tau_j]$,

$$dZ_t^{j,-1} = \sigma_t^{j,-1} d\bar{W}_t^j - \frac{\sigma_t^j}{(1+Z_t^j)}(\sigma_t^{j,-1})' dt$$

with

$$\sigma^{j,-1} := \frac{\sigma^j}{1+Z^j} - \frac{Z^j \sigma^j}{(1+Z^j)^2}.$$

d. Deduce from 2c and 3c that, on $[0,\tau_j]$,

$$\bar{W}_t^{j-1} = \bar{W}_t^j + \int_0^t \left(\alpha_s^j - \alpha_s^{j-1}\right)' ds$$

with

$$(\alpha^j - \alpha^{j-1})(\sigma^{j,-1})' = -\sigma^j(\sigma^{j,-1})'/(1+Z^j).$$

e. Show by using 1c that, for all $i \in \{j,\dots,\kappa-1\}$,

$$\delta\rho^i = \left(\sum_{k=i+1}^{\kappa} \prod_{r=i}^{k-1} \frac{B(\tau_{r+1})}{B(\tau_r)}\right)^{-1} - \left(1 + \sum_{k=i+1}^{\kappa-1} \prod_{r=k}^{\kappa-1} \frac{B(\tau_r)}{B(\tau_{r+1})}\right)^{-1},$$

and deduce that knowing ρ^i for all $i \in \{j,\dots,\kappa-1\}$ implies knowing
$B(\tau_l)/B(\tau_i)$ for all $l, i \in \{j,\dots,\kappa\}$.
f. Deduce that one can compute σ^j and Z^j in terms of $(\rho^i, \lambda^i)_{j\le i \le \kappa-1}$.

4. *Conclusion:*

a. Deduce from the above an algorithm to construct a model that is consistent
with the forward swap rate observed on the market and such that (2.24) holds
for each $j \le \kappa - 1$, given a family $(\lambda^j)_{j \le \kappa-1}$.

b. Why does the construction of a log normal model can be useful to price swaptions?

2.18 (Doubling strategies) The aim of this exercise is to show that, in continuous time, imposing constraints on the strategies is crucial in order to avoid trivial arbitrages.

Let W be a one-dimensional Brownian motion. Define J on $[0, T)$ by

$$J_t := \int_0^t \gamma_s dW_s \, , \ t \in [0, T) \, , \ \text{where } \gamma_s := \frac{1}{\sqrt{T - s}} \text{ pour } s \in [0, T) \, .$$

1. Let Z be defined on $[0, T)$ by $Z_t := W_{\ln \frac{T}{T-t}}$. Show that the processes J and Z have the same law on $[0, T)$ (hint: use the time change theorem, Theorem 7.7).
2. We recall that, for all $a \in \mathbb{R}$, $\tau_a := \inf\{t \geq 0 \ : \ W_t = a\} < \infty \ \mathbb{P} - $a.s. Deduce from the above that $\theta := \inf\{t \in [0, T] \ : \ J_t = 1\} < T \ \mathbb{P} - $a.s., with the convention $\inf \emptyset = \infty$.
3. Show that the process ψ defined by $\psi_t := \gamma_t \mathbf{1}_{t < \theta}$ is well-defined on $[0, T]$, is adapted and satisfies

$$\int_0^T |\psi_t|^2 dt < \infty \ \ \mathbb{P} - \text{a.s.}$$

4. Let us now consider the Black and Scholes model in which the dynamics of the risky asset is

$$S_t = S_0 + \int_0^t S_s \sigma dW_s \, ,$$

with $\sigma, S_0 > 0$. We assume that the risk-free interest rate is equal to zero. We set $\phi_t := \psi_t / (\sigma S_t)$ for $t \leq T$. Show that ϕ is well-defined on $[0, T]$, is adapted and satisfies

$$\int_0^T |\phi_t|^2 dt < \infty \ \ \mathbb{P} - \text{a.s.}$$

5. Write down the dynamics of the wealth process V induced by the strategy ϕ, and starting from 0 at 0. Show that $V_T = 1 \ \mathbb{P} - $a.s.
6. Deduce that we cannot find a constant c such that

$$\mathbb{P}\left[V_t \geq -c \ \forall \, t \in [0, T]\right] = 1.$$

(hint : show that otherwise one would have $\mathbb{E}[V_T] \leq 0$.)
7. Comment.

2.19 (Viable prices under portfolio constraints) We use the setting of Sect. 2.5. Let G be a bounded $\mathcal{F}_T$-measurable random variable. We assume that $p_K(G)$ and $p_K(-G)$ are finite.

1. Define the notion of viable prices under portfolio constraints.
2. Show that any price in $\,]-p_K(-G), p_K(G)[$ is viable.
3. Show that the buyer (resp. seller) can make an arbitrage if the price p at which the option is sold satisfies $p > p_K(G)$ (resp. $-p > p_K(-G)$).
4. Show that $p_K(G)$ is viable if and only if $p_K(G) = \mathbb{E}^{\mathbb{Q}_0}[\tilde{G}]$.
5. What about $-p_K(-G)$?
6. Deduce the set of viable prices for G.

Corrections

2.1

1. The interval of viable prices at 0 is given by the relative interior of $[-p(-1), p(1)]$ with $p(\pm 1) := \sup\{\mathbb{E}^{\mathbb{Q}}[\pm\beta_T], \ \mathbb{Q} \in \mathcal{M}(\tilde{S})\}$.
2. Use the above, or show that an arbitrage can be made if this condition is violated.
3. $B_t(T) = \beta_T / \beta_t$.

2.2 ρ must satisfy

$$\sum_{i=1}^{k} \mathbb{E}^{\mathbb{Q}}[\beta_{t_i}\rho] = \sum_{i=1}^{k} \mathbb{E}^{\mathbb{Q}}[\beta_{t_i}L_{t_i}].$$

2.3

1. Otherwise, we buy the American option, sell the call, and exercise at T. This leads to an arbitrage.
2. Let $\mathbb{Q}$ be a martingale measure. Then, $C_t \geq \mathbb{E}^{\mathbb{Q}}[\beta_T[S_T - K]^+ \mid \mathcal{F}_t]$. By Jensen's inequality and the assumption, this implies that $C_t > [\mathbb{E}^{\mathbb{Q}}[\beta_T(S_T - K) \mid \mathcal{F}_t]]^+ = [S_t - B_t(T)K]^+$.
3. Then, $C_t^a > [S_t - B_t(T)K]^+ \geq [S_t - K]^+$. The market value is strictly bigger than what would be obtained if exercised.
4. It is never exercised, and therefore provides the same flows as a European call.

2.4

1. Apply Itô's lemma to each $\ln(S^i)$.
2. Yes, it suffices to compute the Laplace transform of a Gaussian distribution.
3. This follows from the definition of β.
4. Apply Theorem 2.4.
5. If $d > n$, then there exists $J \subset \{1, \cdots, d\}$ with $|J| = d - n$ and $(\lambda_j)_{j \in J} \in (\mathbb{R}^n)^{|J|}$ such that $\sigma^j = \sum_{i \in J^c} \lambda_j^i \sigma^i$ for all $j \in J$, with $J^c := \{1, \cdots, d\} \setminus J$. The above

then implies that $\mu^j - r = \sum_{i \in J^c} \lambda_j^i (\mu^i - r)$ for each $j \in J^c$. Hence, $d\tilde{S}^j = \sum_{i \in J^c} (S^j/S^i) \lambda_j^i d\tilde{S}^i$. The dynamics of the assets with index in J can be replicated by using assets in J^c.

6. Apply Farkas' Lemma to find a constant risk premium. Hence, Theorem 2.7 can be applied.

7. It is all the $\mathbb{Q}$ such that

$$dQ/d\mathbb{P} = e^{-\frac{1}{2} \int_0^T |\lambda_s|^2 ds + \int_0^T \lambda_s' dW_s}$$

in which λ is an adapted $\mathbb{P}$ − a.s. square integrable process such that $\mathbb{E}[d\mathbb{Q}/d\mathbb{P}] = 1$ and $\sigma\lambda = \mu - r\mathbf{1}_d \, dt \times d\mathbb{P}$-a.e.

8. We can assume $d \leq n$. If $d < n$ there exists more than one deterministic solution to $\sigma\lambda = \mu - r\mathbf{1}_d \, dt \times d\mathbb{P}$-a.e. Each of them yields a different risk neutral measure, so that the market is not complete. Hence, we must have $d = n$. Since $\sigma'\sigma$ is invertible, there is only one solution λ to $\sigma\lambda = \mu - r\mathbf{1}_d \, dt \times d\mathbb{P}$-a.e., it is given by $(\sigma'\sigma)^{-1}\sigma'(\mu - r\mathbf{1}_d)$. Hence, there is only one martingale measure and the market is complete.

9. The density of $\mathbb{Q}$ is given by the risk premium $\lambda^{\mathbb{Q}} := \sigma^{-1}(\mu - r\mathbf{1}_d)$.

10. By Girsanov theorem, the process defined by $W_t^{\mathbb{Q}} := W_t - \lambda^{\mathbb{Q}}t$ is a $\mathbb{Q}$-Brownian motion. It suffices to rewrite S in the exponential form and in terms of $W^{\mathbb{Q}}$ and use that $\sigma\lambda^{\mathbb{Q}} = \mu - r\mathbf{1}_d$.

11. a. This is the price of the call, as given above but in integral form.

    b. $x \geq -d_2$ is equivalent to $S_0 e^{(r-\frac{1}{2}\sigma^2)T + \sigma\sqrt{T}x} \geq K$.

    c. $f(x) = (2\pi)^{-\frac{1}{2}} e^{-x^2/2}$. In the previous expression, combine the terms $\sigma\sqrt{T}x$ and $-x^2/2$ to obtain a quadratic form in x, so as the recognise the density of a Gaussian distribution: $\int_{-d_2} e^{-\frac{1}{2}\sigma^2 T + \sigma\sqrt{T}x} f(x)dx = \Phi(d_1)$. The second term is clear.

12. Use the same trick as above.

13. Apply the previous result.

2.5

1. a. See the definition.

   b. Apply Theorem 2.4.

   c. The domination follows from the previous question. One then applies the Radon-Nikodym theorem.

   d. In this case, we can apply Girsanov's theorem to find $\mathbb{Q} \sim \mathbb{P}$ such that $\tilde{S}$ is a $\mathbb{Q}$-martingale.

2. a. Apply Girsanov's theorem.

   b. No since in any case $\ln S_t = \ln S_0 + \int_t^T (r - \sigma_s^2/2)ds + \int_0^t \sigma_s dW_s^{\mathbb{Q}}$ in which the first term is deterministic and $\int_0^t \sigma_s dW_s^{\mathbb{Q}}$ follows (under $\mathbb{Q}$) a Gaussian distribution with zero mean and variance $\int_t^T \sigma_s^2 ds$.

   c. Yes, by Corollary 2.18 since the above implies that $p(G) = -p(-G)$.

2.6

1. See Exercise 2.4.
2. Apply Girsanov's theorem.
3. Apply Girsanov's theorem and use $d\mathbb{Q}/d\bar{\mathbb{Q}} = H^{-1}$.
4. Take the log in the exponential formulation of S in terms of $\bar{W}$.
5. a. Use $\{\max_{t\in[0,T]} \bar{W}_t \geq a\} = \{\theta_a \leq T\}$ and $\bar{W}_{\theta_a} = a$.
   b. Use that $\bar{\mathbb{Q}}\left[\theta_a \leq T, \bar{W}_T - \bar{W}_{\theta_a} \leq b - a\right] = \bar{\mathbb{Q}}\left[\theta_a \leq T, -\bar{W}_T + \bar{W}_{\theta_a} \leq b - a\right]$ with $\bar{W}_{\theta_a} = a$.
   c. Using the above, the density f follows by differentiating with respect to a and then b (up to the sign):

   $$f(a,b) = \frac{2(2a-b)}{T\sqrt{2\pi T}} e^{-\frac{(2a-b)^2}{2T}}.$$

   d. For $a \leq 0$, this probability is one since $\bar{W}_0 = 0$.
   e. If $b \geq a$ then $\bar{\mathbb{Q}}[\max_{t\in[0,T]} \bar{W}_t \geq a, \bar{W}_T \geq b] = \bar{\mathbb{Q}}[\bar{W}_T \geq b]$. Moreover, the previous formula applied with $a = b$ implies

   $$\bar{\mathbb{Q}}\left[\max_{t\in[0,T]} \bar{W}_t \geq a\right] = \bar{\mathbb{Q}}\left[\max_{t\in[0,T]} \bar{W}_t \geq a, \bar{W}_T \leq a\right] + \bar{\mathbb{Q}}\left[\max_{t\in[0,T]} \bar{W}_t \geq a, \bar{W}_T \geq a\right]$$

   $$= \bar{\mathbb{Q}}[\bar{W}_T \geq a] + \bar{\mathbb{Q}}[\bar{W}_T \geq a].$$

   Hence, $\bar{\mathbb{Q}}[\max_{t\in[0,T]} \bar{W}_t \geq a, \bar{W}_T \leq b] = 2\bar{\mathbb{Q}}[\bar{W}_T \geq a] - \bar{\mathbb{Q}}[\bar{W}_T \geq b]$.
6. We have

   $$p = e^{-rT}\mathbb{E}^{\bar{\mathbb{Q}}}\left[H^{-1}[S_T - K]^+ \mathbf{1}_{\{\tau > T\}}\right]$$

   $$= e^{-rT}\mathbb{E}^{\bar{\mathbb{Q}}}\left[e^{-\frac{1}{2}\lambda^2 T + \lambda\bar{W}_T}[S_0 e^{\sigma\bar{W}_T} - K]^+ \mathbf{1}_{\{\max_{t\in[0,T]} \bar{W}_t < \bar{A}\}}\right]$$

   with $\lambda := (r - \sigma^2/2)$ and $\bar{A} := \ln(A/S_0)/\sigma$. Hence, we must take

   $$\psi(a,b) := e^{-\frac{1}{2}\lambda^2 T + \lambda b}[S_0 e^{\sigma b} - K]^+ \mathbf{1}_{\{a < \bar{A}\}}.$$

7. Write $[S_0 e^{\sigma b} - K]^+ = (S_0 e^{\sigma b} - K)\mathbf{1}_{\{b \geq \ln(K/S_0)/\sigma\}}$ and use similar computations as in Exercise 2.4.
8. Just differentiate.
9. Obtained from the formula.
10. If the delta explodes, then the hedge cannot be applied in practice.[12]

---

[12] The solution is to hedge and price an option with a slightly higher barrier.

2.7

1. This is the martingale property.
2. Clearly

$$p = S_0^2 \mathbb{E}^{\mathbb{Q}} \left[\frac{S_T^2}{S_0^2} \left[S_T^1/S_T^2 - 1 \right]^+ \right] = S_0^2 \, \mathbb{E}^{\tilde{\mathbb{Q}}} \left[\left[S_T^1/S_T^2 - 1 \right]^+ \right].$$

3. Apply Girsanov's theorem.
4. Argue as in Question 12 of Exercise 2.4 with $S = S^1/S^2$.

2.8 To obtain one unit of the foreign currency at T, we can buy $e^{-r^f T}$ units of the foreign currency at 0, it costs $S_0 e^{-r^f T}$. Hence, $m = S_0 e^{-r^f T}$ if the premium is paid at 0. If the premium is paid at T, then we can borrow the amount $S_0 e^{-r^f T}$ at time 0. The reimbursement at T is $S_0 e^{-r^f T} e^{rT}$ and therefore $m = S_0 e^{(r-r^f)T}$.

   The premium $S_0 e^{(r-r^f)T}$ is the price of the forward contract. It is listed on the market. The relation between S_0, r and r^f is that this formula should match the price of the forward contract.

2.9 We must have $m = e^{-rT'} S_0$. Similarly, $F_t = e^{-r(T'-t)} S_t$. The payoff of the call is $(e^{-r(T'-T)} S_T - K)^+ = e^{-r(T'-T)} (S_T - e^{r(T'-T)} K)^+$. Hence, it suffices to use the Black and Scholes formula with the strike $e^{r(T'-T)} K$.

2.10

1. $W_t^{\mathbb{Q}} = W_t + \lambda t$ with $\lambda = (b-r)/\sigma$.
2. Use the integration by part formula, or apply Itô's lemma to $(T-t) W_t^{\mathbb{Q}}$ (recall that $W_0^{\mathbb{Q}} = 0$).
3. Do the computations using the previous result.
4. $\ln(\bar{S}_T)$ has a Gaussian distribution since $\bar{\sigma}$ is deterministic. Its mean is $\ln \bar{S}_0 + \bar{r}T - \frac{1}{2} \int_0^T \bar{\sigma}(t)^2 dt$ and its variance is $\int_0^T \bar{\sigma}(t)^2 dt$.
5. Because $\ln(\bar{S}_T)$ has a Gaussian distribution. This amounts to having an interest rate $\bar{r}$ and a volatility $(\int_0^T \bar{\sigma}(t)^2 dt/T)^{\frac{1}{2}}$ (but do not forget that the payoff is discounted with r and not $\bar{r}$).
6. We can delta hedge with S.

2.11

1. This is by construction.
2. $\max\{C(t_0, \theta, K), P(t_0, \theta, K)\} = P(t_0, \theta, K) + [C(t_0, \theta, K) - P(t_0, \theta, K)]^+$. Then use the call put parity.
3. This is by construction
4. We have $B_t(\tau) = \mathbb{E}^{\mathbb{Q}}[e^{-\int_t^\tau r_s ds} \mid \mathcal{F}_t]$ in which

$$\int_t^\tau r_s ds = \int_t^\tau \left(r_t + b(t-s) + \rho(W_t^1 - W_s^1) + \nu(W_t^2 - W_s^2) \right) ds.$$

By the same arguments as in 2. of Exercise 2.10, it has a Gaussian distribution, conditionally to $\mathcal{F}_t$. Hence, $B_t(\tau)$ is given by the Laplace transform of a Gaussian distribution.

5. The drif of $(B_t(\theta))_{t \leq \theta}$ has to be $rB(\theta)$ because its discounted value is a martingale. Its volatility comes from r in the previous formula. Hence,

$$dB_t(\theta)/B_t(\theta) = r_t - (\theta - t)(\rho dW_t^1 + v dW_t^2).$$

This gives $S_t/B_t(\theta)$ as the ratio of two log-normal processes.

6. p_0 is the only viable price in this complete market.

7. It suffices to factorise $B_{t_0}(\theta)$ in front of the payoff, use the definition of $\mathbb{Q}_\theta$ and the fact that $\mathbb{E}^{\mathbb{Q}}\left[e^{-\int_0^{t_0} r_s ds} B_{t_0}(\theta)\right] = B_0(\theta)$.

8. Just use the dynamics of $B(\theta)$.

9. We are back to a log normal distribution under $\mathbb{Q}_\theta$, as in the Black and Scholes formula.

2.12

1. a. Apply the martingale representation theorem to $(\mathbb{E}[d\mathbb{Q}/d\mathbb{P} \mid \mathcal{F}_t])_{t \leq T}$.
   b. Apply Girsanov's theorem.
   c. If $\lambda \not\equiv 0$, then S is not a martingale.
   d. Apply the martingale representation theorem and use the dynamics of S.

2. a. Otherwise, arbitrage could be made. It ensures that wealth processes are supermartingale.
   b. $V_\tau^{p,\phi} \geq g_\tau$ implies $p \geq \mathbb{E}[g_\tau]$.
   c. The optional sampling theorem implies that $\mathbb{E}[M_\tau] = 0$ and we know that $p_0 \leq \sup_{\tau \in \mathcal{T}_0} \mathbb{E}[g_\tau]$. Moreover, $\sup_{0 \leq t \leq T}(g_t - M_t) \geq g_\tau - M_\tau$ $\mathbb{P}$ − a.s. for all $\tau \in \mathcal{T}_0$.
   d. It is bounded because g is. The dynamic programming principle implies that it has to be a supermartingale.
   e. Use the Doob-Meyer decomposition.
   f. $A_0 = 0$ and $A \geq 0$.
   g. By 2.c and 2.f,

$$p_0 \leq \inf_{M \in \mathcal{M}_b(0)} \mathbb{E}\left[\sup_{0 \leq t \leq T}(g_t - M_t)\right] \leq \mathbb{E}\left[\sup_{0 \leq t \leq T}(g_t - M_t^*)\right] = p_0.$$

   h. Simulate (an approximation of) $\sup_{0 \leq t \leq T}(g_t - M_t^*)$ and compute the mean of the simulations to obtain an estimate of p_0. A guess M on M^*, that can be simulated, would at least provide an upper-bound.

2.13

1. Apply Itô's lemma.
2. This is the same as above.
3. Given $\mathcal{F}_t$, $\int_t^{t+h} e^{-b(t+h-s)} \sigma dW_s^Q$ has a Gaussian distribution with mean 0 and variance $\int_t^{t+h} e^{-2b(t+h-s)} \sigma^2 ds$.
4. This is the only viable price in this complete market.
5. Just apply the integration by parts formula. The law is Gaussian again.
6. Use the Laplace transform of the Gaussian distribution.
7. Use that $\mathbb{E}^Q \left[\exp \left(- \int_0^\tau r_t dt \right) \right] = B_0(\tau)$.
8. (2.21) provides $\int_0^\tau r_t dt$ in terms of integrals with respect to W^Q, from which λ is inferred.
9. Apply Girsanov's theorem.
10. Apply Girsanov's theorem.
11. Yes, we are back to a formulation similar to the one of the Black and Scholes model.

2.14 The reasoning is the same as in Exercise 2.13, only the computations are a little bit more involved. We only answer questions 6 to 9. First, we can compute the option price explicitly by arguing as in Exercise 2.13 (the fact that two Brownian motions are involved does not change much). We cannot hedge the call with only the bond $B(T)$. The reason is that the martingale representation of the claim and the dynamics of B involve both W^Q and $\bar{W}^Q$, terms that cannot be matched. One needs two bonds with different maturities. Since the dynamics of the bond are explicit, see question 4, it is easy to deduce the hedging strategy in terms of ζ and $\bar{\zeta}$, which can in turn be computed explicitly from the explicit expression of the option's price.

2.15 Questions 1 to 7 are solved by arguments used in Exercises 2.13 and 2.14. We only answer Question 8.

a. Use the formula just obtained in the preceding question.
b. $o(t, x, y) = \sum_{i=1}^N c_i b(t, x, y, t_i)$. Use the formula to show that this map is invertible.
c. Use $o(t, \hat{x}(y), y) = \sum_{i=1}^N c_i b(t, \hat{x}(y), y, t_i)$.
d. This follows from the monotonicity of o and b in x: $[o(t, x_t, y_t) - K]^+ = (o(t, x_t, y_t) - K)\mathbf{1}_{\{x \le \hat{x}(y_t)\}}$ and $[b(t, x_t, y_t) - K_i(y_t)]^+ = (b(t, x_t, y_t) - K_i(y_t))\mathbf{1}_{\{x \le \hat{x}(y_t)\}}$.
e. Use the previous question.
f. If $\rho = 0$, then x and y are independent. Hence, $p(y) := \mathbb{E}^Q[e^{-\int_0^t r_s ds}[b(t, x_t, y_t) - K_i(y_t)]^+ \mid y]$ can be computed explicitly by arguing as in Exercise 2.13. This can serve as a variance reduction technique for the design of a Monte-Carlo estimator: one simulates y to obtain realisations of $p(y)$ and takes the mean to obtain an estimate of the price.

2.16

1. a. The discounted bond price has to be a martingale.
   b. It should be 0.
2. a. $H^\tau > 0$.
   b. Write that $H^\tau = H^\tau B_\tau(\tau) = \beta_\tau B_\tau(\tau)/B_0(\tau)$ and use the dynamics of $B(\tau)$.
   c. It is a $\mathbb{Q}_\tau$-Brownian motion by Girsanov's theorem.
3. a. L_t^τ is an interest rate with linear composition fixed at time t for borrowing money at τ and reimbursing together with the interests at $\tau + \delta$ (forward rate).
   b. Fix $s \le t \le \tau$. Use the equality $\tilde{B}(\tau + \delta) = \mathbb{E}^{\mathbb{Q}}[\beta_{\tau+\delta} \mid \mathcal{F}.]$ and the martingale property of $\tilde{B}(\tau)$:

$$\mathbb{E}^{\mathbb{Q}_{\tau+\delta}}[1 + \delta L_t^\tau \mid \mathcal{F}_s] = \mathbb{E}^{\mathbb{Q}}[\beta_{\tau+\delta} B_t(\tau)/B_t(\tau + \delta) \mid \mathcal{F}_s]/\mathbb{E}^{\mathbb{Q}}[\beta_{\tau+\delta} \mid \mathcal{F}_s]$$

$$= \mathbb{E}^{\mathbb{Q}}[\beta_t B_t(\tau) \mid \mathcal{F}_s]/\tilde{B}_s(\tau + \delta)$$

$$= \tilde{B}_s(\tau)/\tilde{B}_s(\tau + \delta).$$

   c. Since it is a $\mathbb{Q}_{\tau+\delta}$-martingale, there should be no dt terms. The form of the ratio in its definition in terms of zero-coupon bond prices leads to this formulation.
4. a. Because λ^τ is deterministic. This leads to explicit prices for caplets and floorlets.
   b. $\lambda^\tau = (1 + \delta L_t^\tau)(\Gamma_t^\tau - \Gamma_t^{\tau+\delta})$ by the previous questions.
   c. $W^{\tau+\delta}$ and λ^τ gives $\Gamma^\tau - \Gamma^{\tau+\delta}$ which is enough to compute W^τ in terms of $W^{\tau+\delta}$ (look at the expression of each of them in terms of W).
   d. In practice, one fixes a longer maturity τ^*. We fix the dynamics of L^{τ^*} in terms of a Brownian motion called $W^{\tau*}$. We let $\mathbb{Q}_{\tau*}$ be the measure under which it is a Brownian motion. Then the dynamics of $L^{\tau^*-\delta}$ is deduced as above for some Brownian motion $W^{\tau*-\delta}$ (under a measure called $\mathbb{Q}_{\tau*-\delta}$ deduced from $\mathbb{Q}_{\tau*}$ by the above), and we go on backward to get $L^{\tau^*-2\delta}$, $L^{\tau^*-3\delta}$, and so on.

2.17

1. a. The two legs should have the same value.
   b. Use that $\tilde{B}(\tau_{i+1})$ is a martingale with value 1 at t_{i+1}, and the definition of the Libor rate.
   c. Combine a. and b.
2. a. Use similar arguments as in 3.b of Exercise 2.16. For $s \le t \le \tau_j$:

$$\mathbb{E}^{\tilde{p}^j}\left[\rho_t^j \mid \mathcal{F}_s\right] = \mathbb{E}\left[(\sum_{i=j}^{\kappa-1} \beta_{\tau_{i+1}})\frac{B_t(\tau_j) - B_t(\tau_\kappa)}{\delta \hat{N}_t^j} \mid \mathcal{F}_s\right]/(\sum_{i=j}^{\kappa-1} \tilde{B}_s(\tau_{i+1}))$$

$$= \mathbb{E}\left[(\sum_{i=j}^{\kappa-1} \tilde{B}_t(\tau_{i+1}))\frac{B_t(\tau_j) - B_t(\tau_\kappa)}{\delta \hat{N}_t^j} \mid \mathcal{F}_s\right]/(\sum_{i=j}^{\kappa-1} \tilde{B}_s(\tau_{i+1}))$$

$$= \mathbb{E}\left[\tilde{B}_t(\tau_j) - \tilde{B}_t(\tau_\kappa) \mid \mathcal{F}_s\right] / (\sum_{i=j}^{\kappa-1} \tilde{B}_s(\tau_{i+1}))$$

$$= (\tilde{B}_s(\tau_j) - \tilde{B}_s(\tau_\kappa)) / (\sum_{i=j}^{\kappa-1} \tilde{B}_s(\tau_{i+1})).$$

   b. Apply it to $\mathbb{E}[d\bar{\mathbb{P}}^j/d\mathbb{P} \mid \mathcal{F}.] > 0$.
   c. It is a Brownian motion under $\bar{\mathbb{P}}^j$ by Girsanov's theorem.
   d. Because ρ^j is a martingale under $\bar{\mathbb{P}}^j$.
3. a. Argue as above.
   b. It follows from direct computations.
   c. Apply Itô's lemma in the preceding formula.
   d. Compare the representations of $Z^{j,-1}$ in terms of $\bar{W}^j$ and $\bar{W}^{j-1}$.
   e. This is obtained by expanding terms.
   f. Once $B(\tau_l)/B(\tau_i)$ is known for all $l, i \in \{j, \ldots, \kappa\}$ σ^j, we deduce σ^j and Z^j by using their definition. From the previous question, they are given in terms of $(\rho^i, \lambda^i)_{j \leq i \leq \kappa-1}$.
4. a. First compute the dynamics of $\rho^{\kappa-1}$ in terms of a Brownian motion $\bar{W}^{\kappa-1}$, and go backward to deduce how to construct the other Brownian motions $\bar{W}^{\kappa-2}$, $\bar{W}^{\kappa-3}$, and so on.
   b. Because they can be written as calls on the swap rate. It allows to use a Black and Scholes type formula.

2.18

1. By the time change theorem, we can find a Brownian motion $\tilde{W}$ such that $W_{\ln \frac{T}{T-t}} = \int_0^{\ln \frac{T}{T-t}} dW_s = \int_0^t \gamma_s d\tilde{W}_s$. Since γ is deterministic, they have the same law.
2. The first hitting time $\tilde{\tau}_1$ of 1 by Z satisfies by construction: $\ln(T/(T-\tilde{\tau}_1)) = \tau_1 < \infty$ $\mathbb{P}$ − a.s.
3. Since $\theta < 1$ $\mathbb{P}$ − a.s., it is $\mathbb{P}$ − a.s. bounded.
4. The path of S is also $\mathbb{P}$ − a.s. bounded (because continuous).
5. $V = J_{.\wedge\theta}$. Hence, $V_T = 1$ $\mathbb{P}$ − a.s.
6. Otherwise, V would be a supermartingale, and we would therefore have $0 \geq \mathbb{E}[V_T] = 1$.
7. If we do not set a lower bound on admissible wealth processes, or a stronger integrability condition on admissible strategies, the notion of absence of arbitrage does not make sense. In the above, we start from 0 and reach 1 with probability 1.

2.19

1. It should not lead to an arbitrage neither for the seller nor for the buyer, using constrained strategies.
2. If $p < p_K(G)$, the seller cannot super-hedge the claim. Similarly, if $-p < p_K(-G)$ then the buyer cannot super-hedge $-G$.
3. They can strictly super-hedge G and $-G$ respectively.
4. If $p_K(G)$ is viable, then we can find $\phi \in \mathcal{A}_K^b$ such that $V_T^{p_K(G),\phi} \geq G \ \mathbb{P}$ – a.s. but $\mathbb{P}\left[V_T^{p_K(G),\phi} > G\right] = 0$. Since G is bounded, we can assume that $V^{p_K(G),\phi}$ is (by stopping the strategy when reaching the upper-bound). Hence, $\tilde{V}^{p_K(G),\phi}$ is a $\mathbb{Q}_0$-martingale. Discounting and taking expectation under $\mathbb{Q}_0$ implies $p_K(G) = \mathbb{E}^{\mathbb{Q}_0}[\tilde{V}_T^{p_K(G),\phi}] = \mathbb{E}^{\mathbb{Q}_0}[G]$. Conversely, if $p_K(G) = \mathbb{E}^{\mathbb{Q}_0}[\tilde{G}]$ then we cannot find we can find $\phi \in \mathcal{A}_K^b$ such that $V_T^{p_K(G),\phi} \geq G \ \mathbb{P}$ – a.s. and $\mathbb{P}\left[V_T^{p_K(G),\phi} > G\right] > 0$ (again use the martingale property).
5. Similarly, $-p_K(-G)$ is viable if and only if $p_K(-G) = \mathbb{E}^{\mathbb{Q}_0}[-\tilde{G}]$.
6. The set of viable prices is

$$]-p_K(-G), p_K(G)[\ \cup \{p_K(G)\}\mathbf{1}_{\{p_K(G)=\mathbb{E}^{\mathbb{Q}_0}[\tilde{G}]\}} \cup \{-p_K(-G)\}\mathbf{1}_{\{p_K(-G)=\mathbb{E}^{\mathbb{Q}_0}[-\tilde{G}]\}}.$$

Chapter 3
Optimal Management and Price Selection

This chapter is dedicated to the resolution of portfolio management problems and to the study of partial hedging strategies, based on a risk criteria. We shall mainly appeal to convex duality and calculus of variations arguments that turn out to be very powerful in complete markets: they will allow us to find explicit solutions.

As explained in the preceding chapter, there is not a unique way to define a viable price for a derivative when the market is incomplete or when portfolio constraints are added. We shall provide here some risk based criteria which permits to select one price within the interval of viable prices, and apply a partial hedging strategy according to this criteria.

3.1 Optimal Management

Let us consider a financial agent whose preferences are modelled by a *utility function U*, i.e. a concave non-decreasing function. We assume that

U1. the closure of dom(U) is $\mathbb{R}_+$;
U2. U is C^2, strictly increasing and strictly concave;
U3. *Inada's conditions* :

$$\lim_{x \searrow 0} \partial U(x) = +\infty \text{ and } \lim_{x \nearrow \infty} \partial U(x) = 0.$$

The aim of the agent is to maximise the expectation under $\mathbb{P}$ of the utility of her terminal wealth. Given an initial endowment $v > 0$, she must solve the optimisation problem

$$u(v) := \sup_{\phi \in \mathcal{A}_U} \mathbb{E}\left[U(V_T^{v,\phi}) \right] \tag{3.1}$$

© Springer International Publishing Switzerland 2016
B. Bouchard, J.-F. Chassagneux, *Fundamentals and Advanced Techniques in Derivatives Hedging*, Universitext, DOI 10.1007/978-3-319-38990-5_3

where

$$\mathcal{A}_U := \left\{ \phi \in \mathcal{A} \ : \ V^{v,\phi} \geq 0 \ \mathbb{P} - \text{a.s. and } U(V_T^{v,\phi})^- \in \mathbb{L}^1(\mathbb{P}) \right\} .$$

3.1.1   Duality in Complete Markets

We first assume that the market is complete, in the sense that $\mathcal{M}(\tilde{S}) = \{\mathbb{Q}\}$. Then, standard tools of convex duality allow to solve this problem "explicitly".

We first introduce the *Fenchel transform* $\bar{U}$ associated to U:

$$\bar{U}(y) := \inf_{x \in \text{dom}(U)} (xy - U(x)) . \tag{3.2}$$

The condition U3 implies that it is well-defined on $(0, \infty)$ and finite at 0 if and only if U is bounded. Moreover, the conditions U2 and U3 imply that the optimum in (3.2) is reached by a unique point $\hat{x}(y) > 0$ satisfying $\partial U(\hat{x}(y)) = y$. Hence,

$$\bar{U}(y) := I(y)y - U(I(y)) \quad \text{where} \quad I(y) := (\partial U)^{-1}(y) . \tag{3.3}$$

Fix $y > 0$ and $\phi \in \mathcal{A}_U$. By (3.2),

$$U(V_T^{v,\phi}) \leq V_T^{v,\phi} y \beta_T H_T - \bar{U}(y \beta_T H_T) \text{ where } H_T := \frac{d\mathbb{Q}}{d\mathbb{P}} ,$$

which, by the last assertion of Theorem 2.7, implies

$$\mathbb{E}\left[U(V_T^{v,\phi}) \right] \leq \mathbb{E}\left[vy - \bar{U}(y \beta_T H_T) \right] .$$

We conclude that

$$\sup_{\phi \in \mathcal{A}_U} \mathbb{E}\left[U(V_T^{v,\phi}) \right] \leq \inf_{y > 0} \left(vy - \mathbb{E}\left[\bar{U}(y \beta_T H_T) \right] \right) . \tag{3.4}$$

The next lemma will be used to show that equality indeed holds, which will then provide the optimal strategy.

Lemma 3.1 *If* $\mathbb{E}^{\mathbb{Q}}\left[\beta_T I(y \beta_T H_T) \right] < \infty$ *for all* $y \in (0, \infty)$, *then there exists a unique* $\hat{y} > 0$ *such that*

$$\mathbb{E}^{\mathbb{Q}}\left[\beta_T I(\hat{y} \beta_T H_T) \right] = v . \tag{3.5}$$

Proof The conditions U2 and U3 imply that $\mathbb{P} - $a.s. the image of $(0, \infty)$ by $y \mapsto I(y \beta_T H_T)$ contains $(0, \infty)$ and that this application is $\mathbb{P} - $a.s. continuous and strictly decreasing. By using the monotone convergence theorem, one can then check that

$y \in (0, \infty) \mapsto \mathbb{E}^{\mathbb{Q}} [\beta_T I(y\beta_T H_T)]$ is continuous, strictly decreasing and that the image of $(0, \infty)$ by this application contains $(0, \infty)$. $\qquad \square$

By the preceding lemma and (3.3), we have

$$\mathbb{E}\left[U(\hat{V}) \right] = v\hat{y} - \mathbb{E}\left[\bar{U}(\hat{y}\beta_T H_T) \right] \quad \text{with} \quad \hat{V} := I(\hat{y}\beta_T H_T) .$$

On the other hand, (3.5) and Theorem 2.17 imply that there exists $\hat{\phi} \in \mathcal{A}$ such that $V_T^{v,\hat{\phi}} = \hat{V}$. The inequality (3.4) allows then one to conclude that $\hat{\phi}$ is the optimal strategy. One checks that $\hat{\phi} \in \mathcal{A}_U$ by using the inequality $\bar{U}(y) \leq y - U(1)$ and (3.3). One can now state the following existence result, which also characterises the optimal terminal wealth.

Theorem 3.1 *Under the conditions of Lemma 3.1, there exists an optimal manage-ment strategy $\hat{\phi}$ and the corresponding wealth process satisfies*

$$V_T^{v,\hat{\phi}} = I(\hat{y}\beta_T H_T),$$

where $\hat{y}$ is the unique solution of

$$\mathbb{E}^{\mathbb{Q}} [\beta_T I(\hat{y}\beta_T H_T)] = v .$$

In the above, the resolution is "explicit" in the sense that we know exactly what should be the terminal value of the optimal portfolio. It remains to compute $\hat{\phi}$. This can be done explicitly in certain examples, see e.g. Exercise 3.2 below. Moreover, by (2.5), one should have

$$\tilde{V}_T^{v,\hat{\phi}} = v + \int_0^T \hat{\phi}_t' \tilde{\sigma}_t dW_t^{\mathbb{Q}}$$

and the condition $\mathbb{E}^{\mathbb{Q}} \left[\tilde{V}_T^{v,\hat{\phi}} \right] = v$ implies that the $\mathbb{Q}$-supermartingale $\tilde{V}^{v,\hat{\phi}}$ is a $\mathbb{Q}$-martingale.

Remark 3.2 The condition of Lemma 3.1 is only used to show that there exists $\hat{y}$ satisfying $\mathbb{E}^{\mathbb{Q}} [\beta_T I(\hat{y}\beta_T H_T)] = v$. One could assume it upfront.

3.1.2 Extension to Incomplete Markets

The duality approach used above can be extended to incomplete markets. However, it does not really help to determine the optimal strategy in general, it only provides an existence result.

Under suitable conditions (but not very restrictive), see [44] and [54], one can obtain that

$$\sup_{\phi \in \mathcal{A}_U} \mathbb{E}\left[U(V_T^{v,\phi})\right] = \inf_{y>0, \mathbb{Q} \in \mathcal{M}_{\mathrm{loc}}(\tilde{S})} \left(vy - \mathbb{E}\left[\bar{U}(y\beta_T \frac{d\mathbb{Q}}{d\mathbb{P}})\right]\right).$$

One can also prove the existence of optimisers $V_T^{v,\hat{\phi}}$ and $(\hat{y}, \hat{\mathbb{Q}})$, and that they are related by the identity

$$V_T^{v,\hat{\phi}} = I(\hat{y}\beta_T \frac{d\hat{\mathbb{Q}}}{d\mathbb{P}}).$$

Unfortunately, the dual problem is now an optimal control problem on a family of martingale measures, which is in general as difficult to solve as the original one.

3.1.2.1   Logarithmic Utility

Let us now specialise to $U(x) = \ln(x)$, a case for which an explicit resolution is possible, even in incomplete market.

We consider the following market model: the risk-free interest rate r is a predictable process and the dynamics of the risky assets is given by

$$S = S_0 + \int_0^{\cdot} \mathrm{diag}\,[S_s]\, \mu_s ds + \int_0^{\cdot} \mathrm{diag}\,[S_s]\, \sigma_s dW_s,$$

where μ, σ are predictable, taking values in $\mathbb{R}^d$ and $\mathbb{M}^{d,n}$. We assume that r, μ and σ are bounded, for simplicity.

Since the utility function is only defined on $(0, \infty)$, it is natural to restrict ourselves to strategies leading to strictly positive wealth processes. One way to ensure this is to describe the strategy in terms of the proportion π of the wealth invested in the risky assets: π^i is the proportion of the wealth invested in the risky asset S^i.

Then, the dynamics of the wealth process, starting from v, is

$$V^{v,\pi} = v + \int_0^{\cdot} V_s^{v,\pi} \pi_s' \mathrm{diag}\,[S_s]^{-1}\, dS_s + \int_0^{\cdot} (V_s^{v,\pi} - V_s^{v,\pi}\pi_s'\mathbf{1}_d)r_s ds,$$

where $\mathbf{1}_d$ is the vector of $\mathbb{R}^d$ with all entries equal to 1. We restrict ourselves to strategies π which are predictable processes taking values in a closed convex set K and such that

$$\mathbb{E}\left[\int_0^T |\pi_s|^2 ds\right] < \infty.$$

We denote the corresponding set by Π_K. It is easily deduced from Itô's lemma that $V_T^{v,\pi}$ is equal to

$$v \exp\left(\int_0^T \left(r_s + \pi_s'(\mu_s - r_s \mathbf{1}_d) - |\pi_s'\sigma_s|^2/2\right) ds + \int_0^T \pi_s'\sigma_s dW_s\right).$$

This implies that $U(V_T^{v,\pi})$ is given by

$$\ln(V_T^{v,\pi}) = \ln v + \int_0^T \left(r_s + \pi_s'(\mu_s - r_s \mathbf{1}_d) - \|\pi_s'\sigma_s\|^2/2\right) ds$$

$$+ \int_0^T \pi_s'\sigma_s dW_s,$$

so that maximising $\mathbb{E}\left[U(V_T^{v,\pi})\right]$ over Π_K is equivalent to

$$\sup_{\pi \in \Pi_K} \mathbb{E}\left[\int_0^T \left(\pi_s'(\mu_s - r_s \mathbf{1}_d) - \|\pi_s'\sigma_s\|^2/2\right) ds\right].$$

The solution is obtained by maximising inside the expectation:

$$\hat{\pi}_s := \arg\max_{k \in K} \left(k'(\mu_s - r_s \mathbf{1}_d) - \|k'\sigma_s\|^2/2\right), \; s \leq T.$$

If $\hat{\pi}$ is square integrable, which is always the case if K is bounded, then this is the solution.

A particular case is when $\sigma\sigma'$ is $dt \times d\mathbb{P}$-a.e. invertible. If $K = \mathbb{R}^d$, i.e. there is no constraint, then

$$\hat{\pi}_s = (\sigma_s\sigma_s')^{-1}(\mu_s - r_s \mathbf{1}_d), \; s \leq T.$$

3.1.3 Indifference Price

When the market is incomplete, the price of a European option $G \in \mathbb{L}^0$ is not uniquely defined, there exists a whole interval of viable prices, see Chaps. 1 and 2. One way to select one price relies on the notion of *indifference price* that was first suggested by Hodges and Neuberger [36]. One first considers the above optimal management problem:

$$u(v; 0) := \sup_{\phi} \mathbb{E}\left[U(V_T^{v,\phi})\right].$$

Then, one considers the same problem but in the case where the agent sells the option of payoff G at a price p. In this case, his wealth is increased by p at 0 and reduced by G at T. Hence, he computes

$$u(v + p; 1) := \sup_{\phi} \mathbb{E}\left[U(V_T^{v+p,\phi} - G) \right] .$$

The indifference price p^I is defined by

$$p^I := \inf\{ p \in \mathbb{R} \: : \: u(v + p; 1) \geq u(v; 0)\} .$$

It is easy to check that $p^I = p(G) = -p(-G)$ if the market is complete and $|G| \in L_S$. In any case, it always holds that $p^I \leq p(G)$ if $G^- \in L_S$.

One can also define a notion of *marginal indifference price*, also called *Davis' price*, as the price p^D at which the agent is indifferent between not selling the option or selling an infinitesimal quantity:

$$\partial_q u(v + q p^D; q) = 0$$

where

$$u(v + p; q) := \sup_{\phi} \mathbb{E}\left[U(V_T^{v+p,\phi} - qG) \right] .$$

3.2 Loss Function Hedging

We now consider the situation in which an agent has sold a European option of payoff G at a price v lower than the super-hedging price. Perfect hedging is no more possible, but he can prescribe a risk criteria to select a partial hedging strategy. We shall study two approaches. The first one consists in maximising the probability of hedging, this is the so-called *quantile hedging* approach. The second consists in minimising the expectation of a loss function, this is the *shortfall hedging* approach.

We restrict to a complete market setting in which $\mathcal{M}(\tilde{S}) = \{\mathbb{Q}\}$, so that the probabilistic arguments of [29] and [30] lead to an explicit resolution. They can be extended to incomplete markets, but, similarly to the optimal management problems, this only leads to existence results.

3.2.1 Quantile Hedging

3.2.1.1 Maximisation of the Probability of Hedging

We first consider the problem of maximising the probability of hedging a claim G paid at T. Namely, we study

$$\sup_{\phi \in \mathcal{A}_+(v)} \mathbb{P}\left[V_T^{v,\phi} \geq G\right], \tag{3.6}$$

where $\mathcal{A}_+(v)$ is the restriction of $\mathcal{A}_b$ to strategies leading to non-negative wealths, when starting from v. We assume that $0 < v < p(G)$.

The key observation of [29] is that the above can be turned into a classical statistical test, which can be solved by appealing to Neyman-Pearson lemma that is recalled below.

From now on, we assume that $G \in \mathbb{L}^0(\mathbb{R}_+) \setminus \{0\}$. One can always reduce to this case whenever the payoff is bounded from below. Let us also recall the notation $\tilde{G} := \beta_T G$.

We first provide an alternative formulation to (3.6).

Proposition 3.3

$$\sup_{\phi \in \mathcal{A}_+(v)} \mathbb{P}\left[V_T^{v,\phi} \geq G\right] = \sup \left\{\mathbb{E}\left[\varphi\right] : \varphi \in \mathbb{L}^0(\{0,1\}), \mathbb{E}^{\mathbb{Q}}[\tilde{G}\varphi] \leq v\right\}. \tag{3.7}$$

Proof Let $\phi \in \mathcal{A}_+(v)$. Then, $\varphi := \mathbf{1}_{\{V_T^{v,\phi} \geq G\}}$ satisfies $\mathbb{P}\left[V_T^{v,\phi} \geq G\right] = \mathbb{E}[\varphi]$ and $G\varphi \leq V_T^{v,\phi}$ so that $\mathbb{E}^{\mathbb{Q}}[\tilde{G}\varphi] \leq \mathbb{E}^{\mathbb{Q}}[\tilde{V}_T^{v,\phi}] \leq v$. This shows that the left-hand side term in (3.7) is lower than the right-hand side. Conversely, if $\varphi \in \mathbb{L}^0(\{0,1\})$ is such that $\mathbb{E}^{\mathbb{Q}}[\tilde{G}\varphi] \leq v$, then one can find $\phi \in \mathcal{A}_b(v)$ such that $V_T^{v,\phi} \geq G\varphi$, see Theorem 2.17. Since $G\varphi \geq 0$, the $\mathbb{Q}$-surmartingale $\tilde{V}^{v,\phi}$ remains positive and hence $\phi \in \mathcal{A}_+(v)$. Moreover, $V_T^{v,\phi} \geq G$ on $\{\varphi = 1\}$. Since $\varphi \in \mathbb{L}^0(\{0,1\})$, this implies that $\mathbb{P}\left[V_T^{v,\phi} \geq G\right] \geq \mathbb{E}[\varphi]$. $\square$

We now observe that the right-hand side in (3.7) can be interpreted as a statistical test:

$$\sup\left\{\mathbb{E}\left[\varphi\right] , \varphi \in \mathbb{L}^0([0,1]) \text{ s.t. } \mathbb{E}^{\mathbb{Q}_G}[\varphi] \leq v/p(G)\right\}, \tag{3.8}$$

where $\mathbb{Q}_G$ is defined by

$$\frac{d\mathbb{Q}_G}{d\mathbb{P}} := \frac{d\mathbb{Q}}{d\mathbb{P}} \frac{\beta_T G}{\mathbb{E}^{\mathbb{Q}}[\beta_T G]}.$$

The main difference is that one looks for a test in $\mathbb{L}^0(\{0,1\})$ and not in $\mathbb{L}^0([0,1])$.

The solution of the above is provided by the *Neyman and Pearson's lemma.*

Lemma 3.2 (Neyman-Pearson) *Let $\mathbb{P}_0$ and $\mathbb{P}_1$ be two probability measures that are dominated by $\mathbb{P}$. Given $\alpha \in]0, 1[$, the solution to*

$$\sup \left\{ \mathbb{E}^{\mathbb{P}_1} [\xi] \ : \ \xi \in \mathbb{L}^0([0, 1]), \ \mathbb{E}^{\mathbb{P}_0} [\xi] \leq \alpha \right\} ,$$

is given by

$$\hat{\xi} := \mathbf{1}_{\{\frac{d\mathbb{P}_1}{d\mathbb{P}} > \hat{a} \frac{d\mathbb{P}_0}{d\mathbb{P}}\}} + \hat{\gamma} \mathbf{1}_{\{\frac{d\mathbb{P}_1}{d\mathbb{P}} = \hat{a} \frac{d\mathbb{P}_0}{d\mathbb{P}}\}}$$

with

$$\hat{a} := \inf \left\{ a > 0 \ : \ \mathbb{P}_0 \left[\frac{d\mathbb{P}_1}{d\mathbb{P}} > a \frac{d\mathbb{P}_0}{d\mathbb{P}} \right] \leq \alpha \right\}$$

and $\hat{\gamma} \in [0, 1]$ such that $\mathbb{E}^{\mathbb{P}_0} \left[\hat{\xi} \right] = \alpha$.

Remark 3.4 In the above lemma, ξ should be interpreted as the test of $\text{Hyp}_0 : \mathbb{P}_0$ against $\text{Hyp}_1 : \mathbb{P}_1$. If the state of nature ω is such that $\xi(\omega) = p$, then we accept Hyp_0 with probability $1 - p$. The quantity $\mathbb{E}^{\mathbb{P}_0} [\xi]$ corresponds to the probability of rejecting Hyp_0 while Hyp_0 is true (first kind risk), and $\mathbb{E}^{\mathbb{P}_1} [\xi]$ corresponds to the probability of rejecting Hyp_0 while Hyp_0 is actually false (power of the test). The test $\hat{\xi}$ is called UMP (uniformly most powerful) at the level α.

By applying the preceding lemma to (3.8), we then find a solution $\hat{\varphi}$ of the following form.

Theorem 3.5 *Assume that*

$$\hat{c} := \inf \left\{ c > 0 \ : \ \mathbb{E}^{\mathbb{Q}} \left[\beta_T G \mathbf{1}_{\{\frac{d\mathbb{P}}{d\mathbb{Q}} > c \frac{d\mathbb{Q}_G}{d\mathbb{Q}}\}} \right] \leq v \right\}$$

satisfies

$$\mathbb{E}^{\mathbb{Q}} \left[\beta_T G \mathbf{1}_{\{\frac{d\mathbb{P}}{d\mathbb{Q}} > \hat{c} \frac{d\mathbb{Q}_G}{d\mathbb{Q}}\}} \right] = v .$$

then, the solution to (3.6) is given by the strategy $\hat{\phi} \in \mathcal{A}_+(v)$ which satisfies

$$V_T^{v, \hat{\phi}} = G\hat{\varphi}$$

where

$$\hat{\varphi} = \mathbf{1}_{\{\frac{d\mathbb{P}}{d\mathbb{Q}} > \hat{c} \frac{d\mathbb{Q}_G}{d\mathbb{Q}}\}} .$$

In most practical applications $\hat{c} > 0$ is such that

$$\mathbb{E}^{\mathbb{Q}}\left[\beta_T G \mathbf{1}_{\{\frac{d\mathbb{P}}{d\mathbb{Q}} > \hat{c}\frac{d\mathbb{Q}_G}{d\mathbb{Q}}\}}\right] = v,$$

recall that $v < p(G) = \mathbb{E}^{\mathbb{Q}}[\beta_T G]$, so that the optimal strategy $\hat{\phi}$ satisfies

$$V_T^{v,\hat{\phi}} = G\mathbf{1}_{\hat{A}}$$

with $\hat{A} := \{d\mathbb{P}/d\mathbb{Q} > \hat{c}\, d\mathbb{Q}_G/d\mathbb{Q}\}$. This means that the optimal strategy consists in replicating a digital type option whose payoff is G on $\hat{A}$ and 0 on $\hat{A}^c$.

Unfortunately, this is in general not a very good solution. In Sect. 4.3.2 we shall see that hedging digital options might be difficult in practice.

When $\hat{c} > 0$ does not satisfy $\mathbb{E}^{\mathbb{Q}}\left[\beta_T G \mathbf{1}_{\{\frac{d\mathbb{P}}{d\mathbb{Q}} > \hat{c}\frac{d\mathbb{Q}_G}{d\mathbb{Q}}\}}\right] < v$, the above cannot be used anymore. However, the same reasoning applies to a relaxed criteria, called *success ratio* maximisation,

$$\sup_{\phi \in \mathcal{A}_+(v)} \mathbb{E}\left[\frac{V_T^{v,\phi}}{G} \wedge 1\right]. \tag{3.9}$$

Here, we use the convention $z/0 = \infty$ for $z \in \mathbb{R}$.

Theorem 3.6 *The solution to (3.9) is given by the strategy $\hat{\phi} \in \mathcal{A}_+(v)$ satisfying*

$$V_T^{v,\hat{\phi}} = G\hat{\varphi}$$

where

$$\hat{\varphi} = \mathbf{1}_{\{\frac{d\mathbb{P}}{d\mathbb{Q}} > \hat{c}\frac{d\mathbb{Q}_G}{d\mathbb{Q}}\}} + \hat{\gamma}\mathbf{1}_{\{\frac{d\mathbb{P}}{d\mathbb{Q}} = \hat{c}\frac{d\mathbb{Q}_G}{d\mathbb{Q}}\}}$$

with

$$\hat{c} := \inf\left\{c > 0 \;:\; \mathbb{E}^{\mathbb{Q}}\left[\beta_T G \mathbf{1}_{\{\frac{d\mathbb{P}}{d\mathbb{Q}} > c\frac{d\mathbb{Q}_G}{d\mathbb{Q}}\}}\right] \leq y\right\}$$

and where $\hat{\gamma} \in [0, 1]$ is such that $\mathbb{E}^{\mathbb{Q}}[\beta_T G\hat{\varphi}] = v$.

When $\hat{\gamma} = 0$, one retrieves the solution of the original quantile hedging problem.

3.2.1.2   Quantile Hedging Price

The above approach also permits to find a price for the claim G which guarantees to be able to cover the claim with a prescribed probability. This is the *quantile hedging price*:

$$p(G; \alpha) := \inf\left\{v \geq 0 \ : \ \exists \ \phi \in \mathcal{A}_+(v) \text{ s.t. } \mathbb{P}[V_T^{v,\phi} \geq G] \geq \alpha\right\},$$

where $\alpha \in [0, 1]$. Clearly, $p(G; 1) = p(G)$ and $p(G; 0) = 0$. Given $\alpha \in (0, 1)$, $p(G, \alpha)$ can be computed by using Theorem 3.5 (or its relaxed version, Theorem 3.6). Indeed, since $0 < v < p(G)$, one can compute

$$\alpha(v) := \sup_{\phi \in \mathcal{A}_+(v)} \mathbb{P}\left[V_T^{v,\phi} \geq G\right].$$

By construction,

$$p(G; \alpha) = \inf\{v \geq 0 \ : \ \alpha(v) \geq \alpha\}.$$

We refer to [29] for various examples of explicit resolutions within Black and Scholes type models, see also Exercise 3.3 below.

3.2.2 Loss Function Hedging

We have seen above that the quantile hedging approach leads to hedging digital type options, which is difficult in practice. This is not the only drawback of this approach, it also does not provide any control on the size of the potential losses.

In order to better control them, one can replace the probability criteria by a risk measure of the form $\mathbb{E}\left[\ell((G - V_T^{x,\phi})^+)\right]$, where ℓ is a *loss function*, i.e. strictly convex and increasing, defined on $\mathbb{R}_+$.

Upon a normalisation of ℓ, one can assume that $\ell(0) = 0$. For the following, we shall also assume that ℓ is C^1 and that $\partial\ell(+\infty) = \infty$, $\partial\ell(0+) = 0$.

As in the preceding section, we restrict to $G \in \mathbb{L}^0(\mathbb{R}_+) \setminus \{0\}$.

3.2.2.1 Minimisation of the Expected Shortfall

We start with the problem of minimising the expected shortfall:

$$\inf_{\phi \in \mathcal{A}_+(v)} \mathbb{E}\left[\ell((G - V_T^{v,\phi})^+)\right], \tag{3.10}$$

with $0 < v < p(G)$. Obviously, the above is 0 if $v \geq p(G)$.

The following Theorem provides an explicit solution. It uses the notation $I := (\partial\ell)^{-1}$.

Theorem 3.7 *There exists an optimal solution $\hat{\phi} \in \mathcal{A}_+(v)$ to (3.10). It satisfies*

$$V_T^{v,\hat{\phi}} = \hat{\varphi}(\hat{c})G$$

where, for all $c > 0$,

$$\hat{\varphi}(c) := \mathbf{1}_{\{G>0\}}\left(1 - \frac{I(c\beta_T d\mathbb{Q}/d\mathbb{P})}{G} \wedge 1\right),$$

and $\hat{c} > 0$ is the unique solution of

$$\mathbb{E}^{\mathbb{Q}}\left[\beta_T \hat{\varphi}(c)G\right] = v .$$

Proof

1. One first observes that

$$\mathbb{E}\left[\ell((G - V_T^{v,\phi})^+)\right] = \mathbb{E}\left[\ell(G(1 - \varphi^{\phi}))\right]$$

where $\varphi^{\phi} := [(V_T^{v,\phi}/G) \wedge 1]\mathbf{1}_{\{G>0\}}$ satisfies $\mathbb{E}^{\mathbb{Q}}\left[\beta_T \varphi^{\phi} G\right] \leq v$. Conversely, if the above constraint holds for $\varphi \in \mathbb{L}^0([0,1])$, then φG can be super-hedged from v, see Theorem 2.17. The discounted value of the portfolio process being a super-martingale under $\mathbb{Q}$, and G being non-negative, the hedging strategy belongs to $\mathcal{A}_+(v)$, see Theorem 2.7. Hence, the above is equivalent to

$$\inf_{\varphi \in \mathbb{L}^0([0,1])} \mathbb{E}\left[\ell((1-\varphi)G)\right] \text{ under the constraint } \mathbb{E}^{\mathbb{Q}}\left[\beta_T \varphi G\right] \leq v. \quad (3.11)$$

2. One checks now that there exists an optimal strategy by using the following technical lemma.

Lemma 3.3 (Komlos' lemma) *Let $(\zeta_n)_n$ be a sequence of random variables that are bounded in $\mathbb{L}^1(\mathbb{P})$. Then, there exists a sequence $(\bar{\zeta}_n)_n$ and a random variable $\bar{\zeta}$ in $\mathbb{L}^1(\mathbb{P})$ such that $\bar{\zeta}_n \to \bar{\zeta}$ $\mathbb{P} - a.s.$ and*

$$\bar{\zeta}_n \in conv\,(\zeta_k,\, k \geq n) \quad \mathbb{P} - a.s.$$

for all $n \geq 1$, where conv denotes the convex envelop of a family (i.e. all elements that can be obtained by forming a convex combination of elements of the family).

Since $\varphi \mapsto \mathbb{E}[\ell(G(1-\varphi))]$ is convex, one deduces from this lemma the existence of a minimising sequence $(\varphi_n)_n$ that converges $\mathbb{P} - a.s.$ to some $\hat{\varphi}$ in $\mathbb{L}^0([0,1])$. One concludes by appealing to Fatou's lemma and using the fact that $\ell \geq 0$.

3. One now checks that $\hat{\varphi}$ has the expected form. Given $\varphi \in \mathbb{L}^0([0,1])$ and $\varepsilon \in [0,1]$, we define

$$\varphi_\varepsilon := \varepsilon\varphi + (1-\varepsilon)\hat{\varphi}$$

and

$$F_\varphi(\varepsilon) := \mathbb{E}\left[\ell((1-\varphi_\varepsilon)G)\right] .$$

Since ℓ is convex, its derivative is non-decreasing. By using a monotone convergence argument, one easily checks that the right-derivative $\partial F_\varphi(0+)$ of F_φ at 0 exists and satisfies

$$\partial F_\varphi(0+) = \mathbb{E}\left[\partial\ell((1-\hat{\varphi})G)(\hat{\varphi}-\varphi)G\right] .$$

Since F_φ is convex, by convexity of ℓ, $\hat{\varphi}$ must satisfy the first order condition: $\partial F_\varphi(0+) \geq 0$ for all $\varphi \in \mathbb{L}^0([0,1])$. This amounts to

$$\mathbb{E}^{\mathbb{Q}_{\hat{\varphi}}}[\hat{\varphi}] \geq \mathbb{E}^{\mathbb{Q}_{\hat{\varphi}}}[\varphi] \tag{3.12}$$

for all $\varphi \in \mathbb{L}^0([0,1])$ such that, see (3.11),

$$\mathbb{E}^{\mathbb{Q}_G}[\varphi] \leq \frac{v}{p(G)} =: \alpha \tag{3.13}$$

with $\mathbb{Q}_{\hat{\varphi}}$ and $\mathbb{Q}_G$ defined by

$$\frac{d\mathbb{Q}_{\hat{\varphi}}}{d\mathbb{P}} = \partial\ell((1-\hat{\varphi})G)G/\mathbb{E}\left[\partial\ell((1-\hat{\varphi})G)G\right] ,$$

$$\frac{d\mathbb{Q}_G}{d\mathbb{P}} = \frac{d\mathbb{Q}}{d\mathbb{P}}\beta_T G/\mathbb{E}^{\mathbb{Q}}[\beta_T G] .$$

As in the preceding section, we end up with a statistical test: one tests the hypothesis $\mathbb{Q}_{\hat{\varphi}}$ against $\mathbb{Q}_G$ at the level α. By Lemma 3.2, the optimal test $\hat{\varphi}$ takes the value 0 if $d\mathbb{Q}_{\hat{\varphi}}/d\mathbb{P} < c\, d\mathbb{Q}_G/d\mathbb{P}$ and the value 1 if $d\mathbb{Q}_{\hat{\varphi}}/d\mathbb{P} > c\, d\mathbb{Q}_G/d\mathbb{P}$, where c is a constant which depends on the level of the test. One can observe that we must have $\hat{\varphi} < 1$ on $\{G > 0\}$ since $\partial\ell(0) = 0$, which implies $d\mathbb{Q}_{\hat{\varphi}}/d\mathbb{P} = 0 < d\mathbb{Q}_G/d\mathbb{P}$ when $\hat{\varphi} = 1$ and $G > 0$. Hence, $d\mathbb{Q}_{\hat{\varphi}}/d\mathbb{Q}_G \leq c$ on $\{G > 0\}$. On $\{G = 0\}$, we have $d\mathbb{Q}_{\hat{\varphi}}/d\mathbb{P} = d\mathbb{Q}_G/d\mathbb{P} = 0$, and we can impose $\hat{\varphi} = 1$, which we will justify below. This is the expected form.

4. It remains to justify the choice $\hat{\varphi} = 1$ on $\{G = 0\}$. To understand this, let us observe that $\hat{c} > 0$ satisfies $\mathbb{E}^{\mathbb{Q}_G}[\hat{\varphi}(\hat{c})] = v/p(G)$. Indeed, $\partial\ell$ is non-decreasing, continuous and satisfies $\partial\ell(+\infty) = \infty$ and $\partial\ell(0+) = 0$, by assumption. Hence, I is non-decreasing, continuous and satisfies $\partial I(+\infty) = \infty$, $\partial I(0+) = 0$. In particular, $\hat{\varphi}((0,\infty)) = [0, \mathbf{1}_{\{G>0\}}]$ $\mathbb{P}$ – a.s. and $c \in (0,\infty) \mapsto \hat{\varphi}(c)$ is $\mathbb{P}$ –

a.s. continuous. Using the monotone convergence theorem, we then obtain that $c \in (0, \infty) \mapsto k(c) := \mathbb{E}^{\mathbb{Q}}[\beta_T G \hat{\varphi}(c)]$ is continuous and satisfies $k((0, \infty)) \supset (0, \mathbb{E}^{\mathbb{Q}}[\beta_T G \mathbf{1}_{\{G>0\}}]) = (0, p(G))$. The fact that the optimum $\hat{c}$ is unique follows from the fact that I is in fact strictly increasing and that $v/p(G) < 1$, so that $\mathbb{P}[I(\hat{c}\beta_T d\mathbb{Q}/d\mathbb{P}) < G] > 0$. $\qquad\square$

3.2.2.2 Short-Fall Hedging Price

As for the quantile hedging criteria, one can now define a minimal price at which the option should be sold so that the shortfall constraint can be satisfied:

$$\inf\left\{v \geq 0 \,:\, \exists\, \phi \in \mathcal{A}_+(v) \text{ s.t. } \mathbb{E}[\ell((G - V_T^{v,\phi})^+)] \leq l\right\}, \; l \in \ell(\mathbb{R}_+)\,.$$

It is deduced from Theorem 3.7 by arguing as in Sect. 3.2.1 above.

3.2.3 Comments

We have restricted ourselves to complete markets, this is the only setting in which explicit optimisers can be obtained, generically. This is obviously very restrictive, in particular because these approaches are essentially useful in incomplete markets, where risks can in general not be fully hedged at a reasonable price (see for instance Sect. 5.2.2) below.

In Chap. 6, we will explain how quantile or shortfall hedging prices can be characterised in terms of Black and Scholes type non-linear partial differential equations, within general Markovian models. Their numerical resolution then allows one to determine the corresponding approximate optimal trading strategy.

3.3 Problems

3.1 (Fenchel transform) The aim of this exercise is to show that the bi-Fenchel's transform of a convex and lower-semicontinuous function is the function itself.

1. Let f be a convex function that is lower-semicontinuous, from $\mathbb{R}$ to $\mathbb{R} \cup \{\infty\}$. We denote by f^* its Fenchel transform[1]:

$$f^*(y) := \sup_{x \in \mathbb{R}}(xy - f(x))\,, \; y \in \mathbb{R}\,.$$

---

[1] See [53] for more on this subject.

Set $C := \{(x, y) \in \mathbb{R} \times (\mathbb{R} \cup \{\infty\}) : y \geq f(x)\}$. Given $(\alpha, p, q) \in \mathbb{R}^3$, define

$$F_{(\alpha,p,q)} := \{(x, y) \in \mathbb{R} \times (\mathbb{R} \cup \{\infty\}) : px + qy \leq \alpha\}$$

and $D := \{(\alpha, p, q) \in \mathbb{R}^3 : C \subset F_{(\alpha,p,q)}\}$.

a. Show that C is convex and closed.
b. By using the Hahn-Banach separation theorem, show that $C = \bigcap_{(\alpha,p,q) \in D} F_{(\alpha,p,q)}$.
c. Show that $C \subset F_{(\alpha,p,q)}$ if and only if $px + qy - \chi_C(x, y) \leq \alpha$ for all (x, y), in which $\chi_C(x, y) = 0 \mathbf{1}_{(x,y) \in C} + \infty \mathbf{1}_{(x,y) \notin C}$.
d. Deduce that $C \subset F_{(\alpha,p,q)}$ if and only if $\chi_C^*(p, q) \leq \alpha$.
e. Deduce that $C = \bigcap_{(p,q) \in \mathbb{R}^2} \{(x, y) \in \mathbb{R} \times (\mathbb{R} \cup \{\infty\}) : px + qy \leq \chi_C^*(p, q)\}$.
f. Compute $\chi_C^*(p, q)$ as a function of f^*.
g. Deduce that $\{(x, y) \in \mathbb{R} \times (\mathbb{R} \cup \{\infty\}) : y \geq (f^*)^*(x)\} = C$ and that $(f^*)^* = f$.

2. Let us now consider a function U from $\mathbb{R}_+$ to $\mathbb{R} \cup \{-\infty\}$, strictly concave, increasing and C^1 on its domain. We set

$$\bar{U}(y) := \inf_{x \in \mathbb{R}} (xy - U(x)) , \; y \in \mathbb{R} .$$

We assume that the conditions U1, U2 and U3 Sect. 3.1 hold for U.

a. Show that $-\bar{U}$ is the Fenchel transform of a function to define in terms of U.
b. Show that the domain of $\bar{U}$ contains $(0, \infty)$. Under which condition does it also contain 0?
c. Show that $-\bar{U}$ is convex.
d. Show that $\inf_{y \in \mathbb{R}} (xy - \bar{U}(y)) = U(x)$.
e. Compute $\bar{U}$ in terms of U and $I := (\partial U)^{-1}$.
f. By using the above and the concavity of U, show that $\bar{U}$ is C^1 in the interior of its domain, and that $\partial \bar{U} = I$.
g. Compute the Fenchel transforms associated to the following concave functions: $\ln(x)$ and x^γ for $\gamma \in (0, 1)$.

3.2 (Optimal management with a CRRA utility function) Within the one dimensional Black and Scholes model,

$$S_t = S_0 + \int_0^t \mu S_s ds + \int_0^t \sigma S_s dW_s$$

with $\mu \in \mathbb{R}$ and $\sigma > 0$, compute the optimal management strategy in the case where the utility function is $U(x) = x^\gamma$, with $\gamma \in (0, 1)$, and the interest rate r is constant.

3.3 (Quantile hedging in the Black and Scholes model) Let us consider the one dimensional Black and Scholes model

$$S_t = S_0 + \int_0^t \mu S_s ds + \int_0^t \sigma S_s dW_s$$

with $\mu \in \mathbb{R}$ and $\sigma > 0$. For simplicity, we take $r = 0$.

1. Given an initial endowment v, we want to maximise the probability of hedging a European call $G := [S_T - K]^+$, with strike $K > 0$:

$$\max_{\phi \in \mathcal{A}_+(v)} \mathbb{P}\left[V_T^{v,\phi} \geq G\right]. \tag{3.14}$$

Let $\mathbb{Q}$ denote the martingale measure and define $\mathbb{Q}_G$ by

$$\frac{d\mathbb{Q}_G}{d\mathbb{Q}} := \frac{G}{\mathbb{E}^{\mathbb{Q}}[G]}.$$

a. Let H be the density of $\mathbb{Q}$ with respect to $\mathbb{P}$. Show that

$$H = (S_T/S_0)^\lambda e^\kappa$$

where $\lambda := \mu/\sigma^2$ and κ is a real number to be defined.
b. With the notations of Theorem 3.5, show that

$$\frac{d\mathbb{P}}{d\mathbb{Q}} > c\frac{d\mathbb{Q}_G}{d\mathbb{Q}} \iff (S_T/S_0)^{-\lambda} > \bar{c}[S_T - K]^+ \iff S_T < b,$$

where $\bar{c}$ and b have to be computed in terms of $c > 0$.
c. Show that one can find $b_1 > 0$ such that the process $\hat{\phi} \in \mathcal{A}_+(v)$ for which $V_T^{v,\hat{\phi}} = [S_T - K]^+ \mathbf{1}_{\{S_T < b_1\}}$ is the solution to (3.14).

2. Given $m \in (0,1)$, we now look for a solution to

$$\inf\left\{v \geq 0 : \exists \phi \in \mathcal{A}_+(v) \text{ s.t. } \mathbb{P}[V_T^{v,\phi} \geq [S_T - K]^+] \geq m\right\}. \tag{3.15}$$

a. Fix $b_2 > 0$ such that $\mathbb{P}[S_T < b_2] = m$. Compute

$$\hat{v} := \mathbb{E}^{\mathbb{Q}}\left[[S_T - K]^+ \mathbf{1}_{\{S_T < b_2\}}\right]$$

in terms of the cumulative function Φ of the standard normal distribution.
b. Show that $\hat{v}$ achieves the infimum in (3.15) and that the optimal strategy is the process $\hat{\phi} \in \mathcal{A}_+(\hat{v})$ which satisfies $V_T^{\hat{v},\hat{\phi}} = [S_T - K]^+ \mathbf{1}_{\{S_T < b_2\}}$.

3. Would the result be different if we were using the success ratio criteria defined in (3.9)?

Corrections

3.1

1. a. Because f is convex and lower-semicontinuous.
   b. By definition of D, $C \subset \bigcap_{(\alpha,p,q)\in D} F_{(\alpha,p,q)}$. Assume now that there is $(x,y) \in \bigcap_{(\alpha,p,q)\in D} F_{(\alpha,p,q)} \setminus C$. Since C is convex, the Hahn-Banach separation theorem implies that we can find $(p',q') \in \mathbb{R}^2 \setminus \{0\}$ such that

   $$\alpha' := \sup\{x'p' + y'q' : (x',y') \in C\} < xp' + yq'.$$

   Then, $(\alpha',p',q') \in D$, and therefore (x,y) cannot belong to $\bigcap_{(\alpha,p,q)\in D}$.
   c. If $C \subset F_{(\alpha,p,q)}$ then $px + qy - \chi_C(x,y) \leq \alpha$ for all (x,y) by definition of χ. Conversely, if $px + qy - \chi_C(x,y) \leq \alpha$ for all (x,y), then $px + qy \leq \alpha$ for all $(x,y) \in C$ and therefore $C \subset F_{(\alpha,p,q)}$.
   d. By the preceding question and definition of χ^*.
   e. By the above, $D := \{(\alpha,p,q) \in \mathbb{R}^3 : \chi_C^*(p,q) \leq \alpha\}$. Then, the required result follows from 1.b.
   f. We have for $(p,q) \neq 0$

   $$\chi_C^*(p,q) = \sup_{(x,y)}(xp + yq - \chi(x,y))$$
   $$= \sup_{(x,y)\in C}(xp + yq)$$
   $$= +\infty\mathbf{1}_{\{q>0\}} + \sup_x(xp - |q|f(x))$$
   $$= +\infty\mathbf{1}_{\{q\geq0\}} + f^*(p/|q|)\mathbf{1}_{\{q<0\}}$$

   and $\chi_C^*(0,0) = 0$.
   g. Hence,

   $$C = \bigcap_{(p,q)\in\mathbb{R}\times(-\infty,0)} \{(x,y) \in \mathbb{R}\times(\mathbb{R}\cup\{\infty\}) : px + qy \leq f^*(p/|q|)\}$$
   $$= \bigcap_{(p,q)\in\mathbb{R}\times(-\infty,0)} \{(x,y) \in \mathbb{R}\times(\mathbb{R}\cup\{\infty\}) : p/|q|x - f^*(p/|q|) \leq y\}$$
   $$= \{(x,y) \in \mathbb{R}\times(\mathbb{R}\cup\{\infty\}) : (f^*)^*(x) \leq y\}.$$

   This shows that $(f^*)^* = f$ by definition of C.

2. a. It is the opposite of the Fenchel's transform of $x \mapsto -U(-x)$.
   b. This follows from U3. It contains 0 if U is bounded from above.
   c. This is the sup of linear maps $y \mapsto xy + U(-x)$, indexed by x.
   d. Since $-U(-\cdot)$ is continuous and convex, we can apply the result of 1.
   e. We use the first order condition of optimality: $\bar{U}(y) = yI(y) - U(I(y))$.
   f. I can be differentiated since $U \in C^2$ by assumption.
   g. This follows from direct computations.

3.2 We follow the steps of Sect. 3.1. We first compute the density of the martingale measure $H_T = d\mathbb{Q}/d\mathbb{P} = \exp(-\frac{1}{2}\lambda^2 T - \lambda W_T)$ with $\lambda := (\mu - r)/\sigma$. We denote by $W^{\mathbb{Q}}$ the corresponding Brownian motion.

Direct computations leads to $I(y) := y^{\frac{1}{\gamma-1}}$. Let $v \geq 0$ be the initial wealth. Then,

$$\hat{y} := I^{-1}(v/\mathbb{E}[H_T I(\beta_T H_T)]) = (v/\mathbb{E}[H_T(\beta_T H_T)^{\frac{1}{\gamma-1}}])^{\gamma-1}$$

verifies $\mathbb{E}[H_T I(\hat{y}\beta_T H_T)] = v$. The optimal terminal wealth is thus given by

$$V_T^{v,\hat{\phi}} = (\hat{y}\beta_T H_T)^{\frac{1}{\gamma-1}} = \hat{y}^{\frac{1}{\gamma-1}} e^{-\frac{1}{\gamma-1}rT} e^{\frac{1}{2}[\frac{\lambda^2}{\gamma-1}+(\frac{1}{\gamma-1})^2]T} e^{-\frac{1}{2}(\frac{\lambda}{\gamma-1})^2 T - \frac{\lambda}{\gamma-1}W_T^{\mathbb{Q}}}.$$

Since $\mathbb{E}^{\mathbb{Q}}[\tilde{V}_T^{v,\hat{\phi}}] = v$, we must have

$$\tilde{V}_T^{v,\hat{\phi}} = v e^{-\frac{1}{2}(\frac{\lambda}{\gamma-1})^2 T - \frac{\lambda}{\gamma-1}W_T^{\mathbb{Q}}},$$

so that $\hat{\phi} = \lambda/(1-\gamma)$.

3.3

1. a. Take $\lambda = -(\mu - r)/\sigma^2$ and compute κ accordingly.
   b. This is an immediate consequence of the preceding question.
   c. We apply Theorem 3.5 and look for

$$c_1 := \inf\left\{c > 0 : \mathbb{E}^{\mathbb{Q}}\left[G\mathbf{1}_{\{\frac{d\mathbb{P}}{d\mathbb{Q}}>c\frac{d\mathbb{Q}_G}{d\mathbb{Q}}\}}\right] \leq v\right\}.$$

But the preceding question implies that it is equivalent to looking for

$$b_1 = \sup\left\{b > 0 : \mathbb{E}^{\mathbb{Q}}\left[G\mathbf{1}_{S_T < b}\right] \leq v\right\}.$$

It is not difficult to check that $b \geq 0 \mapsto \mathbb{E}^{\mathbb{Q}}[G\mathbf{1}_{S_T < b}]$ is continuous and maps $\mathbb{R}_+$ into $[0, \mathbb{E}^{\mathbb{Q}}[G]]$. Hence, $\mathbb{E}^{\mathbb{Q}}[G\mathbf{1}_{S_T < b_1}] = v$. We deduce c_1 as a function of b_1 and conclude with Theorem 3.5.

2. a. Argue as in 12. of Exercise 2.4.
   b. It follows from the preceding questions and the discussion in Sect. 3.2.1.2, that the solution is of the form $[S_T - K]^+ \mathbf{1}_{\{S_T < b\}}$. Minimising the price of this claim is clearly equivalent to minimising over b such that the probability of

hedging G is at least m. By definition of $\hat{v}$ and the martingale representation theorem, there exists $\hat{\phi} \in \mathcal{A}_+(\hat{v})$ such that $V_T^{\hat{v},\hat{\phi}} = [S_T - K]^+ \mathbf{1}_{\{S_T < b_2\}}$, and the probability of hedging G is m.

3. No. It would be the same because the optimum is reached in our case.

Part II
Markovian Models and PDE Approach

Chapter 4
Delta Hedging in Complete Market

In this chapter, we study a class of Markovian models for complete markets. This type of model is the most commonly used in practice. In these models, the underlying price process is solution to a Stochastic Differential Equation.

We give various versions of the so-called Feynman-Kac theorem [27, 40]. In our context, it links the hedging price, given by the conditional expectation of the discounted payoff under the risk neutral measure, to a Partial Differential Equation, the so-called pricing PDE. This link allows in particular to obtain numerical approximation of the option price by using numerical methods for PDEs, see e.g. [45].

We first consider the case where the price function is smooth. In this case, we give explicitly the hedging strategy which is the classical *Delta* hedge, see Sect. 4.2.1. We then study the case where the price function might not be differentiable. In this case, we use the notion of *viscosity* solution for the pricing PDE, which is weaker than the classical one.

4.1 Markovian Models

In this part, we work in a Markovian framework. Let $T > 0$ be a fixed time horizon and $(\Omega, \mathscr{F}, \mathbb{P})$ be a stochastic basis supporting a d-dimensional Brownian motion W. We denote by $\mathbb{F} = (\mathscr{F}_t)_{t \leq T}$ the augmented Brownian filtration. The underlying price process is the solution to the following SDE

$$S_s^{t,x} = x + \int_t^s b(u, S_u^{t,x})du + \int_t^s \sigma(u, S_u^{t,x})dW_u \, , \ s \in [t, T], \qquad (4.1)$$

where the superscript (t, x) means that the price dynamic start from a spot price x at t.

© Springer International Publishing Switzerland 2016
B. Bouchard, J.-F. Chassagneux, *Fundamentals and Advanced Techniques in Derivatives Hedging*, Universitext, DOI 10.1007/978-3-319-38990-5_4

In this chapter, we assume that

$$b : [0, T] \times (0, \infty)^d \mapsto (0, \infty)^d \text{ and } \sigma : [0, T] \times (0, \infty)^d \mapsto \mathbb{M}^{d,d}$$

are continuous and satisfy, for some constant $L > 0$,

$$|b(t, x) - b(t, x')| + |\sigma(t, x) - \sigma(t, x')| \leq L|x - x'| \ \forall (t, x, x') \in [0, T] \times (0, \infty)^{2d}.$$

This implies that (4.1) has a unique strong solution for all $(t, x) \in [0, T] \times (0, \infty)^d$. We moreover assume that this solution is valued in $(0, \infty)^d$ whenever the starting point x belongs to $(0, \infty)^d$.

Example 4.1 Let us assume that the drift coefficient is given by $b(s, x) = \text{diag}[x]\, \bar{b}(s, x)$ and the volatility coefficient by $\sigma(s, x) = \text{diag}[x]\, \bar{\sigma}(s, x)$. This is in particular the case for the Black et Scholes model, for which $\bar{b}$ and $\bar{\sigma}$ are constant. Then, each component $S^{t,x,i}$ of $S^{t,x}$ is given by

$$x^i \exp\left(\int_t^{\cdot} \left(\bar{b}^i(s, S_s^{t,x}) - \frac{1}{2}\|\bar{\sigma}^i(s, S_s^{t,x})\|^2 \right) ds + \int_t^{\cdot} \bar{\sigma}^i(s, S_s^{t,x}) dW_s \right)$$

where $\bar{\sigma}^i$ stands for the i-th row of $\bar{\sigma}$.

In this setting, S has the following flow property:

$$S_{\tau_2}^{t,x} = S_{\tau_2}^{\tau_1, S_{\tau_1}^{t,x}} \quad \text{for all } t \leq \tau_1 \leq \tau_2 \in \mathcal{T}_0. \tag{4.2}$$

Besides, it can be shown that S is a strong Markov process [41]: for all Borel measurable function g with polynomial growth and all stopping time $\tau \in \mathcal{T}_t$,

$$\mathbb{E}\left[g(S_T^{t,x})|\mathscr{F}_\tau\right] = \varphi_g(\tau, S_\tau^{t,x}) \tag{4.3}$$

where

$$\varphi_g(s, x) := \mathbb{E}\left[g(S_T^{s,x})\right], \ (s, x) \in [0, T] \times (0, \infty)^d.$$

For ease of presentation, we assume that the risk free rate is given by a deterministic continuous function $t \in [0, T] \mapsto r_t$, see Remark 4.4 below.

We also consider that the market is complete. More precisely, that σ is invertible and the function (which is well defined)

$$\lambda(t, x) := \sigma(t, x)^{-1}(b(t, x) - r_t x),$$

namely the risk premium, is assumed to be bounded and Lipschitz.

For a market starting at t with an initial condition x, we define the risk neutral measure

$$\frac{d\mathbb{Q}_{t,x}}{d\mathbb{P}} := e^{-\frac{1}{2}\int_t^T \|\lambda(s,S_s^{t,x})\|^2 ds - \int_t^T \lambda(s,S_s^{t,x})' dW_s}.$$

From Girsanov Theorem, see Theorem 2.9, we know that the process

$$W_s^{\mathbb{Q}_{t,x}} := W_s + \int_t^s \lambda(u, S_u^{t,x}) du , \ s \in [t, T],$$

is a $\mathbb{Q}_{t,x}$-Brownian motion.

Using (4.1), we compute that the discounted price process started at t,

$$\tilde{S}_{\cdot}^{t,x} := \beta_{\cdot}^t S_{\cdot}^{t,x} \ \text{where} \ \beta_{\cdot}^t := e^{-\int_t^\cdot r_s ds},$$

has the following dynamics

$$\tilde{S}_s^{t,x} = x + \int_t^s \beta_u^t \sigma(u, S_u^{t,x}) dW_u^{\mathbb{Q}_{t,x}} , \ s \in [t, T].$$

By uniqueness in law, we can assume that $\tilde{S}^{t,x}$ is solution to

$$\tilde{S}_s^{t,x} = x + \int_t^s \beta_u^t \sigma(u, S_u^{t,x}) dW_u^{\mathbb{Q}} , \ s \in [t, T],$$

or, similarly,

$$S_s^{t,x} = x + \int_t^s r_u S_u^{t,x} du + \int_t^s \sigma(u, S_u^{t,x}) dW_u^{\mathbb{Q}} , \ s \in [t, T], \tag{4.4}$$

where $W^{\mathbb{Q}}$ is a $\mathbb{Q}$-Brownian motion possibly on a new probability space $(\hat{\Omega}, \hat{\mathscr{F}}, \mathbb{Q})$. For notational simplicity, we will assume that $(\hat{\Omega}, \hat{\mathscr{F}}) = (\Omega, \mathscr{F})$ and we will use the representation (4.4) in the whole chapter.

4.2 Vanilla Options

In this section, we study the pricing of plain European options. Precisely, we want to characterise the price at time t of a European option whose random terminal payoff at T is given by $g(S_T^{t,x})$, recalling $S_t^{t,x} = x$. We assume that g is bounded from below and has polynomial growth at infinity.

From the Fundamental Asset Pricing Theorem 2.17, the hedging price is given by

$$p(t, x) := \mathbb{E}^{\mathbb{Q}}[\beta_T^t g(S_T^{t,x})]. \qquad (4.5)$$

Using the strong Markov property of S recalled in the previous section, we have that, for all $\tau \in \mathscr{T}_t$,

$$p(t, x) = \mathbb{E}^{\mathbb{Q}}[\beta_\tau^t \mathbb{E}^{\mathbb{Q}}[\beta_T^\tau g(S_T^{t,x})|\mathscr{F}_\tau]] = \mathbb{E}^{\mathbb{Q}}[\beta_\tau^t p(\tau, S_\tau^{t,x})]. \qquad (4.6)$$

In other words, the discounted price process $(\beta_s^t p(s, S_s^{t,x}))_{s \in [t,T]}$ is a $\mathbb{Q}$-martingale. Assuming that p is $C^{1,2}$, we can use Itô's Lemma to compute its dynamics, namely

$$d(\beta_s^t p(s, S_s^{t,x})) = \beta_s^t \left(\mathscr{L}_S p(s, S_s^{t,x}) - r_s p(s, S_s^{t,x}) \right) ds$$
$$+ \beta_s^t \partial p(s, S_s^{t,x}) \sigma(s, S_s^{t,x}) dW_s^{\mathbb{Q}},$$

in which $\mathscr{L}_S$ is the Dynkin operator associated to (4.4):

$$\mathscr{L}_S \varphi(s, x) := \partial_t \varphi(s, x) + r_s \langle x, \partial \varphi(s, x) \rangle + \frac{1}{2} \mathrm{Tr}[\sigma \sigma'(s, x) \partial^2 \varphi(s, x)], \qquad (4.7)$$

where $\partial_t \varphi$ stands for the time derivative of φ, $\partial \varphi$ its Jacobian and $\partial^2 \varphi$ the Hessian with respect to the space variable x.

The martingale property implies that the drift coefficient is equal to 0:

$$\mathscr{L}_S p(s, S_s^{t,x}) = r_s p(s, S_s^{t,x}).$$

Since this is true for all initial condition (t, x), the function p must solve

$$r_t p(t, x) - \mathscr{L}_S p(t, x) = 0 \text{ for all } (t, x) \in [0, T) \times (0, \infty)^d. \qquad (4.8)$$

The boundary condition at the terminal date T is naturally given by

$$\lim_{t' \uparrow T, x' \to x} p(t', x') = g(x) \text{ for all } x \in (0, \infty)^d. \qquad (4.9)$$

The above result, linking the function p, defined in (4.5) as an expectation of a function of a Stochastic Differential Equation to a Partial Differential Equation is known as the *Feynman-Kac Theorem* [27, 40].

4.2.1 Regular Case: Feynman-Kac Formula and Delta-Hedging

The first Theorem of this section justifies the above result in a rigorous way when p is $C^{1,2}$. This is often the case in the applications. But, apart from the case where all the coefficients are regular, this property is not easy to prove a priori. When the coefficients are not regular, it is sometimes possible to rely on the regularising effect of the Brownian Motion. For example, if $\sigma(t,x) = \text{diag}[x]\,\bar{\sigma}(t,x)$ with $\bar{\sigma}$ satisfying a uniform ellipticity condition of the type $\zeta'\bar{\sigma}\bar{\sigma}'\zeta \geq c\|\zeta\|$ for some $c > 0$ independent of $\zeta \in \mathbb{R}^d$, $\ln(S^{t,x}/x)$ has a density $f_{t,x}$ under $\mathbb{Q}$ which is regular in (t,x). This does not require much regularity on r and σ. As

$$p(t,x) = \beta_T^t \int g((x^i e^{z^i})_{i \leq d}) f_{t,x}(z) dz$$

$$= \beta_T^t \int g((e^{y^i})_{i \leq d}) f_{t,x}((y^i - \ln x^i)_{i \leq d}) dy,$$

the derivatives of the function p are deduced from those of the function $(t,x) \mapsto (\beta_T^t, f_{t,x}((\cdot - \ln x^i)_{i \leq d}))$. The regularity of g is not used in this approach.

Theorem 4.2 (Feynman-Kac) *Assume that p is $C^{1,2}([0,T) \times (0,\infty)^d) \cap C^{0,0}([0,T] \times (0,\infty)^d)$. Then it is a solution of* (4.8) *and* (4.9).

Proof The boundary condition follows naturally from the continuity assumption, as $p(T,\cdot) = g$ by definition. Assume now that (4.8) is not satisfied at $(t,x) \in [0,T) \times (0,\infty)^d$, for example

$$r_t p(t,x) - \mathcal{L}_s p(t,x) < 0.$$

The case of the reverse inequality is treated using symmetric arguments. We will work toward a contradiction of the martingale property (4.6). Indeed, by continuity, we have

$$rp - \mathcal{L}_s p < 0,$$

on an open neighbourhood B of (t,x). Let τ be the exit time of this neighbourhood by the process $(s, S_s^{t,x})_{s \geq t}$. It follows from Itô's Lemma that

$$p(t,x) = \beta_\tau^t p(\tau, S_\tau^{t,x}) - \int_t^\tau \beta_s^t \left(\mathcal{L}_s p(s, S_s^{t,x}) - r_s p(s, S_s^{t,x}) \right) ds$$

$$- \int_t^\tau \beta_s^t \partial p(s, S_s^{t,x}) \sigma(s, S_s^{t,x}) dW_s^{\mathbb{Q}}$$

$$< \beta_\tau^t p(\tau, S_\tau^{t,x}) - \int_t^\tau \beta_s^t \partial p(s, S_s^{t,x}) \sigma(s, S_s^{t,x}) dW_s^{\mathbb{Q}}.$$

Taking expectation, we obtain

$$p(t, x) < \mathbb{E}^{\mathbb{Q}}[\beta_\tau^t p(\tau, S_\tau^{t,x})],$$

which contradicts the martingale property (4.6).   □

We now give a reciprocal result, illustrating the notion of *delta-hedging*.

Theorem 4.3 (Verification) *Assume that there exists a function φ in $C^{1,2}([0, T) \times (0, \infty)^d) \cap C^{0,0}([0, T] \times (0, \infty)^d)$ solution to (4.8) and (4.9). If, moreover,*

$$\mathbb{E}^{\mathbb{Q}}\left[\int_t^T \|\partial\varphi(s, S_s^{t,x})\sigma(s, S_s^{t,x})\|^2 ds\right] < \infty, \ \forall \ (t, x) \in [0, T) \times (0, \infty)^d.$$

Then, $\varphi = p$, and

$$p(t, x) + \int_t^T \partial p(s, S_s^{t,x}) d\tilde{S}_s^{t,x} = \beta_T^t g(S_T^{t,x}), \tag{4.10}$$

for all $(t, x) \in [0, T) \times (0, \infty)^d$.

Proof Applying Itô's Lemma and using that φ is solution to (4.8) and (4.9), we obtain straightforwardly

$$\varphi(t, x) + \int_t^T \partial\varphi(s, S_s^{t,x})\beta_s^t \sigma(s, S_s^{t,x}) dW_s^{\mathbb{Q}} = \beta_T^t g(S_T^{t,x}),$$

which is equivalent to (4.10) from (4.4). The integrability condition imposed in the statement of the theorem implies that the $\mathbb{Q}$-expectation of the stochastic integral in the left hand side is equal to zero. We thus deduce that $\varphi(t, x) = \mathbb{E}^{\mathbb{Q}}[\beta_T^t g(S_T^{t,x})] = p(t, x)$.   □

Note from (4.10) that it is possible to hedge the payoff $g(S_T^{t,x})$ starting with an initial wealth $p_{t,x} := p(t, x)$ and holding exactly $\phi_s := \partial p(s, S_s^{t,x})$ units of risky asset in the portfolio at each date s. This technique is known as *delta-hedging* . The *delta* is the derivative of the hedging price with respect to the underlying, i.e. $\partial p(s, S_s^{t,x})$. Observe that, though the volatility $\sigma(s, S_s^{t,x})$ is not constant, delta-hedging allows to cover perfectly the risk associated to the contingent claim. This comes from the fact that it depends on time and the assets value only. We will see that this not the case anymore in stochastic volatility models in which the volatility depends on other sources of randomness, see Chaps. 5, 6 and 8.

Remark 4.4 (Stochastic interest rate) A similar reasoning allows to deal with the case of stochastic interest rates. A simple example is when the interest rate is a deterministic function of the assets (one of them being possibly a zero-coupon bond with maturity T). In this case, the interest rate at time s is given by $r_s^{t,x} = \rho(s, S_s^{t,x})$ where ρ is a function assumed to be non-negative and continuous. The previous results can be extended straightforwardly by replacing r_t by $\rho(t, x)$ and correcting the definition of the Dynkin operator $\mathscr{L}_S$ into

$$\mathscr{L}_S \varphi := \partial_t \varphi + \rho \langle x, \partial \varphi \rangle + \frac{1}{2} \mathrm{Tr}[\sigma \sigma' \partial^2 \varphi]$$

in (4.8).

Let us stress that this characterisation is unique in most of the models used in practice. Indeed, there exists only one solution to (4.8) and (4.9) in the class of function with polynomial growth, which is the class where p belongs typically. In particular, this is the case under the integrability condition of Theorem 4.3. Obviously, such a property is fundamental when one wants to approximate numerically the function p. Without uniqueness, one cannot guarantee the convergence of the numerical approximation to the correct solution. We refer the interested reader to the book [45] for a clear presentation of finite difference methods, which are an industry standard.

Let us give another example for which uniqueness is guaranteed. The argument below uses only a polynomial growth condition. The fact that p has polynomial growth can be directly obtained from standard estimates on solutions of SDEs, when g has itself polynomial growth.

Theorem 4.5 (Uniqueness) *Assume that $\sigma(s, x) = diag\,[x]\,\bar{\sigma}(s, x)$ for all $(s, x) \in [0, T] \times (0, \infty)^d$, where $\bar{\sigma}$ is bounded and Lipschitz-continuous. Let w_1 and w_2 be two solutions to (4.8) and (4.9) in $C^{1,2}([0, T) \times (0, \infty)^d) \cap C^{0,0}([0, T] \times (0, \infty)^d)$. Assume moreover that there exists $q > 0$ such that*

$$\sup_{(t,x) \in [0,T] \times (0,\infty)^d} |w_i(t, x)| / (1 + |x|^q) < \infty, \quad i = 1, 2.$$

Then $w_1 = w_2$ on $[0, T] \times (0, \infty)^d$.

Proof To alleviate the notations, we only consider the case $d = 1$. The proof in the general case follows from the same arguments. To simplify, we also assume $\min_{[0,T]} r =: \underline{r} > 0$. To get rid of this condition, one has simply to multiply $w_1(t, \cdot)$ and $w_2(t, \cdot)$ by e^{ct} where $c > |\underline{r}|$. In this case, $\tilde{w}_i := e^c w_i$ is solution to $(r_t + c)\varphi - \mathscr{L}_S \varphi = 0$ with $\min_{[0,T]}(r + c) > 0$, for $i = 1, 2$. Proving $w_1 = w_2$ is equivalent to prove $\tilde{w}_1 = \tilde{w}_2$.

Assume that $\sup_{[0,T] \times (0,\infty)}(w_1 - w_2) > 0$ where w_1 and w_2 have polynomial growth with degree $q \geq 1$. Let $\varepsilon > 0$. We can then find $(t_\varepsilon, x_\varepsilon) \in [0, T] \times (0, \infty)$ such that

$$\sup_{[0,T] \times (0,\infty)} (w_1 - w_2 - f_\varepsilon) = (w_1 - w_2 - f_\varepsilon)(t_\varepsilon, x_\varepsilon) > 0 \qquad (4.11)$$

where

$$f_\varepsilon(t, x) := \eta L(t, x) + \varepsilon x^{-1} \text{ with } L(t, x) := e^{-Ct}(1 + |x|^{2q})$$

for $C > 0$ a constant that will be set later on, and $\eta, \varepsilon > 0$ close enough to 0.

Note that the penalising function f_ε guarantees that the maximum is finite and is also attained. Remark also that we cannot have $t_\varepsilon = T$ since $w_1(T, \cdot) = g = w_2(T, \cdot)$. This would contradict (4.11).

Since $t_\varepsilon < T$ and $x_\varepsilon > 0$, the first order and second order optimality conditions imply

$$\partial_t(w_1 - w_2 - f_\varepsilon)(t_\varepsilon, x_\varepsilon) \leq 0 , \ \partial(w_1 - w_2 - f_\varepsilon)(t_\varepsilon, x_\varepsilon) = 0$$
$$\text{and } \partial^2(w_1 - w_2 - f_\varepsilon)(t_\varepsilon, x_\varepsilon) \leq 0.$$

Since w_1 and w_2 are solutions to (4.8) and $(w_1 - w_2 - f_\varepsilon)(t_\varepsilon, x_\varepsilon) \geq 0$, we deduce that

$$0 \geq r_{t_\varepsilon} f_\varepsilon(t_\varepsilon, x_\varepsilon) - \mathscr{L}_S f_\varepsilon(t_\varepsilon, x_\varepsilon). \tag{4.12}$$

We can find $C > 0$ such that, for all $(s, x) \in [0, T] \times (0, \infty)$,

$$e^{Cs}(r_s L(s, x) - \mathscr{L}_S L(s, x)) \geq (\underline{r} + C)(1 + |x|^{2q}) - 2qx^{2q}r_s$$
$$-\frac{1}{2}2q(2q - 1)x^{2q}\bar\sigma^2(s, x)$$
$$\geq 1,$$

as $\bar b, \bar\sigma$ are bounded. We thus have

$$r_{t_\varepsilon} L(t_\varepsilon, x_\varepsilon) - \mathscr{L}_S L(t_\varepsilon, x_\varepsilon) \geq e^{-CT} > 0.$$

It follows from (4.12) that

$$-\eta e^{-CT} \geq r_{t_\varepsilon} \frac{\varepsilon}{x_\varepsilon} - \left(-r_{t_\varepsilon} x_\varepsilon \frac{\varepsilon}{x_\varepsilon^2} + \frac{1}{2}\bar\sigma(t_\varepsilon, x_\varepsilon)^2 x_\varepsilon^2 \frac{2\varepsilon}{x_\varepsilon^3} \right).$$

To contradict the previous inequality and thus conclude the proof, it is enough to show that

$$|\varepsilon(x_\varepsilon)^{-1}| \to 0 \text{ when } \varepsilon \to 0,$$

and then to take the limits in the previous inequality.

To the contrary, let us assume that $\varepsilon(x_\varepsilon)^{-1}$ has a limit $c \in (0, \infty]$, along a subsequence. The growth condition on w_1 and w_2 combined with (4.11) implies that the sequence $(t_\varepsilon, x_\varepsilon)$ is bounded. We can thus assume that it converges to (t_0, x_0). It follows that for some point $(\bar{t}_0, \bar{x}_0) \in [0, T] \times \mathbb{R}_+$,

$$
\begin{aligned}
(w_1 - w_2 - \eta L)(\bar{t}_0, \bar{x}_0) &= \max_{[0,T] \times \mathbb{R}_+} (w_1 - w_2 - \eta L) \\
&> (w_1 - w_2 - \eta L)(t_0, x_0) - c \\
&= \lim_{\varepsilon \to 0}(w_1 - w_2 - f_\varepsilon)(t_\varepsilon, x_\varepsilon).
\end{aligned}
$$

But it is clear that one can approximate $(\bar{t}_0, \bar{x}_0)$ by a sequence of points $(\bar{t}_\varepsilon, \bar{x}_\varepsilon)_\varepsilon \subset [0, T] \times (0, \infty)$ such that $\varepsilon(\bar{x}_\varepsilon)^{-1}$ goes to 0. For $\varepsilon > 0$ small enough, the above inequalities lead to

$$
\begin{aligned}
\lim_{\varepsilon \to 0}(w_1 - w_2 - f_\varepsilon)(\bar{t}_\varepsilon, \bar{x}_\varepsilon) &= (w_1 - w_2 - \eta L)(\bar{t}_0, \bar{x}_0) \\
&> \lim_{\varepsilon \to 0}(w_1 - w_2 - f_\varepsilon)(t_\varepsilon, x_\varepsilon),
\end{aligned}
$$

which contradicts (4.11). $\qquad\square$

Remark 4.6 (Comparison principle) If w_1 (resp. w_2) is a only a super-solution (resp. a sub-solution), i.e. (4.8) and (4.9) are satisfied with $\geq$ (resp. with $\leq$) and $w_1(T, \cdot) \geq w_2(T, \cdot)$, the above arguments show that $w_1 \geq w_2$ in the whole domain. This result is called a *comparison principle*: a super-solution greater than a sub-solution on the parabolic boundary[1] of the domain is greater on the whole domain. To obtain such result, one needs to restrict the class of functions under consideration. Above, we only considered function with polynomial growth. In the previous case, we did not use any information associated to the boundary $x = 0$ but this might be required in the general case.

4.2.2 Non-smooth Case: Price Characterisation Using Viscosity Solutions

We now turn to the case where the price function p is not $C^{1,2}$ but only continuous. In this case, it is no more a classical solution to the PDE (4.8). Nevertheless, using a weaker notion of solution, it is still possible to characterise the price function as a solution to (4.8). In our setting, it is very natural to use the notion of viscosity solutions.

---

[1]It is the boundary obtained when excluding the part corresponding to $t = 0$.

We now describe in a general setting this notion, as it will be used several times in the text. For an exhaustive presentation of viscosity solutions, we refer to the seminal survey paper [20].

4.2.2.1 Definitions

Let F be an operator

$$(t, x, v, q, a, A) \in [0, T] \times \mathcal{O} \times \mathbb{R} \times \mathbb{R} \times \mathbb{R}^d \times \mathbb{S}^d \mapsto F(t, x, v, q, a, A) \in \mathbb{R},$$

where $\mathbb{S}^d$ stands for the space of d-dimensional symmetric matrices and $\mathcal{O}$ is an open set of $\mathbb{R}^d$.

We are interested in the solutions to

$$F(t, x, \varphi(t, x), \partial_t \varphi(t, x), \partial \varphi(t, x), \partial^2 \varphi(t, x)) = 0. \tag{4.13}$$

Equation (4.8) corresponds to

$$F(t, x, v, q, a, A) = r_t v - q - r_t \langle a, x \rangle - \frac{1}{2} \mathrm{Tr}[\sigma \sigma'(t, x) A],$$

with $\mathcal{O} = (0, \infty)^d$.

F is said to be *elliptic* if it is non-increasing with respect to $q \in \mathbb{R}$ and $A \in \mathbb{S}^d$, in the symmetric matrices sense.

To motivate the notion of viscosity solutions, let us assume first that $v \in C^{1,2}$ is a classical solution to (4.13) on $[0, T) \times \mathcal{O}$. Let φ be a $C^{1,2}$ function and $(t, x) \in [0, T) \times \mathcal{O}$ a point where $v - \varphi$ reaches a local maximum. Up to modifying φ by a constant, we can always assume that $(v - \varphi)(t, x) = 0$. In this case, the first order and second order optimality conditions read

$$\partial_t (v - \varphi)(t, x) \le 0, \ \partial(v - \varphi)(t, x) = 0 \text{ and } \partial^2(v - \varphi)(t, x) \le 0.$$

Since F is elliptic, we have

$$0 = F(t, x, \varphi(t, x), \partial_t v(t, x), \partial v(t, x), \partial^2 v(t, x))$$
$$\ge F(t, x, \varphi(t, x), \partial_t \varphi(t, x), \partial \varphi(t, x), \partial^2 \varphi(t, x)).$$

Reciprocally, if (t, x) is a point where $v - \varphi$ reaches a local minimum, then

$$0 \le F(t, x, \varphi(t, x), \partial_t \varphi(t, x), \partial \varphi(t, x), \partial^2 \varphi(t, x))$$

using similar arguments.

This leads to the following definition of viscosity solutions.

Definition 4.7 (Viscosity solution) A lower semi-continuous (resp. upper semi-continuous) function v is a *super-solution* (resp. *sub-solution*) in the viscosity sense to (4.13) on $[0, T) \times \mathcal{O}$ if, for all function $\varphi \in C^{1,2}$ and $(t, x) \in [0, T) \times \mathcal{O}$ such that $0 = \min_{[0,T) \times \mathcal{O}}(v - \varphi) = (v - \varphi)(t, x)$ (resp. $0 = \max_{[0,T) \times \mathcal{O}}(v - \varphi) = (v - \varphi)(t, x)$),

$$F(t, x, \varphi(t, x), \partial_t \varphi(t, x), \partial \varphi(t, x), \partial^2 \varphi(t, x)) \geq 0 \quad (\text{resp.} \leq 0). \tag{4.14}$$

A continuous function v is a viscosity solution, if it is both a super- and sub-solution.

Remark 4.8 The smooth function φ is called *test function* for v at (t, x). It will be clear from context if (t, x) is a maximum or minimum point of $v - \varphi$.

The previous reasoning shows that a classical solution is also a viscosity solution. Reciprocally, if a function $v \in C^{1,2}$ is a viscosity solution then it is also a classical solution. To prove this, it is enough to notice that at any point (t, x), $v - v$ reaches a local maximum and local minimum, and then to apply the previous definition with $\varphi := v$. Viscosity solutions are thus indeed an extension of classical solution. This new notion of solution is very powerful as it does not require smoothness conditions on the function under study.

Let us observe that the notion of super-solution is defined for functions v that are lower semi-continuous. This implies that $v - \varphi$ is also lower semi-continuous and thus allows to consider its minimum. In a symmetric way, the notion of sub-solution is restricted to upper semi-continuous function.

Remark 4.9 In Definition 4.7, it is always possible to assume that φ is $C_b^{1,2}$ and that the maximum or minimum is strict. Indeed, if $v - \varphi$ reaches a maximum at (t, x) then $(t', x') \mapsto (v - \varphi)(t', x') - |t - t'|^2 - \|x - x'\|^4$ reaches a strict maximum at (t, x). Moreover, the derivatives of $(t', x') \mapsto \varphi(t', x') + |t - t'|^2 + \|x - x'\|^4$ at (t, x) inserted into the operator F are the same as the ones of φ. Similarly, as soon as v has polynomial growth, we can always assume that its maximum or minimum is local and that the test function has polynomial growth.

Remark 4.10 (Discontinuous solutions) For functions that may not be continuous, it is still possible to define the notion of *discontinuous viscosity solutions*. In this case, a locally bounded function v is a discontinuous viscosity *super-solution* (resp. *sub-solution*) if its lower semi-continuous envelop (resp. upper semi-continuous envelop) is a *super-solution* (resp. *sub-solution*) in the sense of Definition 4.7. In most of the proofs below, we will assume that the functions we work with, are at least continuous on $[0, T)$. However, the results remain valid using discontinuous solutions if this is not the case. We refer to [12] to understand how to adapt the proof in the discontinuous case.

Remark 4.11 (Comparison and uniqueness) As in the case of classical solution, it is possible to obtain uniqueness results and comparison principles. These can be obtained quite generally in the class of discontinuous solutions, see Theorem 4.5 and Remark 4.6 below. Typically, if w_1 (resp. w_2) has polynomial growth and is a super-solution (resp. sub-solution) to (4.13), and if $w_1 \geq w_2$ on $(\{T\} \times \mathscr{O}) \cup ([0, T] \times \partial \mathscr{O})$, then $w_1 \geq w_2$ on $[0, T] \times \mathscr{O}$. Sometimes, it is enough to assume that $w_1 \geq w_2$ on $\{T\} \times \mathscr{O}$ as in the case of Theorem 4.5 with $\mathscr{O} = (0, \infty)$. The proofs of such results rely on very deep and advanced techniques such as Ishii's Lemma. They are far beyond the scope of this book. We refer to [20] for a complete presentation.

Remark 4.12 (Numerical approximation) Let us note that finite difference schemes converge to the viscosity solution whenever it is unique, see [5] for details. It is thus possible to use the same numerical methods to solve the PDEs independently of the notion of solution considered (classical or viscosity).

4.2.2.2 Feynman-Kac Formula in the Viscosity Sense

We now show how to extend Theorem 4.2 to the case where the price function p defined in (4.5) is only continuous.

Theorem 4.13 (Feynman-Kac) *Assume that p is $C^{0,0}([0, T] \times (0, \infty))$, then it is a viscosity solution to (4.8) and (4.9).*

Proof By continuity of p, we have that the boundary condition $p(T, \cdot) = g$ holds true by definition. We now show that p is a super-solution to (4.8). The sub-solution property is shown similarly. Let $\varphi \in C^{1,2}$ and $(t, x) \in [0, T) \times (0, \infty)^d$ be such that

$$\min_{[0,T) \times (0,\infty)^d} (p - \varphi) = (p - \varphi)(t, x) = 0. \tag{4.15}$$

Assume that

$$r_t \varphi(t, x) - \mathscr{L}_S \varphi(t, x) < 0.$$

Then, by continuity,

$$r\varphi - \mathscr{L}_S \varphi < 0,$$

on an open neighbourhood B of (t, x). Let τ be the exit time of B by the process $(s, S_s^{t,x})_{s \geq t}$. Itô's Lemma yields

$$\varphi(t, x) = \beta_\tau^t \varphi(\tau, S_\tau^{t,x}) - \int_t^\tau \beta_s^t \left(\mathscr{L}_S \varphi(s, S_s^{t,x}) - r_s \varphi(s, S_s^{t,x}) \right) ds$$

$$- \int_t^\tau \beta_s^t \partial \varphi(s, S_s^{t,x}) \sigma(s, S_s^{t,x}) dW_s^{\mathbb{Q}}$$

$$< \beta_\tau^t \varphi(\tau, S_\tau^{t,x}) - \int_t^\tau \beta_s^t \partial \varphi(s, S_s^{t,x}) \sigma(s, S_s^{t,x}) dW_s^{\mathbb{Q}}.$$

Taking expectation, we obtain

$$\varphi(t,x) < \mathbb{E}^{\mathbb{Q}}[\beta_\tau^t \varphi(\tau, S_\tau^{t,x})].$$

But, (4.15) implies $\varphi(t,x) = p(t,x)$ and $\varphi(\tau, S_\tau^{t,x}) \leq p(\tau, S_\tau^{t,x})$. We thus deduce from the previous inequality

$$p(t,x) < \mathbb{E}^{\mathbb{Q}}[\beta_\tau^t p(\tau, S_\tau^{t,x})].$$

This contradicts the martingale property of (4.6). □

The previous proof is similar to the proof of Theorem 4.2. The only difference is that Itô's Lemma is applied to the test function φ rather than to p. Moreover, one should note that it is generally easy to show that p is continuous. For example, this is the case as soon as g is continuous with polynomial growth.

The uniqueness result of Theorem 4.5 can be extended to the viscosity solution case. As already mentioned, the proof is much more technical, see [20].

4.2.3 Tangent Process, Malliavin Derivatives and Delta-Hedging

We saw in Chap. 2 that the hedging strategy may be obtained in some case by using the Clark-Ocone formula, see Theorem 2.32.

This formula is well-suited to the case where the price dynamics is given by the solution to a SDE with regular coefficients: the trajectory is then Malliavin differentiable. To fully grasp the following result, it is useful to compute the dynamics of the Malliavin derivative of an Euler scheme using the chain rule given in Proposition 2.29. The next result is then obtained by passing to the limit, as the Euler scheme approximates the true solution in our context, see [48].

Proposition 4.14 *Assume that σ has linear growth and is C_b^1 in space, uniformly with respect to time. Then, $S_s^{t,x} \in \mathbb{D}_{1,2}$ for all $s \in [t,T]$. Moreover, the matrix process $DS^{t,x} = (D^j S^{t,x,i})_{i,j\leq d}$ verifies*

$$D_u S_s^{t,x} = \sigma(u, S_u^{t,x})\mathbf{1}_{\{u\leq s\}} + \int_t^s r_v D_u S_v^{t,x} dv$$

$$+ \sum_{i=1}^d \int_t^s \partial\sigma^{\cdot i}(v, S_v^{t,x}) D_u S_v^{t,x} dW_v^{\mathbb{Q},i}$$

where $\sigma^{\cdot i}$ is the i-th column of σ, $\partial\sigma^{\cdot i}$ its Jacobian matrix with respect to the space variable.

The Malliavin derivative of S can be expressed in term of the tangent process (or gradient) of $S^{t,x}$, which is defined as the sensitivity of $S^{t,x}$ with respect to its initial condition. The following result is proved in e.g. [51] (under a slightly stronger condition).

Proposition 4.15 *Let us assume that the conditions of Proposition 4.14 are satisfied and that σ has Lipschitz-continuous first derivatives. Then, the map $x \in (0, \infty)^d \mapsto S^{t,x}_s$ is $\mathbb{P} - a.s.$ C^1_b, for all $s \in [t, T]$. Besides, the tangent (matrix) process $\partial_x S^{t,x} = \{(\frac{\partial S^{t,x,i}_s}{\partial x^j})_{i,j}\}$ is solution on $[t, T]$ to:*

$$\partial_x S^{t,x}_s = I_d + \int_t^s r_v \partial_x S^{t,x}_v dv + \sum_{i=1}^d \partial \sigma^{\cdot i}(v, S^{t,x}_v) \partial_x S^{t,x}_v dW^{\mathbb{Q},i}_v ,$$

where I_d stands for the d-dimensional identity matrix.

Propositions 4.14 and 4.15 show that the Malliavin derivative and the tangent process follow the same dynamics. This straightforwardly implies the following:

$$D_s S^{t,x}_T = \partial_x S^{t,x}_T (\partial_x S^{t,x}_s)^{-1} \sigma(s, S^{t,x}_s). \tag{4.16}$$

We can then deduce from the Clark-Ocone formula, Theorem 2.32, that, for $g \in C^1_b$, the replication strategy $(\Delta^{t,x}_s)_{t \le s \le T}$ is given by

$$\Delta^{t,x}_s = \beta^s_T \mathbb{E}^{\mathbb{Q}} \left[\partial g(S^{t,x}_T) \partial_x S^{t,x}_T | \mathscr{F}_s \right] (\partial_x S^{t,x}_s)^{-1} . \tag{4.17}$$

Indeed, we compute

$$\beta^t_T g(S^{t,x}_T) = \mathbb{E}^{\mathbb{Q}} \left[\beta^t_T g(S^{t,x}_T) \right] + \int_t^T \beta^t_T \mathbb{E}^{\mathbb{Q}} \left[\partial g(S^{t,x}_T) D_s S^{t,x}_T | \mathscr{F}_s \right] dW^{\mathbb{Q}}_s$$

$$= \mathbb{E}^{\mathbb{Q}} \left[\beta^t_T g(S^{t,x}_T) \right]$$

$$+ \int_t^T \mathbb{E}^{\mathbb{Q}} \left[\partial g(S^{t,x}_T) \beta^t_T \partial_x S^{t,x}_T | \mathscr{F}_s \right] (\partial_x S^{t,x}_s)^{-1} \sigma(s, S^{t,x}_s) dW^{\mathbb{Q}}_s$$

$$= \mathbb{E}^{\mathbb{Q}} \left[\beta^t_T g(S^{t,x}_T) \right]$$

$$+ \int_t^T \mathbb{E}^{\mathbb{Q}} \left[\partial g(S^{t,x}_T) \beta^t_T \partial_x S^{t,x}_T | \mathscr{F}_s \right] (\beta^t_s \partial_x S^{t,x}_s)^{-1} d\tilde{S}^{t,x}_s.$$

To obtain the replication strategy, we then only have to compute $\mathbb{E}^{\mathbb{Q}} \left[\partial g(S^{t,x}_T) \beta^t_T \partial_x S^{t,x}_T | \mathscr{F}_s \right]$. This amounts to compute

$$(t, x, y) \mapsto \psi(t, x, y) := \mathbb{E}^{\mathbb{Q}} \left[\partial g(S^{t,x}_T) Y^{t,x,y}_T \right]$$

with $Y^{t,x,y}$ solution on $[t, T]$ to

$$Y_s^{t,x,y} = y + \sum_{i=1}^{d} \int_t^s \partial \sigma^{\cdot i}(v, S_v^{t,x}) Y_v^{t,x,y} dW_v^{\mathbb{Q},i}.$$

Indeed, it is easy to check, using the flow property of solutions to SDEs with Lipschitz coefficients, that

$$\mathbb{E}^{\mathbb{Q}} \left[\partial g(S_T^{t,x}) \beta_T^t \partial_x S_T^{t,x} | \mathscr{F}_s \right] = \psi(s, S_s^{t,x}, \beta_s^t \partial_x S_s^{t,x}).$$

In particular,

$$\psi(t, x, y) = \mathbb{E}^{\mathbb{Q}} \left[\psi(\tau, S_\tau^{t,x}, Y_\tau^{t,x,y}) \right], \ \tau \in \mathscr{T}_t.$$

The arguments used in Sects. 4.2.2 and 4.2.1 allow then to characterise each component of ψ as the solution of a PDE (which can then be solved numerically).

From the results of Sect. 4.2.2, we observe that this is interesting only if the price function is not smooth. We will thus only give a characterisation of ψ in terms of viscosity solutions. Clearly, if ψ is smooth, it is then solution also in the classical sense. In this case, differentiating (4.8) in x allows to get the PDE satisfied by ψ. The following result is stated only in the unidimensional case for notational simplicity.

Proposition 4.16 *Let us assume that the assumptions of Proposition 4.15 hold true, that g is C_b^1 and $d = 1$. Then the function $(t, x, y) \in [0, T] \times (0, \infty) \times (0, \infty) \mapsto \psi(t, x, y)$ is continuous and is a viscosity solution on $[0, T) \times (0, \infty) \times (0, \infty)$ to*

$$0 = -\partial_t \psi - r_t x \partial_x \psi$$
$$- \frac{1}{2} \left[\sigma^2 \partial_{xx}^2 \psi + 2\sigma (\partial_x \sigma) y \partial_{xy}^2 \psi + (\partial_x \sigma)^2 y^2 \partial_{yy}^2 \psi \right].$$

Besides,

$$\psi(T, x, y) = \partial g(x) y,$$

for all $(x, y) \in (0, \infty) \times (0, \infty)$.

The proof is similar to the one of Theorem 4.13.

Remark 4.17 Whenever g is not C_b^1, the process $\Delta \in \mathscr{A}$ satisfying

$$\beta_T^t g(S_T^{t,x}) = \mathbb{E}^{\mathbb{Q}} \left[\beta_T^t g(S_T^{t,x}) \right] + \int_t^T \Delta_s d\tilde{S}_s^{t,x}$$

may be approximated by using the previous results in the case where g is the limit in $\mathbb{L}^2(\mathbb{Q})$ of a sequence $(g^n)_{n \geq 1}$ of C_b^1 functions. This is, for example, the case of the call or put option (and even the digital option if $S_T^{t,x}$ has a density), and any of their combinations. In this case, the process $\Delta^n \in \mathscr{A}$ associated to the hedging of $g^n(S_T^{t,x})$ can be computed as above. Since

$$\beta_T^t g(S_T^{t,x}) - \beta_T^t g^n(S_T^{t,x}) = \mathbb{E}^{\mathbb{Q}}\left[\beta_T^t g(S_T^{t,x})\right] - \mathbb{E}^{\mathbb{Q}}\left[\beta_T^t g^n(S_T^{t,x})\right]$$
$$+ \int_t^T \beta_s^t (\Delta_s - \Delta_s^n)\sigma(s, S_s^{t,x})dW_s^{\mathbb{Q}},$$

the $\mathbb{L}^2(\mathbb{Q})$ convergence of the payoffs function implies the $\mathbb{L}^2(\mathbb{Q})$ convergence of the hedging strategy.

4.3   Barrier Options

We now consider the pricing of a barrier option whose payoff at T is given by

$$g(S_T^{t,x})\mathbf{1}_{\{\theta_{t,x}>T\}} + h(\theta_{t,x}, S_{\theta_{t,x}}^{t,x})\mathbf{1}_{\{\theta_{t,x}\leq T\}},$$

where

$$\theta_{t,x} := \inf\{s \geq t \ : \ S_s^{t,x} \notin \mathscr{O}\},$$

and $\mathscr{O}$ is an open domain of $(0, \infty)^d$. This means that the options pays $g(S_T^{t,x})$ if the process $S_s^{t,x}$ has not left the domain $\mathscr{O}$ before T, a *recovery* $h(\theta_{t,x}, S_{\theta_{t,x}}^{t,x})$ else.

In the following, we assume that g and h are continuous with polynomial growth. Moreover, $\mathscr{O}$ is C_b^2 is the sense that there exists a C_b^2 function $\mathrm{d}_{\mathscr{O}}$ such that

$$\mathrm{d}_{\mathscr{O}} > 0 \text{ on } \mathscr{O}, \mathrm{d}_{\mathscr{O}} = 0 \text{ on } \partial\mathscr{O}, \text{ and } \mathrm{d}_{\mathscr{O}} < 0 \text{ on } \mathscr{O}^c,$$

and $(\sigma, \mathscr{O})$ satisfies the non-characteristic boundary condition

$$\|\partial\mathrm{d}_{\mathscr{O}}\sigma\| \geq 1 \text{ on } [0, T] \times \partial\mathscr{O}. \tag{4.18}$$

Obviously, the constant 1 is arbitrary and can be obtained by multiplying $\mathrm{d}_{\mathscr{O}}$ by a well chosen constant as soon as the lower bound of the norm is uniformly strictly positive.

Example 4.18 A typical example is the *up-and-out* call on an underlying asset for which: $d = 1, \mathscr{O} = (0, B), h \equiv 0, g(x) = [x - K]^+$ with $B > K > 0, \mathrm{d}_{\mathscr{O}}(x) = B - x$. The case of the *down-and-out* call paying $g(S_T^{t,x})\mathbf{1}_{\{\theta_{t,x}\leq T\}}$ at T is obtained by the parity relation: $g(S_T^{t,x})\mathbf{1}_{\{\theta_{t,x}>T\}} + g(S_T^{t,x})\mathbf{1}_{\{\theta_{t,x}\leq T\}} = g(S_T^{t,x})$. For a *up-or-down-and-out* call: $d = 1, \mathscr{O} = (B_1, B_2), h \equiv 0, g(x) = [x - K]^+$ with $B_2 > K > B_1$, the function $\mathrm{d}_{\mathscr{O}}$ is any function C^2 satisfying $\mathrm{d}_{\mathscr{O}}(x) = |B_i - x|$ on a neighbourhood of $B_i, i = 1, 2$.

4.3.1 Pricing Equation with Dirichlet Boundary Condition

We know that the option price is given by

$$p(t,x) := \mathbb{E}^{\mathbb{Q}}\left[\beta_T^t g(S_T^{t,x})\mathbf{1}_{\{\theta_{t,x}>T\}} + \beta_{\theta_{t,x}}^t h(\theta_{t,x}, S_{\theta_{t,x}}^{t,x})\mathbf{1}_{\{\theta_{t,x}\leq T\}}\right]. \tag{4.19}$$

We first establish the property (4.6) in our setting. In order to do so, we use again the flow and strong Markov properties of the process S. For $\tau \in \mathscr{T}_t$ given:

$$
\begin{aligned}
p(t,x) &= \mathbb{E}^{\mathbb{Q}}\left[\beta_\tau^t \mathbb{E}^{\mathbb{Q}}\left[\beta_T^\tau g(S_T^{t,x})\mathbf{1}_{\{\theta_{t,x}>T\}}|\mathscr{F}_\tau\right]\mathbf{1}_{\{\theta_{t,x}>\tau\}}\right] \\
&+ \mathbb{E}^{\mathbb{Q}}\left[\beta_\tau^t \mathbb{E}^{\mathbb{Q}}\left[\beta_{\theta_{t,x}}^\tau h(\theta_{t,x}, S_{\theta_{t,x}}^{t,x})\mathbf{1}_{\{\theta_{t,x}\leq T\}}|\mathscr{F}_\tau\right]\mathbf{1}_{\{\theta_{t,x}>\tau\}}\right] \\
&+ \mathbb{E}^{\mathbb{Q}}\left[\beta_{\theta_{t,x}}^t h(\theta_{t,x}, S_{\theta_{t,x}}^{t,x})\mathbf{1}_{\{\theta_{t,x}\leq\tau\}}\right] \\
&= \mathbb{E}^{\mathbb{Q}}\left[\beta_\tau^t p(\tau, S_\tau^{t,x})\mathbf{1}_{\{\theta_{t,x}>\tau\}} + \beta_{\theta_{t,x}}^t h(\theta_{t,x}, S_{\theta_{t,x}}^{t,x})\mathbf{1}_{\{\theta_{t,x}\leq\tau\}}\right].
\end{aligned}
\tag{4.20}
$$

This states that the stopped discounted price process $(\beta_{s\wedge\theta_{t,x}}^t p(s \wedge \theta_{t,x}, S_{s\wedge\theta_{t,x}}^{t,x}))_{s\in[t,T]}$ is a $\mathbb{Q}$-martingale. The discounted price behaves like a martingale in the interior of the domain $\mathscr{O}$ and is equal to h on the boundary $\partial\mathscr{O}$ of $\mathscr{O}$. The same arguments as the ones used in Sect. 4.2 allow to prove that p is solution to a PDE with Dirichlet boundary condition.

The main difficulty comes from the possible discontinuity at $\{T\} \times \partial\mathscr{O}$ where p might be equal to h or g. By example, for the *up-and-out* call defined above, on $p(T-, B-) = B - K$ whereas $p(T, B) = 0$. We will only sketch the proof of this result. Because of this possible discontinuity, uniqueness of viscosity solutions for this equation is ill-posed. For numerics, it also causes problems as the derivative of the value function is infinite at the discontinuity point. Moreover, setting-up a delta-hedging strategy as in the previous section might be difficult. Indeed, we observe that the number of assets to hold in the portfolio might explode at T near a discontinuity point, see Exercise 2.6.

To tackle this problem, one solution is to regularise the payoff function. In our setting, this amounts to consider a function g which is continuous and satisfies $g = h(T, \cdot)$ on $\mathscr{O}^c$. In the case of the *up-and-out* call, one can use: $g(x) = [x - K]^+\mathbf{1}_{\{x\leq B_\varepsilon\}} + (B_\varepsilon - K)(B_\varepsilon - B)^{-1}(x - B)\mathbf{1}_{\{B_\varepsilon<x\leq B\}}$ with $K < B_\varepsilon < B$ near B. If $\mathbb{P}\left[S_T^{t,x} \in \partial\mathscr{O}\right] = 0$, it is possible to obtain the convergence of the price function, by using the same arguments as in Remark 4.17. We discuss this point below when we study the case where p is smooth in Sect. 4.3.2.

Proposition 4.19 *The function p is continuous on $([0, T] \times \bar{\mathscr{O}}) \setminus (\{T\} \times \partial\mathscr{O})$. If $g = h(T, \cdot)$ on $\mathscr{O}^c$, then it is also continuous on $\{T\} \times \partial\mathscr{O}$. In particular, it is equal to h on $[0, T) \times \partial\mathscr{O}$ and g on $\{T\} \times \bar{\mathscr{O}}$. It is a viscosity solution to (4.8) on $[0, T) \times$*

$\mathscr{O}$. *Moreover, if* σ *satisfies the assumptions of Theorem 4.5, then* p *is the unique continuous function with polynomial growth satisfying these properties.*

Sketch of proof for Proposition 4.19

1. The viscosity solution property inside the domain follows from the same arguments as in the proof of Theorem 4.13, if p is continuous inside the domain. Indeed, if $(t, x) \in [0, T) \times \mathscr{O}$, then we can choose an open neighbourhood B such that $B \subset [0, T) \times \mathscr{O}$. In this case, the exit time τ verifies $\tau < \theta_{t,x}$, by definition, and Eq. (4.20) rewrites $p(t, x) = \mathbb{E}^{\mathbb{Q}}\left[\beta_\tau^t p(\tau, S_\tau^{t,x})\right]$, as in the setting of Theorem 4.13.

2. We now discuss the continuity property at the boundary $\{T\} \times \bar{\mathscr{O}}$. By definition, $\theta_{t',x'} \wedge T \to T$ when $t' \to T$. Classical estimates on SDE solutions imply that $S_{\theta_{t',x'} \wedge T}^{t',x'} \to x$ in $\mathbb{L}^q(\mathbb{Q})$, for all $q \geq 1$, when $(t', x') \to (T, x)$, and so $\mathbb{Q} - $ a.s. up to a subsequence. We thus have

$$G_{t',x'} := \beta_T^\tau g(S_T^{t',x'})\mathbf{1}_{\{\theta_{t',x'} > T\}} + \beta_{\theta_{t',x'}}^\tau h(\theta_{t',x'}, S_{\theta_{t',x'}}^{t',x'})\mathbf{1}_{\{\theta_{t',x'} \leq T\}}$$

$$= \beta_{T \wedge \theta_{t',x'}}^\tau g(S_{T \wedge \theta_{t',x'}}^{t',x'})\mathbf{1}_{\{S_{T \wedge \theta_{t',x'}}^{t',x'} \in \mathscr{O}\}}$$

$$+ \beta_{T \wedge \theta_{t',x'}}^\tau h(T \wedge \theta_{t',x'}, S_{T \wedge \theta_{t',x'}}^{t',x'})\mathbf{1}_{\{S_{T \wedge \theta_{t',x'}}^{t',x'} \notin \mathscr{O}\}}.$$

Since $\mathscr{O}$ and $\mathscr{O}^c \setminus \partial\mathscr{O}$ are open set, if $x \notin \partial\mathscr{O}$, the $\mathbb{Q} - $ a.s. convergence of $S_{\theta_{t',x'} \wedge T}^{t',x'}$ to x implies the $\mathbb{Q} - $ a.s.-convergence of $G_{t',x'}$ to $G_{T,x}$. We can thus conclude using the dominated convergence theorem and the polynomial growth property of h and g. When $h(T, \cdot) = g$ on $\partial\mathscr{O}$, we can use the same arguments for $x \in \partial\mathscr{O}$.

3. The continuity on $[0, T) \times \partial\mathscr{O}$ is more delicate to obtain. One possible way is to prove that $(t, x) \mapsto \theta_{t,x} \wedge T$ is continuous in probability, and then to pass to the almost-sure limit on a subsequence. In order to do that, we observe that the regularity of $d_{\mathscr{O}}$ allows us to write

$$d_{\mathscr{O}}(S^{t',x'}) = d_{\mathscr{O}}(x') + \int_{t'}^{\cdot} \mathscr{L}_s d_{\mathscr{O}}(s, S_s^{t',x'}) ds$$

$$+ \int_t^{\cdot} (\partial d_{\mathscr{O}}\sigma)(s, S_s^{t',x'}) dW_s^{\mathbb{Q}}.$$

To simplify the arguments, we now assume that the condition (4.18) holds true on the whole domain with $\|\partial d_{\mathscr{O}}\sigma\| = 1$. The general case is obtained by localisation to guarantee $\|\partial d_{\mathscr{O}}\sigma\| > 0$ uniformly on a neighbourhood and using a time change argument to obtain $\|\partial d_{\mathscr{O}}\sigma\| = 1$, see e.g. Theorem 7.7 or [52].

In this case, $\bar{W}^{\mathbb{Q}} := \int_t^{\cdot} (\partial d_{\mathscr{O}}\sigma)(s, S_s^{t',x'}) dW_s^{\mathbb{Q}}$ is a $\mathbb{Q}$-Brownian motion. As $d_{\mathscr{O}} \in C_b^2$, the term $\mathscr{L}_s d_{\mathscr{O}}(s, S_s^{t',x'})$ is integrable, and by Girsanov Theorem there exist a measure $\bar{\mathbb{Q}}_{t',x'} \sim \mathbb{Q}$ and a $\bar{\mathbb{Q}}_{t',x'}$-Brownian motion $\bar{W}^{\bar{\mathbb{Q}}_{t',x'}}$ such that

$$d_{\mathscr{O}}(S_s^{t',x'}) = d_{\mathscr{O}}(x') + \int_{t'}^{s} \mathscr{L}_s d_{\mathscr{O}}(u, S_u^{t',x'})du + \bar{W}_s^{\mathbb{Q}}$$

$$= d_{\mathscr{O}}(x') + \bar{W}_s^{\bar{\mathbb{Q}}_{t',x'}}.$$

We thus have to study the continuity in y of the first passage time at 0, τ_y, of the Brownian motion starting at y, i.e. $y + W$, under the measure $\mathbb{P}$ for which W is a Brownian motion. But saying that $\tau_y > \varepsilon$ amounts to say that $\min_{[0,\varepsilon]} W > -y$. Since $\min_{[0,\varepsilon]} W$ has same law as $-|W_\varepsilon|$, see [52], $\mathbb{P}\left[\min_{[0,\varepsilon]} W > -y\right] = \mathbb{P}\left[|W_\varepsilon| \leq y\right] \to 0$ when $y \to 0$. We thus deduce that $y \in \mathbb{R}_+ \mapsto \tau_y$ is continuous in probability.

4. Uniqueness for smooth functions is obtained as in the proof of Theorem 4.5. If the maximum is reached on $[0, T) \times \partial\mathscr{O} \cup \{T\} \times \bar{\mathscr{O}}$, we straightforwardly have a contradiction using the fact that $w_1 = w_2$ on this domain. Uniqueness is still true for viscosity solution but the arguments are quite involved, see [20].

$\square$

4.3.2 Delta-Hedging, Exploding Behaviour and Regularisation Techniques

We first start by a verification theorem which corresponds to Theorem 4.3 in our setting.

Theorem 4.20 (Verification) *Say that a function* $\varphi \in C^{1,2}([0, T) \times \mathscr{O}) \cap C^{0,0}([0, T] \times \bar{\mathscr{O}} \setminus \{T\} \times \partial\mathscr{O})$ *is solution to* (4.8) *on* $[0, T) \times \mathscr{O}$ *such that* $\varphi(T, \cdot)\mathbf{1}_{(\partial\mathscr{O})^c} = g\mathbf{1}_{\mathscr{O}} + h(T, \cdot)\mathbf{1}_{\mathscr{O}^c}$. *Moreover assume that*

$$\mathbb{Q}\left[S_T^{t,x} \in \partial\mathscr{O}\right] = 0 \text{ and } \mathbb{E}^{\mathbb{Q}}\left[\int_t^T \|\partial p(s, S_s^{t,x})\sigma(s, S_s^{t,x})\|^2 ds\right] < \infty, \qquad (4.21)$$

for all $(t, x) \in [0, T) \times (0, \infty)^d$. *Then,* $\varphi = p$ *on* $[0, T) \times \bar{\mathscr{O}}$. *Moreover,*

$$p(t, x) + \int_t^{T \wedge \theta_{t,x}} \partial p(s, S_s^{t,x})d\tilde{S}_s^{t,x}$$

$$= \beta_{T \wedge \theta_{t,x}}^t \left(g(S_T^{t,x})\mathbf{1}_{\{\theta_{t,x} > T\}} + h(\theta_{t,x}, S_{\theta_{t,x}}^{t,x})\mathbf{1}_{\{\theta_{t,x} \leq T\}}\right),$$

$$(4.22)$$

for all $(t, x) \in [0, T) \times (0, \infty)^d$.

Proof By Itô's Lemma up to $\theta_{t,x} \wedge T_\varepsilon$ with $T_\varepsilon := T - \varepsilon, \varepsilon > 0$, we obtain

$$\varphi(t,x) + \int_t^{T_\varepsilon \wedge \theta_{t,x}} \beta_s^t \partial\varphi(s, S_s^{t,x}) d\tilde{S}_s^{t,x} = \beta_{T_\varepsilon \wedge \theta_{t,x}}^t \varphi(T_\varepsilon \wedge \theta_{t,x}, S_{T_\varepsilon \wedge \theta_{t,x}}^{t,x}).$$

On $\{\theta_{t,x} < T\}$, $\varphi(T_\varepsilon \wedge \theta_{t,x}, S_{T_\varepsilon \wedge \theta_{t,x}}^{t,x}) = h(\theta_{t,x}, S_{\theta_{t,x}}^{t,x})$ for $\varepsilon > 0$ small enough $\mathbb{Q} - $ a.s. On $\{\theta_{t,x} > T\}$, $S_{T_\varepsilon \wedge \theta_{t,x}}^{t,x} \to S_T^{t,x} \in \mathcal{O}$. Since $\mathcal{O}$ is an open set, we have $S_{T_\varepsilon \wedge \theta_{t,x}}^{t,x} \in \mathcal{O}$ for all $\varepsilon > 0$ small enough $\mathbb{Q} - $ a.s. and then $T_\varepsilon \wedge \theta_{t,x} = T_\varepsilon$ for all $\varepsilon > 0$ small enough $\mathbb{Q} - $ a.s. Passing to the limit, we obtain $\varphi(T_\varepsilon \wedge \theta_{t,x}, S_{T_\varepsilon \wedge \theta_{t,x}}^{t,x}) \to g(S_T^{t,x})$ on this event. The case $\{\theta_{t,x} = T\}$ cannot happen since $\mathbb{Q}\left[S_T^{t,x} \in \partial\mathcal{O}\right] = 0$. $\square$

When a smooth enough solution exists and $S_T^{t,x}$ has a density, at least in a neighbourhood of $\partial\mathcal{O}$, the previous result yields the delta-hedging strategy. These conditions, as well as the integrability condition on the term to the right-hand side of (4.21) are satisfied in the Black & Scholes model for standard *up-and-out* options.

As already mentioned in the previous section, putting in place the delta-hedging strategy is problematic. Let us consider the example of an *up-and-out* call with strike $K > 0$ and barrier $B > K$. We have $p(T-, B-) = B - K > p(T-, B+) = 0$. But, in this case, the function p is $C^{1,2}$ on $[0,T) \times (0, B)$ and continuous up to the boundary away from the singularity point $\{(T, B)\}$. We thus have

$$\frac{1}{\varepsilon} \int_{B-\varepsilon}^{B+\varepsilon} \partial p(t,x) dx = \frac{p(t, B + \varepsilon) - p(t, B - \varepsilon)}{\varepsilon} \to -\infty$$

when $(t, \varepsilon) \to (T, 0)$ with $\varepsilon > 0$. This happens only if $\partial p(t,x)$ goes to $-\infty$ when (t,x) is close to (T, B), see Exercise 2.6. The number of assets to short-sell to hedge the option can become really large, which is not realistic in practice.

We have already explained in the previous section a possible way to solve this problem: change the original payoff into a more regular one, with a non exploding behaviour. The market practice is actually slightly different. It consists in hedging the *up-and-out* option with barrier B following the hedging strategy of the option with boundary $B + \varepsilon$ with $\varepsilon > 0$. With this shift, the product is cancelled when $S^{t,x} = B$, i.e. before the level $B + \varepsilon$ at which the hedging strategy may explode. The size of the delta is controlled by this parameter $\varepsilon > 0$. Of course, the *up-and-out* option with boundary B_ε is more expensive than the one with boundary B. But, even if the loss is taken by the option's seller, it cannot be compared to the potential loss $B - K$ that could be realised if the underlying price process reaches the level B at T.

To conclude, let us note that the arguments of Sect. 4.2.3 cannot be applied generally to barrier options. The best that can be obtained, under some regularity conditions, is

$$\partial p(t,x) = \mathbb{E}^{\mathbb{Q}}\left[\partial p(\theta_{t,x}, S_{\theta_{t,x}}^{t,x}) \beta_{\theta_{t,x}}^t \partial_x S_{\theta_{t,x}}^{t,x}\right].$$

Nevertheless, the very strong assumptions allowing to replace ∂p by the Jacobian of h and g in the expectation do not hold in practice for the standard options.

4.4 American Options

We study now the pricing and hedging of American Options with payoff $g(S^{t,x})$.
Following Sect. 2.4.4, the hedging price is given by:

$$p(t,x) := \sup_{\tau \in \mathscr{T}_t} \mathbb{E}^{\mathbb{Q}} \left[\beta_\tau^t g(S_\tau^{t,x}) \right]. \tag{4.23}$$

Moreover, Remark 2.41 states that the discounted price $\beta^t p(\cdot, S^{t,x})$ is a supermartingale and a martingale up to the first time where $p(\cdot, S^{t,x}) = g(S^{t,x})$. Since $p \geq g$ by definition, we deduce that p must solve the quasi-variational inequality

$$\min\{r - \mathscr{L}_S p , \, p - g\} = 0 \quad \text{on} \quad [0, T) \times (0, \infty)^d. \tag{4.24}$$

Indeed, the previous equation reflects the fact that the drift $\beta^t (\mathscr{L}_S p - rp)(\cdot, S^{t,x})$ of $\beta^t p(\cdot, S^{t,x})$ is equal to zero when $p(\cdot, S^{t,x}) > g(S^{t,x})$. This is the martingale property. Moreover, it is always non-positive: This is the supermartingale property. This can happen only when the price is equal to the payoff.

To simplify the arguments, we will assume in this section that

$$g \text{ is uniformly Lipschitz, non-negative and bounded,} \tag{4.25}$$

$$(r, \sigma) \text{ is Hölder-}1/2 \text{ in time and Lipschitz in space.}$$

4.4.1 Dynamic Programming Principle

To obtain rigorously the pricing equation, we need the property (4.6) adapted to our context. This is what we call the dynamic programming principle. It links the price functional at time t to the same functional at a later date via a control problem. Here, this control problem is an optimal stopping problem.

Theorem 4.21 (dynamic programming) *The function p is continuous, bounded, and satisfies the dynamic programming principle: Let $(t, x) \in [0, T) \times (0, \infty)^d$ and $\theta \in \mathscr{T}_t$, then*

$$p(t,x) = \sup_{\tau \in \mathscr{T}_t} \mathbb{E}^{\mathbb{Q}} \left[\beta_{\tau \wedge \theta}^t \left(p(\theta, S_\theta^{t,x}) \mathbf{1}_{\theta < \tau} + g(S_\tau^{t,x}) \mathbf{1}_{\theta \geq \tau} \right) \right]. \tag{4.26}$$

Proof

1. We first show the inequality $\leq$. Let $\tau \in \mathscr{T}_t$, then:

$$\mathbb{E}^{\mathbb{Q}}\left[\beta_\tau^t g(S_\tau^{t,x})\right] = \mathbb{E}^{\mathbb{Q}}\left[\mathbf{1}_{\theta<\tau}\beta_\theta^t \mathbb{E}^{\mathbb{Q}}\left[\beta_\tau^\theta g(S_\tau^{t,x})|\mathscr{F}_\theta\right] + \beta_\tau^t g(S_\tau^{t,x})\mathbf{1}_{\theta\geq\tau}\right].$$

The Markov property of S and the definition of p in (4.23) imply that

$$\mathbb{E}^{\mathbb{Q}}\left[\beta_\tau^\theta g(S_\tau^{t,x})|\mathscr{F}_\theta\right] \leq p(\theta, S_\theta^{t,x}) \quad \text{on } \{\tau > \theta\}.$$

The stopping time $\tau \in \mathscr{T}_t$ being arbitrary, combining these two equations gives the first inequality

$$p(t,x) = \sup_{\tau\in\mathscr{T}_t} \mathbb{E}^{\mathbb{Q}}\left[\beta_\tau^t g(S_\tau^{t,x})\right]$$

$$\leq \sup_{\tau\in\mathscr{T}_t} \mathbb{E}^{\mathbb{Q}}\left[\beta_{\tau\wedge\theta}^t \left(p(\theta, S_\theta^{t,x})\mathbf{1}_{\theta<\tau} + g(S_\tau^{t,x})\mathbf{1}_{\theta\geq\tau}\right)\right].$$

2. In the sequel, we use the fact that there exists a continuous function $c > 0$ such that, for all $\tau \in \mathscr{T}_0$,

$$|p(t_1,x_1) - p(t_2,x_2)| + |J(t_1,x_1,\tau) - J(t_2,x_2,\tau)|$$

$$\leq c(\|x_1\| + \|x_2\|)(|t_1 - t_2|^{\frac{1}{2}} + \|x_1 - x_2\|),$$

where

$$J(t_i,x_i,\tau) := \mathbb{E}^{\mathbb{Q}}\left[\beta_{t_i\vee\tau}^{t_i} g(S_{t_i\vee\tau}^{t_i,x_i})\right].$$

This is obtained by observing that

$$|p(t_1,x_1) - p(t_2,x_2)| \leq \sup_{\tau\in\mathscr{T}_0} |J(t_1,x_1,\tau) - J(t_2,x_2,\tau)|,$$

where

$$|J(t_1,x_1,\tau) - J(t_2,x_2,\tau)| \leq \mathbb{E}^{\mathbb{Q}}\left[|\beta_{t_1\vee\tau}^{t_1} g(S_{t_1\vee\tau}^{t_1,x_1}) - \beta_{t_2\vee\tau}^{t_2} g(S_{t_2\vee\tau}^{t_2,x_2})|\right],$$

and then by using standard estimates based on the regularity assumptions (4.25). For $n \geq 1$, we set $t_i^n := t + (T-t)i/n$, $i \leq n$. Let also $\varepsilon > 0$. The functions J and p being continuous in space, uniformly with respect to the other variables, we can find a countable partition of $(0,\infty)^d$ namely $(B_k^\varepsilon)_{k\geq 1}$, points $(x_k^\varepsilon)_{k\geq 1}$ such that $x_k^\varepsilon \in B_k^\varepsilon$, and stopping times $(\tau_{ki}^{\varepsilon,n})_{k,i\geq 1}$ such that

$$
\begin{aligned}
J(t_i^n, \cdot, \tau_{ki}^{\varepsilon,n}) &\geq J(t_i^n, x_k^\varepsilon, \tau_{ki}^{\varepsilon,n}) - \varepsilon \\
&\geq p(t_i^n, x_k^\varepsilon) - 2\varepsilon \\
&\geq p(t_i^n, \cdot) - 3\varepsilon \text{ on } B_k^\varepsilon.
\end{aligned}
$$

Let θ_n be the first t_i^n after θ. Given $\tau \in \mathscr{T}_t$, we define the new stopping time

$$
\bar{\tau}^{\varepsilon,n} = \tau \mathbf{1}_{\tau < \theta_n} + \mathbf{1}_{\tau \geq \theta_n} \sum_{i=1}^{n} \sum_{k\geq 1} \tau_{ki}^{\varepsilon,n} \mathbf{1}_{\{S_{\theta_n}^{t,x} \in B_k^\varepsilon\}}.
$$

The previous inequalities imply that

$$
J(\theta_n, S_{\theta_n}^{t,x}, \bar{\tau}^{\varepsilon,n}) \geq p(\theta_n, S_{\theta_n}^{t,x}) - 3\varepsilon. \tag{4.27}
$$

Moreover, if each $\tau_{k,i}^{\varepsilon,n}$ depends only on $\sigma(W_{\cdot \wedge t_i^n} - W_{t_i^n})$, then we can show that

$$
J(\theta_n, S_{\theta_n}^{t,x}, \bar{\tau}^{\varepsilon,n}) = \mathbb{E}^{\mathbb{Q}}\left[\beta_{\bar{\tau}^{\varepsilon,n}}^{\theta_n} g(S_{\bar{\tau}^{\varepsilon,n}}^{t,x}) | \mathscr{F}_{\theta_n} \right]. \tag{4.28}
$$

This is generally not true because of the dependence between $\mathscr{F}_{\theta_n}$ and $\bar{\tau}^{\varepsilon,n}$, but one can always reduce to this case. We then have,

$$
\begin{aligned}
J(t, x, \bar{\tau}^{\varepsilon,n}) &= \mathbb{E}^{\mathbb{Q}}\left[\beta_\tau^t g(S_{\bar{\tau}^{\varepsilon,n}}^{t,x}) \right] \\
&= \mathbb{E}^{\mathbb{Q}}\left[\beta_\tau^t g(S_\tau^{t,x}) \mathbf{1}_{\theta_n > \tau} \right] \\
&\quad + \mathbb{E}^{\mathbb{Q}}\left[\mathbf{1}_{\theta_n \leq \tau} \beta_{\theta_n}^t \mathbb{E}^{\mathbb{Q}}\left[\beta_{\bar{\tau}^{\varepsilon,n}}^{\theta_n} g(S_{\bar{\tau}^{\varepsilon,n}}^{t,x}) | \mathscr{F}_{\theta_n} \right] \right] \\
&= \mathbb{E}^{\mathbb{Q}}\left[\beta_\tau^t g(S_\tau^{t,x}) \mathbf{1}_{\theta_n > \tau} + \mathbf{1}_{\theta_n \leq \tau} \beta_{\theta_n}^t J(\theta_n, S_{\theta_n}^{t,x}, \bar{\tau}^{\varepsilon,n}) \right] \\
&\geq \mathbb{E}^{\mathbb{Q}}\left[\beta_\tau^t g(S_\tau^{t,x}) \mathbf{1}_{\theta_n > \tau} + \mathbf{1}_{\theta_n \leq \tau} \beta_{\theta_n}^t p(\theta_n, S_{\theta_n}^{t,x}) \right] - O(\varepsilon),
\end{aligned}
$$

where we used (4.28), (4.27) and the fact that β^t is bounded, since r is itself bounded. This implies

$$
\begin{aligned}
p(t, x) &\geq J(t, x, \bar{\tau}^{\varepsilon,n}) \\
&\geq \mathbb{E}^{\mathbb{Q}}\left[\beta_\tau^t g(S_\tau^{t,x}) \mathbf{1}_{\theta_n > \tau} + \mathbf{1}_{\theta_n \leq \tau} \beta_{\theta_n}^t p(\theta_n, S_{\theta_n}^{t,x}) \right] - O(\varepsilon).
\end{aligned}
$$

Since $\varepsilon > 0$ is arbitrary, we deduce

$$
p(t, x) \geq \mathbb{E}^{\mathbb{Q}}\left[\beta_\tau^t g(S_\tau^{t,x}) \mathbf{1}_{\theta_n > \tau} + \mathbf{1}_{\theta_n \leq \tau} \beta_{\theta_n}^t p(\theta_n, S_{\theta_n}^{t,x}) \right].
$$

Finally, when $n \to \infty$, we have $\theta_n \downarrow \theta$ and $S_{\theta_n}^{t,x} \to S_\theta^{t,x}$ $\mathbb{Q}$ − a.s. The continuity of p and g, an application of the dominated convergence theorem, and the fact that $0 \leq g \leq p$ imply

$$p(t,x) \geq \mathbb{E}^\mathbb{Q} \left[\beta_\tau^t g(S_\tau^{t,x}) \mathbf{1}_{\theta \geq \tau} + \mathbf{1}_{\theta < \tau} \beta_\theta^t p(\theta, S_\theta^{t,x}) \right],$$

which proves the Proposition as $\tau \in \mathcal{T}_t$ is arbitrary.								$\square$

4.4.2   Associated Quasi-variational Inequalities

We can now show that p is a viscosity solution to (4.24).

Proposition 4.22 *The function p is a continuous viscosity solution to (4.24). Moreover, $p(T, \cdot) = g$ on $(0, \infty)^d$.*

Proof

1. The proof of the super-solution property is not more difficult than the one of Theorem 4.13. If the super-solution property of (4.24) is not satisfied in a point $(t, x) \in [0, T) \times (0, \infty)^d$ for a test function φ, we then have

$$(r\varphi - \mathscr{L}_S \varphi)(t, x) < 0,$$

as $p \geq g$ by definition. This is still true in a neighbourhood B of (t, x). Following the arguments of the proof of Theorem 4.13, we can find a stopping time $\theta < T$ such that

$$p(t, x) < \mathbb{E}^\mathbb{Q} \left[\beta_\theta^t p(\theta, S_\theta^{t,x}) \right].$$

But the dynamic programming principle in Theorem 4.21 implies that

$$p(t, x) \geq \mathbb{E}^\mathbb{Q} \left[\beta_\theta^t p(\theta, S_\theta^{t,x}) \right],$$

for $\theta < T$ (replace the supremum by the term obtained for $\tau \equiv T$ in (4.26)).
2. We now prove the sub-solution property. Let $(t, x) \in [0, T) \times (0, \infty)^d$ and φ be a test function at this point:

$$0 = \max_{[0,T) \times (0,\infty)^d} (p - \varphi) = (p - \varphi)(t, x). \tag{4.29}$$

Assume that at (t, x)

$$\min\{ r - \mathscr{L}_S , \ \varphi - g \} > \eta, \tag{4.30}$$

with $\eta > 0$. Then, this stays true on a neighbourhood of B. Moreover, up to changing φ by $\tilde{\varphi}(s, y) := \varphi(s, y) + |s - t|^2 + \|x - y\|^4$ and η, we can assume that

$$p \leq \varphi - \eta \quad \text{on} \quad \partial B. \tag{4.31}$$

Indeed, the derivatives of order 1 in time and order 1 and 2 in space of $\tilde{\varphi}$ and φ are equal at (t, x). This means that (4.30) still holds for $\tilde{\varphi}$ at (t, x), and thus on a neighbourhood of (t, x). The only difference is that (t, x) is now a strict maximum of $p - \tilde{\varphi}$.

Let θ be the first exit time of $(s, S_s^{t,x})_{s \geq t}$ from B. We use Itô's Lemma and (4.30) to obtain

$$\varphi(t, x) \geq \mathbb{E}^{\mathbb{Q}} \left[\beta_{\tau \wedge \theta}^t \left(\varphi(\theta, S_\theta^{t,x}) \mathbf{1}_{\theta < \tau} + \varphi(\tau, S_\tau^{t,x}) \mathbf{1}_{\theta \geq \tau} \right) \right].$$

Using (4.30) and (4.31), we then obtain

$$p(t, x) = \varphi(t, x) \geq \mathbb{E}^{\mathbb{Q}} \left[\beta_{\tau \wedge \theta}^t \left(p(\theta, S_\theta^{t,x}) \mathbf{1}_{\theta < \tau} + g(S_\tau^{t,x}) \mathbf{1}_{\theta \geq \tau} \right) \right] + \eta \min_{[t,T]} \beta^t.$$

Since $\tau \in \mathcal{T}_t$ is arbitrary and $\eta \min_{[t,T]} \beta^t > 0$, this contradicts the dynamic programming principle stated in Theorem 4.21. $\qquad \square$

Let us conclude this section with a uniqueness result.

Proposition 4.23 (Uniqueness) *Assume that σ satisfies the conditions of Theorem 4.5. Then p is the unique continuous viscosity solution of (4.24). Moreover, $p(T, \cdot) = g$ on $(0, \infty)^d$.*

Sketch of proof. Once again, we will only discuss the case of smooth solutions. Assume that w_1 and w_2 are two classical solutions. Then, if $w_1 = g$ we have $w_1 = g \leq w_2$. At a maximum point of $w_1 - w_2$ (which is positive), we must then have $rw_i - \mathcal{L}_S w_i = 0$ for $i = 1, 2$. We can then follow the proof of Theorem 4.5. $\qquad \square$

4.4.3 Delta-Hedging in the Smooth Case

We now consider the regular case. But since p is generally equal to g on a subset of $[0, T) \times (0, \infty)^d$, it is not reasonable to impose that it is $C^{1,2}$ on the whole domain $[0, T) \times (0, \infty)^d$.

Theorem 4.24 (Verification) *We assume that σ satisfies the conditions of Proposition 4.23. Let $w \geq g$ be a continuous bounded function on $[0, T] \times (0, \infty)$. Let us assume that it is $C^{1,2}$ on $\mathcal{D} := \{(t, x) \in [0, T) \times (0, \infty)^d : w(t, x) > g(x)\}$ and that it is solution to*

$$r - \mathcal{L}_S w = 0 \text{ on } \mathcal{D} \text{ and } w(T, \cdot) = g \text{ on } (0, \infty)^d.$$

Let us assume moreover

$$\mathbb{E}^{\mathbb{Q}}\left[\int_t^{\hat{\tau}_{t,x}} \|(\partial w\sigma)(s, S_s^{t,x})\|^2 ds\right] < \infty$$

where

$$\hat{\tau}_{t,x} := \inf\{s \geq t : w(s, S_s^{t,x}) = g(S_s^{t,x})\},$$

for all $(t, x) \in \mathscr{D}$. *Then,* $w = p$. *Moreover,*

$$w(t,x) + \int_t^{\hat{\tau}_{t,x}\wedge\cdot} \partial w(s, S_s^{t,x}) d\tilde{S}_s^{t,x} \geq \beta_{\hat{\tau}_{t,x}\wedge\cdot}^t g(S_{\hat{\tau}_{t,x}\wedge\cdot}^{t,x}) \qquad (4.32)$$

and

$$w(t,x) + \int_t^{\hat{\tau}_{t,x}} \partial w(s, S_s^{t,x}) d\tilde{S}_s^{t,x} = \beta_{\hat{\tau}_{t,x}}^t g(S_{\hat{\tau}_{t,x}}^{t,x}). \qquad (4.33)$$

Finally,

$$p(t,x) = \mathbb{E}^{\mathbb{Q}}\left[\beta_{\hat{\tau}_{t,x}}^t g(S_{\hat{\tau}_{t,x}}^{t,x})\right].$$

Proof The fact that $w = p$ comes from Proposition 4.23. Equations (4.32) and (4.33) are obtained by applying Itô's Lemma:

$$w(t,x) + \int_t^{\hat{\tau}_{t,x}\wedge\cdot} \partial w(s, S_s^{t,x}) d\tilde{S}_s^{t,x} = \beta_{\hat{\tau}_{t,x}\wedge\cdot}^t w(\hat{\tau}_{t,x}\wedge\cdot, S_{\hat{\tau}_{t,x}\wedge\cdot}^{t,x})$$

$$\geq \beta_{\hat{\tau}_{t,x}\wedge\cdot}^t g(S_{\hat{\tau}_{t,x}\wedge\cdot}^{t,x}),$$

since $w \geq g$, with equality holding true at $\hat{\tau}_{t,x}$. Taking expectation on both sides, we obtain

$$p(t,x) = w(t,x) = \mathbb{E}^{\mathbb{Q}}\left[\beta_{\hat{\tau}_{t,x}}^t g(S_{\hat{\tau}_{t,x}}^{t,x})\right].$$

$\square$

When the price function p is C^1 in space up to the boundary of $\mathscr{D}$, it is possible to apply the argument of Sect. 4.2.3 to obtain

$$\partial p(t,x) = \mathbb{E}^{\mathbb{Q}}\left[\beta_{\hat{\tau}_{t,x}}^t \partial g(S_{\hat{\tau}_{t,x}}^{t,x})(\partial_x S^{t,x})_{\hat{\tau}_{t,x}}\right].$$

It is possible to estimate $\mathscr{D}$ by solving (4.24) numerically using finite difference methods, see e.g. [45], and then to simulate $\hat{\tau}_{t,x}$ to compute the delta by Monte Carlo methods. It is also possible to estimate it by solving a PDE. The PDE is obtained by considering the right-hand side term of the equation as the price function of a barrier option paying $\partial g(S^{t,x})\partial_x S^{t,x}$ the first time $\hat{\tau}_{t,x}$ where $S^{t,x}$ leaves $\mathscr{D}$. We are then dealing with a problem similar to the one considered in Sect. 4.3.

4.5 Problems

4.1 (Call in the Black and Scholes model: verification arguments) Let $K, r, \sigma > 0$ be fixed. For $(t, x) \in [0, T] \times (0, \infty)$, we define

$$f(t,x) := x\Phi(d(t,x)) - Ke^{-r(T-t)}\Phi(d(t,x) - \sigma\sqrt{T-t})$$

where Φ if the cumulative distribution function of the standard normal distribution and

$$d(t,x) := \frac{1}{\sigma\sqrt{T-t}}\left(\ln(x/K) + (r + \frac{1}{2}\sigma^2)(T-t)\right).$$

1. Computing its derivatives, check that f is solution on $[0, T) \times (0, \infty)$ to .

$$rf - \partial_t f - rx\partial_x f - \frac{1}{2}\sigma^2 x^2 \partial_{xx}^2 f = 0.$$

2. Show that f is continuous and check that $f(T, x) = [x - K]^+$ for all $x > 0$.
3. We now consider the process S given by

$$dS_t = S_t\mu dt + \sigma S_t dW_t$$

and the initial condition $S_0 > 0$. Here μ is a constant. Show that $|\partial_x f(t, S_t)| \le 1$ $\mathbb{P}$ − a.s. for all $t \le T$.
4. Using 1. and 2., show that

$$f(0, S_0) + \int_0^T \partial_x f(s, S_s)d\tilde{S}_s = \beta_T[S_T - K]^+.$$

5. Show that $f(0, S_0) + \int_0^t \partial_x f(s, S_s)d\tilde{S}_s \ge 0$ $\mathbb{P}$ − a.s. for all $t \le T$.
6. Deduce that f is the hedging price of the call option with exercise price K in the Black and Scholes model.

4.2 (Model with dividends) In some models, it is assumed that S pays a dividend whose cumulated value is a predictable RCLL[2] process C. In this case, the dynamics of the risky assets read

$$S_t = S_0 + \int_0^t b_s ds + \int_0^t \sigma_s dW_s - C_t \quad, t \in \mathbb{T}.$$

The dividend value must be subtracted to the asset price to avoid any arbitrage opportunity.

If we hold a quantity ϕ_t^i of asset i at time t, we receive then a dividend equal to $\phi_t^i dC_t^i$ on the period $[t, t+dt]$.

1. Show that the dynamics of the wealth process read

$$dV_t = \left[r_t \left(V_t - \phi_t' S_t \right) \right] dt + \phi_t' dC_t + \phi_t' dS_t .$$

2. Deduce that

$$dV_t = \left(r_t V_t + \phi_t' (b_t - r_t S_t) \right) dt + \phi_t' \sigma_t dW_t .$$

3. Deduce also that

$$d\tilde{V}_t = \phi_t' d\tilde{Z}_t .$$

where

$$\tilde{Z}_t = S_0 + \int_0^t (\tilde{b}_s - r_s \tilde{S}_s) ds + \int_0^t \tilde{\sigma}_s dW_s \quad, t \in \mathbb{T}.$$

4. How the notion of martingale measure must be modified in this context?
5. We assume now that $b_t = \mu S_t$ and $\sigma_t = \nu S_t$ for some constants $\mu, \nu > 0$ and that $dC_t = \delta S_t dt$ where $\delta \geq 0$ is also a constant. Give the price at the date 0 of a European call with strike $K > 0$ in this model (in terms of the cumulative distribution function Φ of the centered reduced normal law).

4.3 (Double barrier Option) We consider the Black and Scholes Model in dimension one where

$$dS_s^{t,x} = S_s^{t,x} \mu ds + \sigma S_s^{t,x} dW_s, \ S_t^{t,x} = x,$$

with $\mu \in \mathbb{R}$, $\sigma > 0$ and W is a one-dimensional Brownian Motion. We assume that the risk free interest rate is constant equal to r.

---

[2]Right continuous with left limits.

We consider the barrier option with payoff

$$g(S_T^{t,x})\mathbf{1}_{A_{t,x}}$$

where g is a continuous function whose support is in the open interval $(\underline{B}, \bar{B})$, $\underline{B} < \bar{B}$, and

$$A_{t,x} := \{\min_{[t,T]} S^{t,x} \geq \underline{B} \text{ and } \max_{[t,T]} S^{t,x} \leq \bar{B}\}.$$

Write down the PDE associated to the hedging price on the domain $[0, T) \times (\underline{B}, \bar{B})$. What are the boundary conditions on $[0, T) \times \{\underline{B}, \bar{B}\}$ and $\{T\} \times (0, \infty)$?

4.4 (Bermudan options) We consider the Black and Scholes Model in one dimension where

$$dS_s^{t,x} = S_s^{t,x}\mu ds + \sigma S_s^{t,x} dW_s, \ S_t^{t,x} = x,$$

with $\mu \in \mathbb{R}$, $\sigma > 0$ and W a one dimensional Brownian Motion. We assume that the risk free interest rate is constant equal to r.

Let g be a bounded non-negative continuous function and let $0 = t_0 < t_1 < t_2 < \cdots < t_\kappa = T$ be a finite sequence of dates. We consider the Bermudan Option with payoff $g(S^{t,x})$. This option gives the right to receive the payoff $g(S_{t_i}^{t,x})$ at t_i if one chooses to exercise the option at this date.[3] We denote p^{BE} the super-replication price of this option.

1. Justify the following dual formulation for the price:

$$p^{BE}(t,x) = \sup_{\tau \in \mathscr{T}_t^{BE}} \mathbb{E}^{\mathbb{Q}}\left[\beta_\tau^t g(S_\tau^{t,x})\right]$$

   where $\mathscr{T}_t^{BE}$ is the set of stopping times with values in $[t, T] \cap \{t_i, \ i = 0, \ldots, \kappa\}$.
2. Justify quickly the dynamic programming principle

$$p^{BE}(t,x) = \sup_{\tau \in \mathscr{T}_t^{BE}} \mathbb{E}^{\mathbb{Q}}\left[\beta_{\tau \wedge \theta}^t \left(p^{BE}(\theta, S_\theta^{t,x})\mathbf{1}_{\theta < \tau} + g(S_\tau^{t,x})\mathbf{1}_{\theta \geq \tau}\right)\right]$$

   for all $\theta \in \mathscr{T}_t$.
3. Assume that p^{BE} is smooth on each interval $[t_i, t_{i+1})$, $i \leq \kappa - 1$, show that it is a solution to

$$rp^{BE} - \mathscr{L}_S p^{BE} = 0 \text{ on } [t_i, t_{i+1}) \times (0, \infty), \ i = 0, \cdots, \kappa - 1,$$

---

[3]Similarly to American Options, Bermudan Options can only be exercised once. The main difference with American Option is that the exercise time is restricted to be in $\{t_i, \ i = 0, \ldots, \kappa\}$.

with boundary conditions

$$\lim_{t'\uparrow t_{i+1}, x'\to x} p^{BE}(t', x) = \max\{g(x), p^{BE}(t_{i+1}, x)\},$$

for all $x \in (0, \infty)$, $i = 0, \cdots, \kappa - 1$, with the convention

$$p^{BE}(t_\kappa, \cdot) = g.$$

4. What should be the rational exercise time for this option?

4.5 (Asian Option: arithmetic mean)   We consider the same model as in the previous exercise but now we suppose that the volatility process $(\sigma_t)_{t\geq 0}$ is random, positive, and that σ_t and σ_t^{-1} are uniformly bounded in (t, ω). We are interested in the pricing and hedging of the Asian Option whose payoff at maturity $T > 0$ is

$$G = (Y_T - K)^+ \text{ where } Y_T = \int_0^T S_u g(u) du$$

and $g : [0, T] \longrightarrow \mathbb{R}$ is a given deterministic function. We denote by $\mathbb{Q}$ the risk-neutral probability, W the Brownian motion under $\mathbb{Q}$, and by $\mathbb{E}_t^\mathbb{Q} = \mathbb{E}^\mathbb{Q}[.|\mathscr{F}_t]$ the associated conditional expectation operator.

1. a. Check that

$$\mathbb{E}_t^\mathbb{Q}\left[e^{-r(T-t)}Y_T\right] = \hat{\theta}(t)S_t + e^{-r(T-t)}\int_0^t g(u)S_u du,$$

   where $\hat{\theta}: [0, T] \longrightarrow \mathbb{R}$ is a deterministic function to precise.
   b. We consider the self-financing portfolio strategy consisting in holding $\hat{\theta}(t)$ units of risky assets at each dates $t \in [0, T]$. We note by X_t the portfolio value at time t. Give the dynamics of the process $\tilde{X}$ defined by $\tilde{X}_t = e^{-rt}X_t$ for $t \in [0, T]$ in terms of W.
   c. With the notation $\tilde{S}_t = e^{-rt}S_t$, $t \in [0, T]$, show that

$$\hat{\theta}(0)S_0 + \int_0^T \hat{\theta}(u)d\tilde{S}_u = e^{-rT}Y_T.$$

   Give the financial interpretation of $\hat{\theta}(t)$ and of $\hat{\theta}(0)S_0$.
   d. Show that there exists an initial capital $\widehat{X}_0$ such that

$$\widehat{X}_t := e^{rt}\widehat{X}_0 + e^{rt}\int_0^t \hat{\theta}(u)d\tilde{S}_u$$

satisfies

$$e^{-rT}\widehat{X}_T = \widehat{X}_0 + \int_0^T \hat{\theta}(u)d\tilde{S}_u = e^{-rT}(Y_T - K).$$

2. We note $Z_t := \frac{\widehat{X}_t}{S_t}, t \in [0, T]$.

   a. Show that

   $$\mathbb{E}^{\mathbb{Q}}\left[e^{-rT}G\right] = S_0\mathbb{E}^{\hat{\mathbb{Q}}}\left[Z_T^+\right]$$

   where $\hat{\mathbb{Q}}$ is a probability measure equivalent to $\mathbb{Q}$ to be determined.

   b. Write down the dynamics of the Brownian Motion B associated to $\hat{\mathbb{Q}}$ by the Girsanov Theorem.

   c. Applying Ito's formula, give the dynamics of the process $Z_t = \frac{\widehat{X}_t}{S_t}, t \in [0, T]$ in terms of B.

   d. We assume that $(\sigma_t)_t$ is constant equal to σ. Use the representation obtained in a. to propose a pricing PDE for the option using only time, a variable z and $\hat{\theta}$.

3. We assume now that $g(t) = 1$ for all $t \in [0, T]$ and that the volatility process is constant: $\sigma_t = \sigma > 0$ for all $t \geq 0$. We denote $K = e^{-k}$ and introduce the double Laplace transform:

   $$L(\lambda, \mu) := \iint \mathbb{E}^{\mathbb{Q}}\left[e^{-rt}(Y_t - e^{-k})^+\right]e^{-\mu t}e^{-\lambda k}dkdt$$

   for $\lambda, \mu > 0$.

   a. Show that

   $$L(\lambda, \mu) = \frac{1}{\mu\lambda(1 + \lambda)}\mathbb{E}^{\mathbb{Q}}\left[e^{-r\theta}Y_\theta^{1+\lambda}\right]$$

   where θ is a random variable independent from the Brownian Motion W and with exponential distribution with parameter μ, i.e. $\mathbb{P}[\theta \leq t] = (1 - e^{-\mu t})\mathbf{1}_{\mathbb{R}_+}(t)$.

   b. Check that for all $t \geq 0$:

   $$S_t^{1+\lambda} = S_0^{1+\lambda}e^{(1+\lambda)(r+\frac{\sigma^2}{2}\lambda)t}L_t$$

   where

   $$L_t := e^{(1+\lambda)\sigma W_t - \frac{1}{2}(1+\lambda)^2\sigma^2 t}.$$

c. Check that $\tilde{\mathbb{Q}}$ defined by $d\tilde{\mathbb{Q}} := L_t d\mathbb{Q}$ is a probability measure equivalent to $\mathbb{Q}$ on $\mathscr{F}_t$ for all $t \geq 0$, and give the distribution of the Brownian Motion W under $\tilde{\mathbb{Q}}$.

d. Show that

$$L(\lambda, \mu) = \frac{S_0^{(1+\lambda)}}{\lambda(1+\lambda)} \mathbb{E}^{\tilde{\mathbb{Q}}}\left[\int_0^\infty e^{-\beta t}\xi_t^{1+\lambda}dt\right]$$

where

$$\xi_t := \frac{Y_t}{S_t}, \; t \geq 0, \; \text{and} \; \beta := r + \mu - (1+\lambda)(r + \frac{\sigma^2}{2}\lambda).$$

e. Write down the dynamics of ξ and deduce that ξ is a Markov process.

f. We admit that the function

$$f(\xi_0) := \mathbb{E}^{\tilde{\mathbb{Q}}}\left[\int_0^\infty e^{-\beta u}\xi_u^{1+\lambda}\,du\right]$$

is of class C^2. Using Itô's formula, find a linear ODE satisfied by f.

g. How can all this be used to price the asian option?

4.6 (Probabilistic representation of greeks) Let $(\Omega, \mathscr{F}, \mathbb{P})$ be a probability space supporting a one-dimensional standard Brownian motion $(W_t)_{t\geq 0}$ and let T be a positive real number. We consider X as the solution of the following SDE

$$X_t = 1 + \int_0^t rX_s ds + \int_0^t \sigma(X_s)dW_s, \quad 0 \leq t \leq T,$$

for some $r > 0$. The function $\sigma : \mathbb{R} \to \mathbb{R}$ is twice differentiable with first and second derivatives bounded by $L > 0$.

1. For $\epsilon \geq 0$, we denote by X^ϵ the solution of the following SDE

$$X_t^\epsilon = 1 + \int_0^t (r+\epsilon)X_s^\epsilon ds + \int_0^t \sigma(X_s^\epsilon)dW_s, \quad 0 \leq t \leq T. \tag{4.34}$$

a. Explain why there is a unique solution to (4.34).

b. Show that, for $\epsilon \geq 0$ and $p \geq 2$,

$$\mathbb{E}\left[\sup_{t\in[0,T]} |X_t^\epsilon - X_t|^p\right] \leq C_p \epsilon^p,$$

where C_p is a positive constant which depends on L, T and p.

c. Show that

$$\sigma(X_s^\epsilon) = \sigma(X_s) + \xi_s(X_s^\epsilon - X_s)$$

where ξ is a bounded stochastic process.

d. For ξ found in the previous question, show that, for all $s \in [0, T]$,

$$|\xi_s - \sigma'(X_s)| \leq C|X_s^\epsilon - X_s| \,,$$

where σ' denotes the first derivative of σ.

2. Let Z be the unique solution to

$$Z_t = \int_0^t (rZ_s + X_s)ds + \int_0^t \sigma'(X_s)Z_s dW_s \,. \tag{4.35}$$

We admit that the following holds true

$$\mathbb{E}\left[\sup_{t \in [0,T]} |Z_t|^p\right] \leq C_p \,, \ p \geq 2 \,, \tag{4.36}$$

for some positive constant C_p, depending on L, T and p.

a. For $0 \leq t \leq T$ and $\epsilon > 0$, we define $\Gamma_t^\epsilon := \frac{X_t^\epsilon - X_t}{\epsilon}$, show that

$$\lim_{\epsilon \to 0} \mathbb{E}\left[\sup_{t \in [0,T]} |\Gamma_t^\epsilon - Z_t|^2\right] = 0 \,.$$

b. Let $g : \mathbb{R} \mapsto \mathbb{R}$ be a twice differentiable function with first and second derivatives bounded by L. We are interested in the sensitivity of $\mathbb{E}\left[e^{-rT}g(X_T)\right]$ with respect to a small perturbation of the parameter r. We thus want to estimate

$$\rho := \lim_{\epsilon \to 0} \frac{\mathbb{E}\left[e^{-(r+\epsilon)T}g(X_T^\epsilon)\right] - \mathbb{E}\left[e^{-rT}g(X_T)\right]}{\epsilon} \,. \tag{4.37}$$

Show that

$$\rho = \mathbb{E}\left[e^{-rT}\left(g'(X_T)Z_T - Tg(X_T)\right)\right] \,,$$

where g' denotes the first derivative of g.

Corrections

4.1

1. Direct computations.
2. If $x < K$, then $\lim_{t \to T} d(t,x) = -\infty$ and $f(t,x) \to 0$. If $x > K$, $\lim_{t \to T} d(t,x) = \infty$ and $f(t,x) \to x - K$. If $x = K$, $\lim_{t \to T} d(t,x) = 0$ and $f(t,x) \to 0$. Combining the three cases, we get $f(t,x) \to [x - K]^+$. The continuity follows easily.
3. We compute $\partial_x f(t,x) = \Phi(t,x) \in [0,1]$, for all $(t,x) \in [0,T] \times (0,+\infty)$.
4. Introduce $\tilde{f}(t,x) := e^{-rt} f(t, e^{rt}x)$ and observe that

$$\partial_t \tilde{f}(t, \tilde{S}_t) = e^{-rt}\left(-rf(t,S_t) + \partial_t f(t,S_t) + r\partial_x f(t,S_t)\right), \ \ \partial_x \tilde{f}(t, \tilde{S}_t) = \partial_x f(t,S_t).$$

Applying Ito's formula to $\tilde{f}(t, \tilde{S}_t)$ and using the PDE satisfied by f, we obtain

$$f(0,S_0) + \int_0^T \partial_x f(s, S_s) d\tilde{S}_s = \tilde{f}(T, \tilde{S}_T),$$

and we conclude observing that $\tilde{f}(T, \tilde{S}_T) = \beta_T [S_T - K]^+$.
5. Let $\mathbb{Q}$ denote the risk-neutral probability, $\tilde{S}$ is a $\mathbb{Q}$ martingale. From 3, taking the conditional $\mathbb{Q}$-expectation at time t in the equality of 4, we obtain $f(0,S_0) + \int_0^t \partial_x f(s, S_s) d\tilde{S}_s \geq 0$ $\mathbb{Q}$ − a.s. for all $t \leq T$. It holds also $\mathbb{P}$ − a.s. as $\mathbb{P}$ and $\mathbb{Q}$ are equivalent.
6. It follows from the above.

4.2

1. Denote α_t the amount of money on the cash account, $d\alpha_t = r_t \alpha dt + \phi_t' dC_t$ (the last term takes the dividends into account). Moreover the strategy is self-financing so $dV_t = d\alpha_t + \phi_t' dS_t$. To conclude one notes that $\alpha_t = V_t - \phi_t' S_t$.
2. Use the dynamics of S.
3. Apply Ito's formula to $(\tilde{V}_t)_{t \in [0,T]}$ with $\tilde{V} = e^{-rt}V_t$ and use 2.
4. $\tilde{Z}$ is a martingale under the risk neutral probability measure (that we denote $\mathbb{Q}$).
5. From 3–4 we can use the usual approach to recover the hedging price (an hedging portfolio). In particular, we have $V_0 = \mathbb{E}^{\mathbb{Q}}\left[[\tilde{S}_T - e^{-rT}K]_+\right]$ with $d\tilde{S}_t = -\delta\tilde{S}_t dt + v\tilde{S}_t dW_t^{\mathbb{Q}}$ and compute

$$V_0 = S_0 e^{-\delta T}\Phi(d_+) - Ke^{-rT}\Phi(d_-) \text{ with } d_\pm = \frac{\ln(\frac{S_0}{K}) + (r - \delta \pm \frac{1}{2}\sigma^2)T}{\sigma\sqrt{T}}.$$

4.3

Let p denote the pricing function. We apply Proposition 4.19.

1. Within the domain, p satisfies the usual pricing PDE namely

$$\partial_t p + r x \partial_x p + \frac{1}{2}\sigma^2 x^2 \partial_{xx}^2 p = rp.$$

2. The boundary condition is $p = 0$ on $[0, T) \times \{\underline{B}, \bar{B}\}$ and $p = g$ on $\{T\} \times (0, \infty)$.

4.4

1. Work in discrete time by setting $G_i := g(S_{t_i})$, $i \leq \kappa$ and use the arguments of Sect. 1.5.
2. Follow the arguments of the proof of Theorem 4.21.
3. Using the previous question, observe that for $t \in [t_i, t_{i+1})$ and any θ valued in (t, t_{i+1}),

$$p^{BE}(t, x) = \mathbb{E}^{\mathbb{Q}}\left[\beta_\theta^t p^{BE}(\theta, S_\theta^{t,x})\right].$$

Apply then the arguments of the introduction of Sect. 4.2.

To prove the boundary condition in time, set $\theta = t_{i+1}$, the programming principle yields

$$\mathbb{E}^{\mathbb{Q}}\left[\beta_{t_{i+1}}^t p^{BE}(t_{i+1}, S_{t_{i+1}}^{t,x})\right] \vee \mathbb{E}^{\mathbb{Q}}\left[\beta_{t_{i+1}}^t g(S_{t_{i+1}}^{t,x})\right]$$
$$\leq p^{BE}(t, x) \leq$$
$$\mathbb{E}^{\mathbb{Q}}\left[\beta_{t_{i+1}}^t p^{BE}(t_{i+1}, S_{t_{i+1}}^{t,x}) \vee g(S_{t_{i+1}}^{t,x})\right],$$

and the proof is concluded by letting $t \uparrow t_{i+1}$, $x' \to x$.
4. $\tau^* = \inf\{t \in \mathscr{T}_0 \,|\, p(t, S_t) = g(S_t)\}$.

4.5

1. a. Observing that $\mathbb{E}_t^{\mathbb{Q}}[Y_T] = \int_0^t S_u g(u)du + \int_t^T \mathbb{E}_t^{\mathbb{Q}}[S_u]\, g(u)du$ and $\mathbb{E}_t^{\mathbb{Q}}[S_u] = e^{r(u-t)}S_t$, we get $\hat{\theta}(t) := \int_t^T e^{-r(T-u)}g(u)du$.
   b. From the definition of self-financing strategies, we have

$$dX_t = r(X_t - \hat{\theta}(t)S_t)dt + \hat{\theta}(t)dS_t.$$

Applying Ito's formula to $e^{-rt}X_t$, we obtain

$$d\tilde{X}_t = -r\hat{\theta}(t)\tilde{S}_t dt + \hat{\theta}(t)e^{-rt}dS_t$$
$$= \hat{\theta}(t)\sigma_t \tilde{S}_t dW_t.$$

   c. Apply Ito's formula to $(\hat{\theta}(t)\tilde{S}_t)_{t\in[0,T]}$. Comparing with the dynamics of $\tilde{X}$, we get that $\hat{\theta}$ is an hedging strategy for Y_T with $\hat{\theta}(0)S_0$ the initial wealth.

   d. The terminal payoff $-K$ corresponds to selling K zero-coupon, whose hedging strategy uses no risky asset obviously. We thus have $\hat{X}_0 = \hat{\theta}(0)S_0 - e^{-rT}K$.

2. a. $\mathbb{E}^{\mathbb{Q}}\left[e^{-rT}G\right] = S_0\mathbb{E}^{\mathbb{Q}}\left[\frac{\tilde{S}_T}{S_0}[Z_T]^+\right]$.

   b. $dB_t = dW_t - \sigma_t dt$. and thus $\frac{d\tilde{\mathbb{Q}}}{d\mathbb{Q}} := \frac{\tilde{S}_t}{S_0}$.

   c. $dZ_t = (\hat{\theta}(t) - Z_t)\sigma_t dB_t$.

   d. From c. Z is a Markov process and then $\mathbb{E}^{\mathbb{Q}}\left[e^{-rT}G\right] = v(0, Z_0)$ where v is solution to $\partial_t v + \frac{1}{2}\sigma^2(\hat{\theta}(t) - z)^2\partial^2_{zz}v = 0$ and $v(T, z) = S_0[z]^+$.

3. a. We first observe that $L(\lambda, \mu) = \frac{1}{\mu}\int \mathbb{E}^{\mathbb{Q}}\left[e^{-r\theta}[Y_\theta - e^{-k}]^+\right]e^{-\lambda k}dk$, then using Fubini's theorem,

$$L(\lambda, \mu) = \frac{1}{\mu}\mathbb{E}^{\mathbb{Q}}\left[e^{-r\theta}\int_{-\ln(Y_\theta)}^{+\infty}(Y_\theta - e^{-k})e^{-\lambda k}dk\right] = \frac{1}{\mu\lambda(1+\lambda)}\mathbb{E}^{\mathbb{Q}}\left[e^{-r\theta}Y_\theta^{1+\lambda}\right].$$

   b. Direct application of Itô's formula.

   c. L is indeed a positive martingale with expectation equal to 1. For $t \in [0, T]$, $W_T = \tilde{W}_t + (1+\lambda)\sigma t$ with $\tilde{W}$ a Brownian motion under $\tilde{\mathbb{Q}}$.

   d. Observe that

$$L(\lambda, \mu) = \frac{S_0^{1+\lambda}}{\mu\lambda(1+\lambda)}\mathbb{E}^{\mathbb{Q}}\left[\frac{S_\theta^{1+\lambda}}{S_0^{1+\lambda}}e^{-r\theta}\xi_\theta^{1+\lambda}\right],$$

the rest follows by simple algebra.

   e. We compute $d\xi_t = \{1 - \xi_t(\lambda\sigma^2 + r)\}dt - \xi_t\sigma d\tilde{W}_t$. Thus ξ is solution to a linear SDE and is a strong Markov process.

   f. Using the Markov property of ξ, we have

$$f(\xi_0) = \mathbb{E}^{\tilde{\mathbb{Q}}}\left[\int_0^t e^{-\beta u}\xi_u^{1+\lambda}\,du + e^{-\beta t}f(\xi_t)\right],$$

and, by applying Itô's formula, we get

$$0 = \mathbb{E}^{\tilde{\mathbb{Q}}}\left[\int_0^t e^{-\beta u}\left(\xi_u^{1+\lambda} - \beta f(\xi_u) + f'(\xi_u)\{1 - \xi_u(\lambda\sigma^2 + r)\} + \frac{1}{2}f''(\xi_u)\xi_u^2\sigma^2\right)du\right].$$

Dividing by t and letting $t \to 0$, we obtain the ODE

$$\xi^{1+\lambda} - \beta f(\xi) + f'(\xi)\{1 - \xi(\lambda\sigma^2 + r)\} + \frac{1}{2}f''(\xi)\xi^2\sigma^2 = 0.$$

4. Solve the ODE and invert the Laplace transform.

4.6

1. a. SDE with Lipschitz coefficients.
   b. We set $\Delta X = X^\epsilon - X$ and get

$$\beta_t \Delta X_t = \epsilon \int_0^t \beta_s X_s^\epsilon ds + \int_0^t \beta_s \big(\sigma(X_s^\epsilon) - \sigma(X_s)\big) dW_s \ .$$

We then compute

$$\sup_{0 \le u \le t} |\Delta X_u|^p \le 2^p e^{rT} \Big(\sup_{0 \le u \le t} | \int_0^u \beta_s \big(\sigma(X_s^\epsilon) - \sigma(X_s)\big) dW_s |^p + \epsilon^p \sup_{0 \le u \le t} | \int_0^u X_s^\epsilon ds |^p \Big).$$

Using Burkholder-Davis-Gundy inequality, the Lipschitz continuity of σ

$$\mathbb{E}\left[\sup_{0 \le u \le t} |\Delta X_u|^p \right] \le C_p \big(\mathbb{E}\left[| \int_0^t |\Delta X_s|^2 ds |^{\frac{p}{2}} \right] + \epsilon^p \big).$$

Using Hölder inequality (if $p > 2$), the fact that $|\Delta X_s| \le \sup_{0 \le u \le s} |\Delta X_u|$ and Fubini's Theorem, we obtain

$$\mathbb{E}\left[\sup_{0 \le u \le t} |\Delta X_u|^p \right] \le C_p \big(\int_0^t \mathbb{E}\left[\sup_{0 \le u \le s} |\Delta X_s|^p \right] ds + \epsilon^p \big) \ .$$

The proof is concluded by using Gronwall's Lemma.
   c. Let $\theta(\lambda) = X_s + \lambda(X_s^\epsilon - X_s)$, for $\lambda \in [0, 1]$. Applying the fundamental theorem of calculus to the function $\lambda \mapsto \sigma(\theta(\lambda))$, we compute

$$\sigma(X_s^\epsilon) - \sigma(X_s) = \sigma(\theta(1)) - \sigma(\theta(0)) = \int_0^1 \sigma'(\theta(\lambda))(X_s^\epsilon - X_s) d\lambda$$

and then $\xi_s = \int_0^1 \sigma'(X_s + \lambda(X_s^\epsilon - X_s)) d\lambda$. It is bounded since σ is C_b^1.
   d. We compute

$$|\xi_s - \sigma'(X_s)| = | \int_0^1 \big(\sigma'(X_s + \lambda(X_s^\epsilon - X_s)) - \sigma'(X_s)\big) d\lambda |$$

$$\le C \int_0^1 \lambda |X_s^\epsilon - X_s| d\lambda$$

leading to the result.

2. a. Using 1.c, we obtain

$$\Gamma_t^\epsilon = \int_0^t (r\Gamma_s^\epsilon + X_s^\epsilon)ds + \int_0^t \xi_s\Gamma_s^\epsilon dW_s$$

so that $Y^\epsilon := \Gamma^\epsilon - Z$ solves

$$Y_t^\epsilon = \int_0^t \{rY_s^\epsilon + (X_s^\epsilon - X_s)\}ds + \int_0^t \left(\xi_s Y_s^\epsilon + \{\xi_s - \sigma'(X_s)\}Z_s\right)dW_s.$$

Applying Burkholder-Davis-Gundy inequality, using 1.c, 1.d, we compute

$$\mathbb{E}\left[\sup_{u\le t}|Y_u^\epsilon|^2\right] \le C\Big(\mathbb{E}\left[\int_0^t |Y_s^\epsilon|^2 ds\right]$$
$$+ \mathbb{E}\left[\int_0^T |X_s^\epsilon - X_s|^2|Z_s|^2 ds + \int_0^T |X_s^\epsilon - X_s|^2 ds\right]\Big). \quad (4.38)$$

Then, Cauchy-Schwarz inequality implies

$$\mathbb{E}\left[\int_0^T |X_s^\epsilon - X_s|^2|Z_s|^2 ds\right] \le C\mathbb{E}\left[\sup_{s\le T}|X_s^\epsilon - X_s|^4\right]^{\frac12}\mathbb{E}\left[\sup_{s\le T}|Z_s|^4\right]^{\frac12}$$
$$\le C\epsilon^2,$$

where we use 1.a and (4.36) for the last inequality. Inserting the previous bound back into Eq. (4.38) and using 1.a again, we obtain

$$\mathbb{E}\left[\sup_{u\le t}|Y_u^\epsilon|^2\right] \le C\mathbb{E}\left[\int_0^t |Y_s^\epsilon|^2 ds + \epsilon^2\right].$$

Observing that $\sup_{u\le s}|Y_u^\epsilon|^2 \ge |Y_s^\epsilon|^2$ and applying Fubini Theorem, we deduce that

$$\mathbb{E}\left[\sup_{u\le t}|Y_u^\epsilon|^2\right] \le C\Big(\mathbb{E}\left[\int_0^t \sup_{u\le s}|Y_u^\epsilon|^2 ds\right] + \epsilon^2\Big).$$

The proof is concluded by using Gronwall's Lemma and letting ϵ go to 0.
 b. We have

$$\rho = -Te^{-rT}\mathbb{E}\left[g(X_T)\right] + e^{-rT}\tilde\rho \quad \text{with } \tilde\rho = \lim_{\epsilon\to 0}\mathbb{E}\left[\frac{g(X_T^\epsilon) - g(X_T)}{\epsilon}\right]. \quad (4.39)$$

Hence,

$$\mathbb{E}\left[\frac{g(X_T^\epsilon) - g(X_T)}{\epsilon} - g'(X_T)Z_T\right] = \mathbb{E}\left[\tilde{g}_T^\epsilon \Gamma_T^\epsilon - g'(X_T)Z_T\right]$$

$$= \mathbb{E}\left[\tilde{g}_T^\epsilon(\Gamma_T^\epsilon - Z_T)\right] + \mathbb{E}\left[\left(\tilde{g}_T^\epsilon - g'(X_T)\right)Z_T\right]$$

with $\tilde{g}_T^\epsilon = \int_0^1 g'(X_s + \lambda(X_T^\epsilon - X_T))d\lambda$.

Since g is C_b^2, we get

$$|\tilde{g}_T^\epsilon - g'(X_T)| \le \frac{L}{2}|X_T^\epsilon - X_T| \quad \text{and} \quad |\tilde{g}_T^\epsilon| \le L$$

for some $L > 0$. We then use Cauchy-Schwarz inequality to obtain

$$|\mathbb{E}\left[\frac{g(X_T^\epsilon) - g(X_T)}{\epsilon} - g'(X_T)Z_T\right]| \le C\left(\mathbb{E}\left[|\Gamma_T^\epsilon - Z_T|^2\right]^{\frac{1}{2}} + \mathbb{E}\left[|X_T^\epsilon - X_T|^2\right]^{\frac{1}{2}}\right),$$

and the proof is then concluded by using 1.b and 2.a.

Chapter 5
Super-Replication and Its Practical Limits

In complete markets, the notion of viable price is very satisfying as it leads to the definition of a unique no-arbitrage price. Moreover, this price is the solution of a replication problem, which can be characterised quite simply. In particular, we have seen in Chap. 4 that this price is given as the unique solution of a linear PDE in a Markovian framework.

In incomplete markets, the notion of viable price leads to a range of prices. Selecting a price can then be done by using various criterions. A conservative approach in term of risk taking is to choose the most expensive one: the super-hedging price. It has been characterised in probabilistic terms in Chap. 2.

In a complete market, uniqueness of the no-arbitrage price is also lost when the hedging portfolio is constrained. Actually, incompleteness and portfolio constraints can essentially be studied together in Markovian settings, as non-hedgeable sources of risk can be viewed as assets that cannot be traded dynamically.

We pursue here the study of Sect. 2.5, but in a Markovian setting. We will first show that the super-replication price with constraints of a European contingent claim is the solution to a non-linear PDE. In some cases, one can show that this non-linear equation is equivalent to a linear one with a modified terminal condition. In terms of hedging, this means that one can hedge an option with portfolio constraints by considering the un-constrained strategy associated to a modified (face-lifted) payoff.

Finally, we will use the results obtained for complete markets with portfolio constraints to characterise the super-replication price in some incomplete markets. The main result is typically negative: the super-replication price is too expensive to be used in practice. We refer to Sect. 6.2 for the study of alternative approaches.

© Springer International Publishing Switzerland 2016

B. Bouchard, J.-F. Chassagneux, *Fundamentals and Advanced Techniques in Derivatives Hedging*, Universitext, DOI 10.1007/978-3-319-38990-5_5

5.1 Hedging with Portfolio Constraints

We have seen in the previous part that option prices can be characterised by a PDE
when the price of the underlying assets is solution to a SDE. In Sect. 4.2.3, we have
also shown how to delta-hedge this contingent claim. Sometimes, this delta-hedging
approach cannot be used in practice because the delta is too large. This is typically
the case for digital options or barrier options, see Sect. 4.3.2. One can then decide
to use a delta-hedging approach but with a delta constrained to belong to a bounded
set.

More generally, it is also possible that portfolio constraints are imposed by
market regulators, for example by forbidding short-selling practice.

We continue here the study started in Sect. 2.5, in a Markovian setting. In
particular, we will characterise the super-replication price with portfolio constraints
as the solution of a non-linear PDE.

5.1.1 Framework

As previously, the market is constituted of d risky assets whose price is given by

$$S_s^{t,x} = x + \int_t^s b(u, S_u^{t,x})du + \int_t^s \sigma(u, S_u^{t,x})dW_u , \ s \in [t, T], \tag{5.1}$$

and the risk free interest rate $(r_t)_{0\le t\le T}$ is a deterministic continuous function. We
assume that the assets price process takes values in $\mathbb{R}^d$. We also assume that σ has a
bounded inverse, that b and σ are Lipschitz in space and $\frac{1}{2}$-Hölder in time, and that
the same holds true for the function

$$(t, x) \mapsto \lambda(t, x) := \sigma^{-1}(t, x)(b(t, x) - r_t x)$$

which is bounded as well.

We study here the super-replication of a contingent claim with bounded and
lower-semicontinuous payoff g under portfolio constraints given by K, a closed
convex set containing 0.

From Theorem 2.42, we know that the super-replication price $p_K(t, x)$ of g under
portfolio constraints is given by

$$p_K(t, x) = \sup_{v\in\hat{\mathcal{K}}_b} J(t, x, v)$$

with

$$J(t, x, v) := \mathbb{E}\left[\mathcal{E}_T^{t,x,v}\left(\beta_T^t g(S_T^{t,x}) - \int_t^T \beta_s^t \delta_K(v_s)ds\right)\right]$$

and for $v \in \hat{\mathcal{K}}_b, t \leq s \leq T$,

$$\mathcal{E}_s^{t,x,v} = e^{-\frac{1}{2}\int_t^s \|\lambda_u^{t,x,v}\|^2 du - \int_t^s \lambda_u^{t,x,v} dW_u},$$

$$\lambda_s^{t,x,v} = \lambda(s, S_s^{t,x}) - \sigma^{-1}(s, S_s^{t,x})v_s.$$

The set $\hat{\mathcal{K}}_b$ is the set of progressively measurable processes, which are essentially bounded and take their values in $\hat{K}$. The set $\hat{K}$ is the domain of the support function of K, recall Sect. 2.5.

Equivalently, the super-replication price under constraints rewrites

$$p_K(t,x) = \sup_{v \in \hat{\mathcal{K}}_b} \mathbb{E}^{\mathbb{Q}_{t,x}^v}\left[\beta_T^t g(S_T^{t,x}) - \int_t^T \beta_s^t \delta_K(v_s) ds\right], \qquad (5.2)$$

with $\frac{d\mathbb{Q}_{t,x}^v}{d\mathbb{P}} := \mathcal{E}_T^{t,x,v}$. From Girsanov's theorem, see Theorem 2.9, the process defined by

$$W_s^{\mathbb{Q}_{t,x}^v} := W_s + \int_t^s \lambda_u^{t,x,v} du, \ s \in [t, T],$$

is a $\mathbb{Q}_{t,x}^v$-Brownian motion. Note that

$$S_s^{t,x} = x + \int_t^s (r_u S_u + v_u) du + \int_t^s \sigma(u, S_u^{t,x}) dW_u^{\mathbb{Q}_{t,x}^v}, \ s \in [t, T]. \qquad (5.3)$$

We will show that the function p_K is a viscosity solution of a non-linear PDE. This theoretical result is very important in practice as it allows to use numerical schemes for PDEs to approximate the super-replication price.

This result is based on a dynamic programming principle for the function p_K, whose intuitive formulation is

$$p_K(t,x) = \sup_{v \in \hat{\mathcal{K}}_b} \mathbb{E}\left[\mathcal{E}_\theta^{t,x,v}\left(\beta_\theta^t p_K(\theta, S_\theta^{t,x}) - \int_t^\theta \beta_s^t \delta_K(v_s) ds\right)\right]$$

for all stopping time $\theta \in \mathcal{T}_t$.

In the sequel, we will use the following version.

Theorem 5.1

(i) *For all control $v \in \hat{\mathcal{K}}_b$ and all stopping time $\theta \in \mathcal{T}_t$ such that $S^{t,x}$ is essentially bounded on $[t, \theta]$,*

$$p_K(t,x) \geq \mathbb{E}\left[\mathcal{E}_\theta^{t,x,v}\left(\beta_\theta^t p_K(\theta, S_\theta^{t,x}) - \int_t^\theta \beta_s^t \delta_K(v_s) ds\right)\right].$$

(ii) For all $\epsilon > 0$, there exists a control $v^\epsilon \in \hat{\mathcal{K}}_b$ such that, for all stopping time $\theta \in \mathcal{T}_t$ for which $S^{t,x}$ is essentially bounded on $[t, \theta]$,

$$p_K(t,x) - \epsilon \leq \mathbb{E}\left[\mathcal{E}_\theta^{t,x,v^\epsilon}\left(\beta_\theta^t p_K(\theta, S_\theta^{t,x}) - \int_t^\theta \beta_s^t \delta_K(v_s^\epsilon)ds\right)\right].$$

The proof of this result is very technical, relying on deep results of measure theory. It is largely outside the scope of this book, see [12] and the references therein.

5.1.2   Pricing Equation

We show in this section that p_K is a viscosity solution of a PDE by using the previous dynamic programming principle. This PDE involves the following differential operator which is linked to the constraints:

$$\mathcal{C}_K(p) := \inf_{\|\zeta\|=1, \zeta \in \hat{K}} \delta_K(\zeta) - \zeta'p \,.$$

Moreover, to characterise the terminal condition of the PDE, we introduce the face-lift operator.

Definition 5.2 (Face-lift) For h a measurable function from $\mathbb{R}^d$ to $\mathbb{R}$, the *face-lift* $F_K[h]$ is given by

$$F_K[h](x) := \sup_{y\in\mathbb{R}^d} h(x+y) - \delta_K(y) = \sup_{y\in\hat{K}} h(x+y) - \delta_K(y)\,, \quad x \in \mathbb{R}^d\,.$$

The function $F_K[h]$ is the smallest function above h whose gradient, if it exists, satisfies the constraints.[1] In financial terms, $F_K[h]$ is the cheapest payoff satisfying the delta constraint and allowing to hedge the payoff h at the same time. It is then quite natural that this function is the terminal condition at $T-$ of the super-replication problem. Exercise 5.2 studies some properties of this *face-lift* transform.

In the following, we limit our study to the case where p_K is continuous on $[0, T) \times \mathbb{R}^d$ to simplify. As explained above, there is little chance that p_K is continuous at T except if $F_K[g] = g$. In the general case, it is possible to show that the lower semi-continuous envelop (resp. upper) of p_K is a super-solution (resp. sub-solution) of the same equation. Note that p_K is at least lower semi-continuous as soon as g is lower semi-continuous itself. This is obtained easily from (5.2), having in mind that the supremum of lower semi-continuous functions is lower semi-continuous. The assumption of upper semi-continuity is used only to prove the sub-solution property.

---

[1] See Exercise 5.2 for a clear mathematical statement. It is remarkable that such a function exists.

Theorem 5.3 *Assume that p_K is continuous on $[0, T) \times \mathbb{R}^d$. Then, p_K is a viscosity solution of*

$$\min\{r\varphi - \mathcal{L}_S\varphi \,,\, C_K(\partial\varphi)\} = 0 \quad \text{on } [0, T) \times \mathbb{R}^d \tag{5.4}$$

with terminal condition

$$\liminf_{t'\uparrow T, x'\to x} p_K(t', x') \geq F_K[g](x) \,, \; x \in \mathbb{R}^d.$$

If $F_K[g]$ is upper semi-continuous and σ is bounded, then

$$\limsup_{t'\uparrow T, x'\to x} p_K(t', x') \leq F_K[g](x) \,, \; x \in \mathbb{R}^d.$$

The proof is divided in several steps. We study first the characterisation of the function on the domain $[0, T) \times \mathbb{R}^d$ by proving the super- and sub-solution properties.

Proposition 5.4 *The function p_K is a viscosity super-solution of* (5.4).

Proof We consider the constant control $\nu := y$ for $y \in \hat{K}$ and we introduce for $t < T, h > 0$,

$$\theta^h = \inf\{s \geq t \mid \|S_s^{t,x} - x\| + |\mathcal{E}_s^{t,x,\nu} - 1| \geq 1\} \wedge (t + h) \,.$$

Then $S^{t,x}$ is essentially bounded on $[t, \theta^h]$ and from Theorem 5.1 (i),

$$p_K(t, x) \geq \mathbb{E}\left[\mathcal{E}_{\theta^h}^{t,x,\nu}\left(\beta_{\theta^h}^t p_K(\theta^h, S_{\theta^h}^{t,x}) - \int_t^{\theta^h} \beta_s^t \delta_K(y) ds \right) \right].$$

Let φ be a smooth test function for p_K such that $p_K - \varphi$ reaches a minimum at (t, x), with $(p_K - \varphi)(t, x) = 0$. We thus have

$$\varphi(t, x) = p_K(t, x) \geq \mathbb{E}\left[\mathcal{E}_{\theta^h}^{t,x,\nu}\left(\beta_{\theta^h}^t \varphi(\theta^h, S_{\theta^h}^{t,x}) - \int_t^{\theta^h} \beta_s^t \delta_K(y) ds \right) \right]. \tag{5.5}$$

Applying Itô's formula, we obtain

$$\beta_{\theta^h}^t \varphi(\theta^h, S_{\theta^h}^{t,x}) = \varphi(t, x) + \int_t^{\theta^h} \beta_s^t (\mathcal{L}_S\varphi(s, S_s^{t,x}) - r_s\varphi(s, S_s^{t,x})) ds$$

$$+ \int_t^{\theta^h} \beta_s^t \partial\varphi(s, S_s^{t,x})\sigma(s, S_s^{t,x}) dW_s^{\mathbb{Q}_{t,x}^\nu} + \int_t^{\theta^h} \beta_s^t \partial\varphi(s, S_s^{t,x}) y ds \,.$$

Using (5.5), it comes

$$0 \geq \frac{1}{h}\mathbb{E}\left[\mathcal{E}_{\theta^h}^{t,x,\nu}\left(\int_t^{\theta^h}\beta_s^t\left(\mathcal{L}_S\varphi(s,S_s^{t,x}) + \partial\varphi(s,S_s^{t,x})y - \delta_K(y)\right)ds\right)\right].$$

Letting h go to 0, and applying the dominated convergence theorem, we obtain

$$r_t\varphi(t,x) - \mathcal{L}_S\varphi(t,x) - \partial\varphi(t,x)y + \delta_K(y) \geq 0. \qquad (5.6)$$

Setting $y = 0$ in (5.6), we have $r\varphi(t,x) - \mathcal{L}_S\varphi(t,x) \geq 0$.
Choosing y as $\lambda\zeta$ with $\|\zeta\| = 1$ and $\lambda \to \infty$, we obtain

$$-\partial\varphi(t,x)\zeta + \delta_K(\zeta) \geq 0, \quad \forall\, \zeta \in \hat{K} \text{ s.t. } \|\zeta\| = 1.$$

$\square$

Proposition 5.5 *The function p_K is a viscosity sub-solution of (5.4).*

Proof First, we observe that $\mathcal{C}_K$ is a continuous operator and therefore so is the operator in (5.4).
   Let then φ be a smooth test function such that $(p_K - \varphi)$ reaches a strict local maximum[2] at (t,x) and that, without loss of generality, $(p_K - \varphi)(t,x) = 0$.
   Assume that

$$\min\{r_t\varphi(t,x) - \mathcal{L}_S\varphi(t,x),\, \mathcal{C}_K(\partial\varphi(t,x))\} > 0.$$

We are working toward a contradiction.
   By continuity, we have that

$$\min\{r\varphi - \mathcal{L}_S\varphi,\, \mathcal{C}_K(\partial\varphi)\} > 0 \text{ on } B \qquad (5.7)$$

where B is an open neighbourhood of (t,x), and that there exists $\xi > 0$ such that

$$(p_K - \varphi)(s,y) \leq -\xi \text{ for } (s,y) \in \partial B. \qquad (5.8)$$

The last statement comes from the fact that a strict maximum is reached at (t,x) and that (t,x) does not belong to the compact set ∂B.
   We define then the following exit time:

$$\theta = \inf\{s \geq t \mid (s,S_s^{t,x}) \notin B\} \wedge T.$$

---

[2]See Remark 4.9.

Let $\epsilon > 0$ and ν^ϵ be given by Theorem 5.1 (ii). By applying Itô's Lemma and using (5.7), we obtain

$$\varphi(t,x) \geq \mathbb{E}\left[\mathcal{E}_\theta^{t,x,\nu^\epsilon}\left(\beta_\theta^t \varphi(\theta, S_\theta^{t,x}) - \int_t^\theta \beta_s^t \delta_K(\nu_s^\epsilon)ds\right)\right].$$

Using (5.8), we then deduce that

$$\varphi(t,x) \geq \mathbb{E}\left[\mathcal{E}_\theta^{t,x,\nu^\epsilon}\left(\beta_\theta^t p_K(\theta, S_\theta^{t,x}) - \int_t^\theta \beta_s^t \delta_K(\nu_s^\epsilon)ds + \xi\beta_\theta^t\right)\right].$$

From the definition of ν^ϵ, it follows

$$0 \geq -\epsilon + \xi e^{-\int_0^T r_s ds} + p_K(t,x) - \varphi(t,x).$$

Since $p_K(t,x) = \varphi(t,x)$, letting ϵ go to 0, the previous inequality leads to $0 \geq \xi e^{-\int_0^T r_s ds}$ which is absurd. $\qquad\square$

The two following propositions characterise p_K on the boundary of the domain, i.e. at $\{T\} \times \mathbb{R}$.

Proposition 5.6 *For all $x \in \mathbb{R}^d$,*

$$\liminf_{t' \uparrow T, x' \to x} p_K(t', x') \geq F_K[g](x).$$

Proof Consider a sequence $(t_n, x_n)_{n \geq 1}$, such that $(t_n, x_n) \to (T, x)$, $p_K(t_n, x_n) \to \liminf_{t' \uparrow T, x' \to x} p_K(t', x')$ and $t_n < T$ for all n. We consider the constant control $\nu_s^n = \frac{1}{T-t_n}y$, $s \in [t_n, T]$, $y \in \hat{K}$. By definition of p_K, it follows

$$p_K(t_n, x_n) \geq \mathbb{E}\left[\mathcal{E}_T^{t_n, x_n, \nu^n}\left(\beta_T^{t_n} g(S_T^{t_n, x_n}) - \int_{t_n}^T \beta_s^{t_n} \delta_K(\nu_s^n)ds\right)\right]$$

$$\geq \mathbb{E}^{\mathbb{Q}_{t_n, x_n}^{\nu_n}}\left[\beta_T^{t_n} g(S_T^{t_n, x_n}) - \delta_K(y)\int_{t_n}^T \frac{\beta_s^{t_n}}{T-t_n}ds\right].$$

To conclude, we observe that

$$S_t^{t_n, x_n} = x_n + \int_{t_n}^t (r_s S_s + \nu_s^n)ds + \int_{t_n}^t \sigma(s, S_s^{t_n, x_n})dW_s^{\mathbb{Q}_{t_n, x_n}^{\nu_n}}$$

has same law under $\mathbb{Q}_{t_n, x_n}^{\nu_n}$ as

$$Z_t^n := x_n + \frac{t-t_n}{T-t_n}y + \int_{t_n}^t r_s Z_s^n ds + \int_{t_n}^t \sigma(s, Z_s^n)dW_s^{\mathbb{Q}}$$

under $\mathbb{Q}$, for $t \geq t_n$.

The random variable Z_T^n converges in $L^2(\mathbb{Q})$ to $x + y$ and then a.s. along a subsequence $(Z_T^{n'})$.

Moreover, we have

$$\delta_K(y) \int_{t_{n'}}^{T} \frac{\beta_s^{t_{n'}}}{T - t_{n'}} ds \to \delta_K(y)$$

and, by using Fatou's Lemma,

$$\liminf_{t' \uparrow T, x' \to x} p_K(t', x') = \lim_{n \to \infty} p_K(t_n, x_n) = \lim_{n' \to \infty} p_K(t_{n'}, x_{n'})$$

$$\geq \liminf_{n' \to \infty} \mathbb{E}^{\mathbb{Q}} \left[\beta_T^{t_{n'}} g(Z_T^{n'}) - \delta_K(y) \int_{t_{n'}}^{T} \frac{\beta_s^{t_{n'}}}{T - t_{n'}} ds \right]$$

$$\geq g(x + y) - \delta_K(y) .$$

Since $y \in \hat{K}$ was arbitrarily chosen, we obtain finally

$$\liminf_{t' \uparrow T, x' \to x} p_K(t', x') \geq \sup_{y \in \hat{K}} g(x + y) - \delta_K(y),$$

which concludes the proof. $\qquad\qquad\qquad\qquad\qquad\qquad\qquad\qquad\qquad\square$

We can obtain a complete characterisation of the terminal condition by strengthening slightly the smoothness assumptions on $F_K[g]$.

Proposition 5.7 *Assume that $F_K[g]$ is Lipschitz[3] and that σ is bounded then*

$$\limsup_{t' \uparrow T, x' \to x} p_K(t', x') \leq F_K[g](x) .$$

Proof

1. Let $(t_n, x_n)_{n \geq 0}$ be a sequence such that $(t_n, x_n) \to (T, x)$, $t_n < T$ for all n, and $p_K(t_n, x_n) \to \limsup_{t' \uparrow T, x' \to x} p_K(t', x')$. By definition of $p_K(t_n, x_n)$, there exists a control ν^n such that

$$p_K(t_n, x_n) \leq \mathbb{E}^{\mathbb{Q}_{t_n, x}^{\nu^n}} \left[\beta_T^{t_n} \big(g(S_T^{t_n, x_n}) - \int_{t_n}^{T} (\beta_T^{t_n})^{-1} \beta_s^{t_n} \delta_K(\nu_s^n) ds \big) \right] + \frac{1}{n} .$$

---

[3]This is the case as soon as K is bounded or g is Lipschitz, see Exercise 5.2.

By sub-additivity and 1-homogeneity of δ_K, we observe that

$$\delta_K\Big(\int_{t_n}^T (\beta_T^{t_n})^{-1}\beta_s^{t_n}v_s^n ds\Big) \leq \int_{t_n}^T (\beta_T^{t_n})^{-1}\beta_s^{t_n}\delta_K(v_s^n)ds,$$

and therefore

$$g(S_T^{t_n,x_n}) - \int_{t_n}^T (\beta_T^{t_n})^{-1}\beta_s^{t_n}\delta_K(v_s^n)ds \leq g(S_T^{t_n,x_n}) - \delta_K\Big(\int_{t_n}^T (\beta_T^{t_n})^{-1}\beta_s^{t_n}v_s^n ds\Big)$$

$$\leq F_K[g]\Big(S_T^{t_n,x_n} - \int_{t_n}^T (\beta_T^{t_n})^{-1}\beta_s^{t_n}v_s^n ds\Big).$$

Combining the previous inequalities, we obtain

$$p_K(t_n,x_n) \leq \mathbb{E}^{\mathbb{Q}_{t_n,x_n}^{v^n}}\Big[\beta_T^{t_n}F_K[g]\Big(S_T^{t_n,x_n} - \int_{t_n}^T (\beta_T^{t_n})^{-1}\beta_s^{t_n}v_s^n ds\Big)\Big] + \frac{1}{n}$$

and thus

$$\limsup_{t'\uparrow T, x'\to x} p_K(t',x')$$

$$\leq \limsup_{n\to\infty} \mathbb{E}^{\mathbb{Q}_{t_n,x_n}^{v^n}}\Big[\beta_T^{t_n}F_K[g]\Big(S_T^{t_n,x_n} - \int_{t_n}^T (\beta_T^{t_n})^{-1}\beta_s^{t_n}v_s^n ds\Big)\Big]. \qquad (5.9)$$

Since $F_K[g]$ is Lipschitz, we have

$$\mathbb{E}^{\mathbb{Q}_{t_n,x_n}^{v^n}}\Big[\beta_T^{t_n}\Big(F_K[g]\Big(S_T^{t_n,x_n} - \int_{t_n}^T (\beta_T^{t_n})^{-1}\beta_s^{t_n}v_s^n ds\Big)\Big)\Big]$$

$$\leq$$

$$\beta_T^{t_n}F_K[g](x_n) + C\mathbb{E}^{\mathbb{Q}_{t_n,x_n}^{v^n}}\Big[|\beta_T^{t_n}S_T^{t_n,x_n} - \int_{t_n}^T \beta_s^{t_n}v_s^n ds - \beta_T^{t_n}x_n|\Big],$$

for $C > 0$ that does not depend on n. Since σ is bounded, we compute, using the dynamics of $S^{t_n,x}$ under $\mathbb{Q}_{t_n,x}^{v^n}$ in (5.3), that

$$\mathbb{E}^{\mathbb{Q}_{t_n,x_n}^{v^n}}\Big[|\beta_T^{t_n}S_T^{t_n,x_n} - \int_{t_n}^T \beta_s^{t_n}v_s^n ds - \beta_T^{t_n}x_n|\Big]$$

$$\leq$$

$$C\Big(\sqrt{T-t_n} + |T-t_n| + (1-\beta_T^{t_n})x_n\Big),$$

after possibly increasing the value of C. We thus have

$$\mathbb{E}^{\mathbb{Q}^{y^n}_{t_n,x_n}}\left[\beta_T^{t_n}\left(F_K[g](S_T^{t_n,x_n} - \int_{t_n}^{T}(\beta_T^{t_n})^{-1}\beta_s^{t_n}v_s^n ds))\right)\right]$$

$$\leq$$

$$\beta_T^{t_n}F_K[g](x_n) + C\left(\sqrt{T - t_n} + |T - t_n| + (1 - \beta_T^{t_n})x_n\right).$$

Since $F_K[g]$ is upper-semicontinuous, the proof is concluded by inserting the previous inequality in (5.9) and observing that $\beta_T^{t_n} \to 1$.   □

Remark 5.8 The result of Proposition 5.7 is still true if we assume that $F_K[g]$ is upper semi-continuous only. One needs to use a mollification argument.

5.1.3   Equivalence Property: Hedging a Modified Payoff Without Constraint

We study here constraints of the type

$$K = \prod_{i=1}^{d} I_i \tag{5.10}$$

with I_i an interval of $\mathbb{R}$. This type of constraints has a clear financial interpretation.

For instance, $I_i = [0, \infty)$ means that short selling asset i is not allowed, $I_i = [-m_i, M_i]$ for some $m_i, M_i \geq 0$ means that the trader can only take limited positions in the asset.

Let us consider the precise example of a *digital* option with payoff $\mathbf{1}_{[B,+\infty)}$ on a risky asset. We have seen in Chap. 4 how to compute the price and the hedge of such an option. We also remarked in Sect. 4.3.2, that the discontinuity of the payoff function at B leads to a very large delta near the maturity, if the underlying has a value close to B. In practice, the trader has to comply with a portfolio constraint for this asset, given by $K = [-m, M]$ for $m, M > 0$. If the trader applies the classical delta hedging approach, it might be that the delta constraint will not be satisfied...

The theoretical solution to this problem is given in the previous section. In particular, Theorem 5.3 allows to characterise the price of such option under constraints. The market practice consists simply in replicating an option with a modified payoff which is a well chosen *call spread*, corresponding to the *face-lift* of the digital option payoff. The price p_{CS} of this strategy is solution of the PDE

$$r\varphi - \mathcal{L}_S\varphi = 0 \text{ on } [0, T) \times \mathbb{R}^d , \tag{5.11}$$

with terminal condition

$$\lim_{t' \to T, x' \to x} p_{CS}(t', x') = F_K[\mathbf{1}_{[B,+\infty)}](x), \quad x \in \mathbb{R}^d.$$

The question is then: is this market practice correct? Namely, will the portfolio constraint be satisfied for the solution and is this an optimal solution? In other words, does it hold that $p_{CS} = p_K$?

In the following paragraph, we will give a positive answer for the one-dimensional case. The multidimensional case is more complicated. Nevertheless, we are going to give a sufficient condition for this property to be satisfied.

5.1.3.1 The Case of One Risky Asset

Proposition 5.9 *Assume that σ is a C_b^1 function that does not depend on time. When $d = 1$, super-replicating a payoff g with delta constraints is equivalent to hedging the payoff $F_K[g]$. The super-replication price with constraints is thus solution to*

$$r\varphi - \mathcal{L}_S \varphi = 0 \quad on \ [0, T) \times \mathbb{R}, \tag{5.12}$$

with terminal condition

$$\lim_{t' \to T, x' \to x} p_K(t', x') = F_K[g](x), \quad x \in \mathbb{R}^d.$$

Proof We assume that $F_K[g]$ is a C_b^1 function . The general case is obtained by using an approximation argument, see [17]. We recall, see equation (4.17), that the hedging strategy of $F_K[g]$ is given by

$$\Delta_s^{F_K[g]} = e^{-\int_s^T r_s ds} \mathbb{E}^{\mathbb{Q}_{t,x}} \left[\partial F_K[g](S_T^{t,x}) \frac{\partial_x S_T^{t,x}}{\partial_x S_s^{t,x}} \ \Big| \ \mathcal{F}_s \right], \ t \le s \le T,$$

with $\partial_x S^{t,x}$ defined as the tangent process of $S^{t,x}$. It satisfies

$$\partial_x S_s^{t,x} = 1 + \int_t^s r_u \partial_x S_u^{t,x} du + \int_t^s \partial \sigma(S_u^{t,x}) \partial_x S_u^{t,x} dW_u^{\mathbb{Q}_{t,x}}, \ t \le s \le T,$$

see Sect. 4.2.3. Since σ is a C_b^1 function, we know that $(\beta_s^t \partial_x S_s^{t,x})_{t \le s \le T}$ is a positive martingale whose expectation is equal to 1. We thus introduce $\mathbb{P}_{t,x}^\nabla$ a new equivalent probability given by $\frac{d\mathbb{P}_{t,x}^{\nabla}}{d\mathbb{Q}_{t,x}} = \beta_s^t \partial_x S_T^{t,x}$. We obtain then

$$\Delta_s^{F_K[g]} = \mathbb{E}^{\mathbb{P}_{t,x}^{\nabla}} \left[\partial F_K[g](S_T^{t,x}) \mid \mathcal{F}_s \right], \qquad t \le s \le T.$$

Since $\partial F_K[g]$ takes its values in a convex set K, we have that $\Delta^{F_K[g]}$ takes also its values in K. We deduce that $\mathbb{E}^{\mathbb{Q}_{t,x}}[\beta_T^t F_K[g](S_T^{t,x})] \ge p_K(t,x)$. Besides, from Theorem 5.3, we know that p_K is a super-solution of equation (5.12), for which a comparison theorem holds true. We thus have

$$\mathbb{E}^{\mathbb{Q}_{t,x}} \left[\beta_T^t F_K[g](S_T^{t,x}) \right] \le p_K(t,x),$$

see the Feynman-Kac formula in Theorem 4.2, which concludes the proof. □

5.1.3.2 The Multidimensional Case

Proposition 5.10 *Assume that σ is C_b^1 and that $\partial_{x_j} \sigma^i(x) = 0$ for $1 \le j \ne i \le d$, i.e. the volatility of asset i only depends on its own value. Assume moreover that K is as in (5.10) and that g is bounded. Then, super-replicating the payoff g with delta constraints is equivalent to replicating the payoff $F_K[g]$.*

Proof We do the proof in the case where $F_K[g]$ is a C_b^1 function. The general case is obtained by an approximation result, see [17]. We use the same technique as in the one dimensional case, working with each component separately. Indeed, the tangent process $\partial_x S^{t,x}$ of $S^{t,x}$ satisfies

$$\partial_x (S_s^{t,x})^i = 1 + \int_t^s r_u \partial_x (S_u^{t,x})^i du + \int_t^s \nabla \sigma^{i\cdot}((S_u^{t,x})^i) \partial_x (S_u^{t,x})^i dW_u^{\mathbb{Q}_{t,x}}.$$

Hence,

$$(\Delta_s^{F_K[g]})^i = \mathbb{E}^{\mathbb{P}_{t,x}^{\nabla^i}} \left[\partial_i F_K[g](S_T^{t,x}) \mid \mathcal{F}_s \right], \qquad t \le s \le T,$$

with $\frac{d\mathbb{P}_{t,x}^{\nabla^i}}{d\mathbb{Q}_{t,x}} := \beta_T^t \partial_x (S_s^{t,x})^i$ and where $\partial_i F_K[g]$ is the derivative with respect to the component i. Since $\partial_i F_K[g]$ is valued in I_i a convex set, we have that $(\Delta_s^{F_K[g]})^i$ takes its values in I_i. This implies that $\mathbb{E}^{\mathbb{Q}_{t,x}} \left[\beta_t^T F_K[g](S_T^{t,x}) \right] \ge p_K(s,x)$. The proof is concluded then as in the one dimensional case. □

Remark 5.11 A necessary and sufficient condition allowing to replace the initial hedging problem by the hedging problem without any constraint of the face-lifted payoff is given in [17]. On the applications side, these results are quite interesting for an easy numerical resolution as soon as the condition are satisfied. Moreover, this validates the market practice in the one dimensional case.

5.2 Application to Incomplete Markets

In this section, we use the previous study of the super-replication problem in a complete market with constraints to characterise the super-replication price in an incomplete market. We study here incompleteness due to a larger number of risk factors than hedging assets. This situation can be modelled by a multidimensional process S satisfying equation (5.1), i.e.

$$S_s^{t,x} = x + \int_t^s b(u, S_u^{t,x})du + \int_t^s \sigma(u, S_u^{t,x})dW_u , \ s \in [t, T].$$

The first n components of S represent then the hedging assets values, i.e. assets that are tradable, and the last $d - n$ components represent the values of non-tradable risk factors. We assume that σ is invertible. If this is not the case it is always possible to add other risk factors so that this is true. It then appears clearly that the super-replication problem without constraints in incomplete market with n hedging assets is the same as the problem of super-replication with constraints. The fictitious market to consider here is constituted of the assets and risk factors whose dynamics are given by S, and the set of constraints is $K = \mathbb{R}^n \times \{0\}^{d-n}$.

When σ is bounded, we can characterise the super-replication price p_K of a contingent claim with payoff $g(S^1, \ldots, S^n)$, by using Theorem 5.3. Indeed, p_K is solution to

$$\min\{r\varphi - \mathcal{L}_S\varphi \, , \, \mathcal{C}_K(\partial\varphi)\} = 0 \ \text{on} \ [0, T) \times \mathbb{R}^d , \tag{5.13}$$

with terminal condition

$$\lim_{t'\uparrow T, x'\to x} p_K(t', x') = h(x), \ x \in \mathbb{R}^d ,$$

with $h(x_1, \ldots, x_d) = g(x_1, \ldots, x_n)$. Observe that $F_K[h] = h$ here.

When σ is not bounded, Theorem 5.3 gives a partial characterisation of the terminal condition. We complete this result in Sect. 5.2.1 below in the case $d = 2$, $n = 1$. In Sect. 5.2.2, we study the particular case of a stochastic volatility model, with σ not bounded for $d = 2, n = 1$.

5.2.1   Non-hedgeable Volatility: The Black-Scholes-Barenblatt Equation

We consider a market with one risky asset S^1, one risk factor S^2, and a constant interest rate r. The dynamics of S are given by

$$S_t^1 = S_0^1 + \int_0^t r_s S_s^1 ds + \int_0^t S_s^1 \sigma(S_s^2) dW_s^1 ,$$

$$S_t^2 = S_0^2 + \gamma_1 W_t^1 + \gamma_2 W_t^2 .$$

We assume here that the function σ satisfies $\bar{\sigma} \geq \sigma(\cdot) \geq \underline{\sigma}$, with $\bar{\sigma}, \underline{\sigma}$ two strictly positive constants, and is a continuous function.

The volatility of S^1 depends on the risk factor S^2: this model is a stochastic volatility model, see Chap. 8. The risk factor S^2 is not tradable, the initial market is thus incomplete.

In this market, we are interested in the characterisation of the super-replication price of a contingent claim with payoff $g(S^1)$. To do that, we interpret (S^1, S^2) as a fictitious market and impose the constraint of zero investment in the second asset, i.e. $K = \mathbb{R} \times \{0\}$.

The main result is that the super-hedging price solves the Black-Scholes-Barenblatt equation, (5.14) below.

Proposition 5.12 *Assume that p_K is continuous on $[0, T) \times (0, \infty) \times \mathbb{R}$. Assume moreover that $\inf_{x \in \mathbb{R}} \sigma(x) = \underline{\sigma}$ and $\sup_{x \in \mathbb{R}} \sigma(x) = \bar{\sigma} < \infty$. Then the super-replication price p_K does not depend on s^2 and is a viscosity solution on $[0, T) \times (0, \infty)$ of*

$$r\varphi - \partial_t \varphi - rs^1 \partial_{s^1} \varphi - \frac{1}{2}(s^1)^2 \left(\underline{\sigma}^2 [\partial^2_{s^1 s^1} \varphi]^- + \bar{\sigma}^2 [\partial^2_{s^1 s^1} \varphi]^+ \right) = 0 , \qquad (5.14)$$

with terminal condition $\lim_{t' \uparrow T, s^{1'} \to s^1} p_K(t', s^{1'}) = g(s^1)$, *for all* $s^1 \in (0, \infty)$.

Proof

1. From Theorem 5.3 and by using the constraints' shape, we know that p_K is a viscosity solution to

$$\min \{ r\varphi - \mathcal{L}_S \varphi , \ -\partial_{s^2} \varphi, \partial_{s^2} \varphi \} (t, x) = 0 , \qquad (5.15)$$

with

$$\mathcal{L}_S = \partial_t + rs^1 \partial_{s^1} + \frac{1}{2} [(s^1 \sigma(s^2))^2 \partial^2_{s^1 s^1} + \gamma^2 \partial^2_{s^2 s^2} + 2s^1 \sigma(s^2) \gamma_1 \partial^2_{s^1 s^2}]$$

and $\gamma^2 := \gamma_1^2 + \gamma_2^2$.

In particular, the function p_K is a viscosity super-solution of $-\partial_{s^2}\varphi(t, s^1, s^2) = 0$ and also of $\partial_{s^2}\varphi(t, s^1, s^2) \doteq 0$. As in the regular case, see Exercise 5.5, this implies that p_K does not depend on s^2 and justifies the notation $p_K(t, s^1) = p_K(t, s^1, s^2)$.

Let $(t, s^1) \in [0, T) \times (0, \infty)$ and φ be a smooth test function for p_K such that $p_K - \varphi$ reaches a minimum at (t, s^1), with $(p_K - \varphi)(t, s^1) = 0$. Using the previous result, we obtain that, for all $s^2 \in \mathbb{R}$,

$$r\varphi(t, s^1) - \partial_t \varphi(t, s^1) - rs^1 \partial_{s^1} \varphi(t, s^1) - \frac{1}{2}(s^1 \sigma(s^2))^2 \partial^2_{s^1 s^1} \varphi(t, s^1) \geq 0.$$

Taking the infimum on s^2, we obtain that p_K is a super-solution of (5.14).

2. We now prove the sub-solution property. Let φ be a smooth bounded function and (t, s^1)be a point at which $p_K - \varphi$ reaches a strict maximum (equal to 0). We work toward a contradiction. If the sub-solution property to (5.15) is not satisfied at (t, s^1) then

$$(r\varphi - \mathcal{L}_S\varphi) \geq 0 \text{ on } B \times \mathbb{R}, \tag{5.16}$$

where $B \subset [0, T) \times (0, \infty)$ is an open neighbourhood of (t, s^1). We consider the process $S = (S^1, S^2)$ starting from $s = (s^1, s^2)$ at t. Let θ be the first exit time of $(u, S_u^1)_{u \geq t}$ from B. Let us observe that, since the maximum of $p_K - \varphi$ is strict at (t, s^1), we have $\max_{\partial B}(p_K - \varphi) =: -\zeta < 0$. Moreover, we have $\hat{K} = \{0\} \times \mathbb{R}$ so that the measure change associated to the density $\mathcal{E}^{t,s,\nu}$, $\nu \in \hat{\mathcal{K}}_b$, does not affect the *drift* of S^1. Applying Itô's Lemma and using (5.16), we obtain, for all $\nu \in \hat{\mathcal{K}}_b$,

$$p_K(t, s_1) = \varphi(t, s^1) \geq \mathbb{E}\left[\mathcal{E}^{t,s,\nu}_\theta \beta^t_\theta \varphi(\theta, S^1_\theta)\right]$$
$$\geq \mathbb{E}\left[\mathcal{E}^{t,s,\nu}_\theta \beta^t_\theta p_K(\theta, S^1_\theta)\right] + \zeta e^{-rT}.$$

Since $\delta_K(z) = 0$ if $z \in \hat{K}$, and $\zeta > 0$, this contradicts the dynamic programming principle of Theorem 5.1.

3. To conclude the proof, we have to specify the terminal condition. We know that $\liminf_{t' \uparrow T, s^{1'} \to s^1} p_K(t', s^{1'}) \geq g(s^1)$, see Proposition 5.6. Here, since σ is bounded and the constraint is only on S^2, it is possible to show, using the proof of Proposition 5.7, that $\limsup_{t' \uparrow T, s^{1'} \to s^1} p_K(t', s^{1'}) \leq g(s^1)$ even though $(s^1, s^2) \mapsto s^1 \sigma(s^2)$ is not bounded.

$\square$

The Proposition 5.12 leads to a so-called "robustness" property of the Black-Scholes formula. For a European contingent claim with payoff g, we denote p^g the price given by the Black-Scholes-Barrenblatt equation and $p^g_{BS}(\sigma)$ the price given by the Black-Scholes formula for a (constant) volatility equal to σ, see Exercise 4.1.

Then, an easy application of the comparison theorem shows that

$$\sup_{\underline{\sigma} \leq \sigma \leq \bar{\sigma}} p^g_{BS}(\sigma) \leq p^g ,\qquad(5.17)$$

and that

$$p^g_{BS}(\bar{\sigma}) = p^g \ \text{ if } g \text{ is convex } , \ \ p^g_{BS}(\underline{\sigma}) = p^g \ \text{ if } g \text{ is concave.}$$

Hence, for convex (resp. concave) payoffs, hedging in a stochastic volatility environment reduces to hedging in the Black-Scholes model with the highest (resp. lowest) expected level of volatility, unless the volatility factor can be hedged by another mean (like by trading other liquid options, see Chap. 8).

Exercise 5.3 discusses some properties of the solution to the Black-Scholes-Barenblatt equation. We will also comment again on this point in Sect. 7.3.1.

5.2.2 Non-hedgeable Volatility: The Unbounded Case, Buy-and-Hold Strategy

We go back to the study of the previous model but we now assume that the function σ is not bounded. In this setting, we have

Proposition 5.13 *Assume that* $\inf_{x \in \mathbb{R}} \sigma(x) = 0$, $\sup_{x \in \mathbb{R}} \sigma(x) = +\infty$, g *is bounded. Then the super-replication price of* g *is* $g^c(S^1_0)$, *where* g^c *is the concave envelop of* g.

Proof

1. Applying the same arguments as in the proof of Proposition 5.12, we obtain that p_K is super-solution to

$$-\partial_t \varphi - \frac{1}{2}(s^1 \sigma(s^2))^2 \partial^2_{s^1 s^1} \varphi = 0,$$

for all $s_2 \in \mathbb{R}$. Since $\inf_{x \in \mathbb{R}} \sigma(x) = 0$, $\sup_{x \in \mathbb{R}} \sigma(x) = +\infty$, we deduce that p_K is necessarily a viscosity super-solution of

$$-\partial_t \varphi = 0, \ \text{and} \ -\partial^2_{s^1 s^1} \varphi = 0.$$

As in the regular case, this implies that p_K is decreasing in time and concave in space, thus $p_K(t, s^1) \geq p_K(T, s^1) \geq g^c(s^1)$.

2. Moreover, the *buy-and-hold* strategy with initial value $g^c(S^1_0)$ and constant delta $\partial^- g^c(S^1_0)$, where $\partial^- g^c$ is the left-derivative of the concave function g^c, allows to super-hedge the payoff g. This is obtained directly from the inequality $g(S^1_T) \leq g^c(S^1_T) \leq g^c(S^1_0) + \partial^- g^c(S^1_0)(S^1_T - S^1_0)$ due to the concavity of g^c.

□

To conclude, let us observe that, although the terminal condition is not completely specified, we can characterise the super-replication price of the contingent claim and this by using essentially the super-solution property. This model shows also that when σ is not bounded, we typically have $p_K(T-, .) \neq F_K[g]$.

To conclude, let us consider the case of a call with strike k, i.e. $g(s^1) = [s^1 - k]^+$. Then, $g^c(s^1) = s^1$. The super-replication price is equal to the price of the underlying asset, this is quite an onerous strategy... In incomplete markets, it appears that the super-replication price is very often too expensive to be used in practice.

5.3 Problems

5.1 (Replication with portfolio constraints) In dimension 1, we consider a local volatility model, with $(S_t)_{0 \leq t \leq T}$, the risky asset price, satisfying

$$S_t = S_0 + \int_0^t rS_s ds + \int_0^t \sigma(S_s)dW_s , \quad 0 \leq t \leq T , \qquad (5.18)$$

where W is a Brownian motion under the risk neutral probability . $T > 0$ is a finite time horizon and $r \geq 0$ is the risk-free interest rate.

We assume that S takes values in $(0, \infty)$.

1. Give a sufficient condition on σ implying that there exists a unique solution to the SDE (5.18).
2. Let g be a C_b^∞ function. We denote $(t, x) \mapsto p(t, x)$ the pricing function associated to the European option with payoff g and maturity T, for all initial date $t \in [0, T]$ and initial value of the underlying x.

   a. Recall the risk neutral pricing formula for p.
   b. Give the PDE satisfied by p.

3. We assume that p is C^∞, σ is C_b^2 and $\partial \sigma \sigma$ is Lipschitz.

   a. Give the PDE satisfied by the Delta of the option.
   b. Deduce an expression for the Delta using an expectation.
   c. A trader has to hedge the option with payoff g satisfying investment limits in the risky asset: her strategy must be valued in $[-m, M]$, with $m, M > 0$.
   Give a condition on g guaranteeing that she can use the usual valuation formula and hedging strategy.

5.2 (Some properties of the *Face-lift* transform) In the sequel, K is a closed convex set and $0 \in K$. Recall the notations of Sect. 5.1.2.

1. Show that

   a. If $h(x) \geq g(x)$ for all x, then $F_K[h](x) \geq F_K[g](x)$, for $x \in \mathbb{R}^d$.
   b. If h is a constant function then $F_K[h] = h$.
   c. $F_K[h] \geq h$ and $F_K[h] = F_K[F_K[h]]$.
   d. $F_K[h \vee g] = F_K[h] \vee F_K[g]$.
   e. If h is a Lipschitz function, then $F_K[h]$ is Lipschitz.
   f. If K is bounded, then $F_K[h]$ is Lipschitz.

2. Let h be a lower semi-continuous function bounded from below by $-m_h$, for some $m_h \geq 0$ given.

   Show that there exists a sequence $(h_n)_{n \geq 1}$ of bounded Lipschitz functions, uniformly bounded from below by $-m_h$, converging (simply) to h, and such that $F_K[h_n] \uparrow F_K[h]$.

3. Let h be a Lipschitz function from $\mathbb{R}^d$ to $\mathbb{R}$.

   a. Show that $F_K[h]$ is a viscosity super-solution of

   $$\min\{\varphi - h \, , \, \mathcal{C}_K(\partial\varphi)\} = 0. \tag{5.19}$$

   b. Let v be a C^1 function, super-solution on $\mathbb{R}^d$ of (5.19). Show that

   $$v \geq F_K[v] \geq F_K[h] \, .$$

   c. Deduce that if h is C^1 and $\partial h \in K$, then $F_K[h] = h$.

5.3 (Black-Scholes-Barenblatt PDE) We work in the framework of Sect. 5.2.1.

1. Prove the inequality (5.17).
2. Show that the pricing function $p_{BS}^g(\bar{\sigma})$ is convex if g is convex.
3. Deduce that $p^g = p_{BS}^g(\bar{\sigma})$ in this case.
4. Characterise p^g if g is concave.

5.4 (An upper bound for the hedging price with portfolio constraints) We work in dimension 1. We consider a financial market with a risk-free interest rate set to zero ($r = 0$), and where the price of the risky assets is given by

$$S_s^{t,x} = x + \int_t^s S_u^{t,x} \sigma(S_u^{t,x}) dW_u, \ t \leq s \leq T,$$

when the stock price is x at t, and t is the initial date. The function $\bar{\sigma} : x \mapsto x\sigma(x)$ is bounded, twice continuously differentiable with bounded derivatives. We consider a European option with payoff $g(S_T^{0,x_0})$ at T, where $x_0 > 0$ and g is a bounded function.

1. We assume that the function $(t, x) \in [0, T] \times (0, \infty) \mapsto p(t, x) = \mathbb{E}\left[g(S_T^{t,x})\right]$ is $C_b^{1,2}([0, T) \times (0, \infty)) \cap C^{0,0}([0, T] \times (0, \infty))$.

   a. What is the PDE satisfied by p?

b. Give the hedging strategy of the option in terms of p and its derivative with respect to the space variable x.

2. Let $\pi_N := \{t_i^N := iT/N,\ i = 0,\dots,N\},\ N \in \mathbb{N} \setminus \{0\}$ and $S^{0,x,N}$ be defined by

$$S^{0,x,N}_{t_{i+1}^N} := S^{0,x,N}_{t_i^N} + \bar{\sigma}(S^{0,x,N}_{t_i^N})(W_{t_{i+1}^N} - W_{t_i^N})\,,\ i = 0,\dots,N-1,$$

with $S_0^{0,x,N} = x$.

a. Give the induction formula verified by $Z^{0,x,N} := (\partial S^{0,x,N}_{t_i^N}/\partial x)_{i=0,\dots,N}$.

b. From now on, we assume that the process $Y^{t,x}$ solution to

$$Y^{t,x}_s = 1 + \int_t^s \partial\bar{\sigma}(S_u^{t,x})Y_u^{t,x}dW_u,\ t \le s \le T,$$

satisfies $Y_T^{t,x} = \partial S_T^{t,x}/\partial x$ $\mathbb{P}$-a.s., for all $(t,x) \in [0,T] \times (0,\infty)$.
What can we say about the couples $(S^{0,x_0,N}_{t_i^N}, Z^{0,x_0,N}_{t_i^N})_{i=0,\dots,N}$ and (S^{0,x_0}, Y^{0,x_0})?

3. We now restrict the admissible portfolio strategies to the set of strategies such that the number of risky assets held in portfolio takes its values in $K := [m,M] \ni 0$ where $m < M$. We denote $\mathcal{A}_K$ the set of predictable processes satisfying this constraint and we introduce $p_K(0,x_0)$ given by the infimum of the set Γ defined by

$$\left\{ v \mid \exists\ \phi \in \mathcal{A}_K \text{ s.t. } V_T^{v,\phi} := v + \int_0^T \phi_s dS_s^{0,x_0} \ge g(S_T^{0,x_0})\ \mathbb{P}\text{--a.s.}\right\}.$$

a. Let H be the set of C_b^1 functions above g on $(0,\infty)$. Show that, for all $h \in H$ and $(t,x) \in [0,T] \times (0,\infty)$,

$$I[h](t,x) := \frac{\partial}{\partial x}\mathbb{E}\left[h(S_T^{t,x})\right]$$

is well defined and satisfies

$$I[h](t,x) = \mathbb{E}\left[\partial h(S_T^{t,x})Y_T^{t,x}\right].$$

b. Show that, for all $(t,x) \in [0,T] \times (0,\infty)$, there exists a measure $\mathbb{Q}^{t,x} \sim \mathbb{P}$ such that $d\mathbb{Q}^{t,x}/d\mathbb{P} = Y_T^{t,x}$.

c. We denote now H_K the set of functions $h \in H$ such that $\partial h \in K$. Deduce from the two previous questions that $I[h](t,x) \in K$ for all $h \in H_K$ and $(t,x) \in [0,T] \times (0,\infty)$.

d. We suppose for this question that $(t,x) \in [0,T] \times (0,\infty) \mapsto p_h(t,x) = \mathbb{E}\left[h(S_T^{t,x})\right]$ is $C_b^{1,2}([0,T) \times (0,\infty)) \cap C^{0,0}([0,T] \times (0,\infty))$ for all $h \in H_K$. Show that, for all $h \in H_K$, we can find an hedging strategy of $g(S_T^{0,x_0})$ in $\mathcal{A}_K$

(give it explicitely) starting from $p_h(0, x_0)$ at 0, and deduce that

$$p_K(0, x_0) \leq \inf_{h \in H_K} p_h(0, x_0). \tag{5.20}$$

e. Explain why the previous inequality should be an equality.

4. We consider here the particular case of $g : x \in (0, \infty) \mapsto \mathbf{1}_{x \geq A}$ and $K = [0, 1]$, where $A > 0$. Find a Lipschitz function $\hat{h}$ such that there exists $(h_n)_n \subset H_K$ satisfying $h_n \to \hat{h}$ (simply) and

$$\inf_{h \in H_K} p_h(0, x_0) = p_{\hat{h}}(0, x_0) = \lim_{n \to \infty} p_{h_n}(0, x_0).$$

5. What can you say if (5.20) is an equality and if $p_{\hat{h}}$ is $C_b^{1,2}([0, T) \times (0, \infty)) \cap C^{0,0}([0, T] \times (0, \infty))$?

5.5 (Characterisation of non-decreasing functions in the viscosity sense) Let I be an open interval of $\mathbb{R}$ and h a lower semi-continuous and bounded function from I to $\mathbb{R}$. We assume that h is a viscosity super-solution of

$$\partial \varphi \geq 0 \quad \text{on } I. \tag{5.21}$$

The goal of this exercise is to show that h is a non-decreasing function.

1. For all $\epsilon > 0$, we define $h^\epsilon(x) = h(x) + \epsilon x$. Let $x_0 \in I$ and φ be a smooth test function for h^ϵ at x_0, i.e.

$$(h^\epsilon - \varphi)(x_0) = \min_{x \in I}(h^\epsilon - \varphi)(x) . \tag{5.22}$$

Show that the function φ is increasing on an interval (x_1, x_2) containing x_0.
2. We admit that if w_1 (resp. w_2) is a bounded viscosity super-solution (resp. sub-solution) of (5.21) on (x_1, x_2) and if $w_1 \geq w_2$ on $\{x_1, x_2\}$, then $w_1 \geq w_2$ on $[x_1, x_2]$. Show that the assumption $h^\epsilon(x_1) \geq h^\epsilon(x_2)$ contradicts (5.22).
3. Deduce that h is non-decreasing at x_0.[4]

Corrections

5.1

1. σ Lipschitz continuous.
2. $p = \mathbb{E}\left[e^{-rT} g(S_T)\right].$

---

[4]One can show that the set of points at which h admits a test function is indeed dense. See e.g. [37]. This implies that h is indeed non-decreasing on I.

3. $\partial_t p + rx\partial_x p + \frac{1}{2}\sigma^2(x)\partial^2_{xx}p = rp$.

4. a. Let $\Delta(t,x) = \partial_x p(t,x)$. Differentiating the previous PDE in x, we obtain

$$\partial_t \Delta + (rx + \partial\sigma\sigma(x))\partial_x\Delta + \frac{1}{2}\sigma^2(x)\partial^2_{xx}\Delta = 0 \text{ and } \Delta(T,x) = \partial g(x).$$

(5.23)

b. Set $X^{t,x}_s = x + \int_t^s(rX^{t,x}_u + \partial\sigma\sigma(X^{t,x}_u))du + \int_t^s \sigma(X^{t,x}_u)dW_u$. By the Feynman-Kac formula: $\Delta(t,x) = \mathbb{E}[\partial g(X_T)]$.

c. If $-m \le \partial g \le M$, then $-m \le \Delta(t,x) \le M$ from the previous representation and the trader can then use the usual pricing equation.

5.2

1. a. Follows directly from the definition.

b. $\delta_{\tilde{K}} \ge 0$ $(0 \in K)$, take $y = 0$ in the Definition of $F_K[h]$.

c. For the first point: take $y = 0$ in the Definition of $F_K[h]$. For the second point:

$$F_K[F_K[h]](x) = \sup_{y_2 \in \mathbb{R}^d} F_K[h](x + y_2) - \delta_K(y_2)$$

$$= \sup_{y_2 \in \mathbb{R}^d} \sup_{y_1 \in \mathbb{R}^d} h(x + y_2 + y_1) - \delta_K(y_1) - \delta_K(y_2)$$

$$= \sup_{y_1, y_2 \in \mathbb{R}^d} h(x + y_2 + y_1) - \delta_K(y_1) - \delta_K(y_2)$$

$$\le \sup_{y_1, y_2 \in \mathbb{R}^d} h(x + y_2 + y_1) - \delta_K(y_1 + y_2)$$

$$\le F_K[h](x) \le F_K[F_K[h]](x)$$

d. From 1. $F_K[h \vee g] \ge F_K[h] \vee F_K[g]$. For any $y \in \tilde{K}$,

$$(h\vee g)(x+y)-\delta_K(y) = [h(x+y)-\delta_K(y)]\vee[g(x+y)-\delta_K(y)] \le F_K[h]\vee F_K[g](x)$$

and the result follows by taking the supremum over y.

e. For any $x, x' \in \mathbb{R}^d$ and $y \in \tilde{K}$, we compute

$$h(x + y) - \delta_K(y) \le h(x' + y) - \delta_K(y) + C|x - x'| \le F_K[h](x') + C|x - x'|.$$

Taking the supremum over y leads to $F_K[h](x) \le F_K[h](x') + C|x - x'|$. Observing that x and x' play a symmetric role, we get the result.

f. If K is bounded, δ_K is Lipschitz and $\tilde{K} = \mathbb{R}^d$. In particular,

$$F_K[h](x) = \sup_{z \in \mathbb{R}^d} h(z) - \delta_K(z - x),$$

and the proof follows from the same arguments as the one used in the previous question, by using the Lipschitz continuity of δ_K.

2. We define a sequence of functions $(g_n)_n$ by

$$g_n(x) = \inf_{y \in \mathbb{R}^d} \{h(y) + n|x - y|\} , \quad x \in \mathbb{R}^d ,$$

for $n \geq 1$. It is clear that the sequence $(g_n)_n$ is non-decreasing, that $-m_h \leq g_n \leq h$, and that g_n is n-Lipschitz continuous for all $n \geq 1$.

We now prove that g_n converges simply to h. Fix some $x \in \mathbb{R}^d$. Since h is l.s.c and bounded from below there exists a sequence $(x_n)_n$ in $\mathbb{R}^d$ such that

$$g_n(x) = h(x_n) + n|x - x_n| , \tag{5.24}$$

for all $n \geq 1$. We then have, since h is bounded from below by $-m_h$,

$$n|x - x_n| \leq h(x) - h(x_n) \leq h(x) + m_h ,$$

for all $n \geq 1$. Therefore we get $\lim_{n \to \infty} x_n = x$.

From this last convergence and (5.24) and since h is l.s.c. we get

$$\lim_{n \to \infty} g_n(x) \geq \liminf_{n \to \infty} h(x_n) \geq h(x) .$$

Thus, $g_n(x) \uparrow h(x)$ as $n \uparrow \infty$ for all $x \in \mathbb{R}^d$.

Define now the sequence of functions $(h_n)_n$ by

$$h_n(x) := g_n(x) \wedge n , \quad x \in \mathbb{R}^d ,$$

for all $n \geq 1$. Since g_n is Lipschitz continuous and bounded from below, we get that h_n is bounded and Lipschitz continuous for all $n \geq 1$. Moreover, since $(g_n)_n$ is non-decreasing and converges simply to h, we also get that $(h_n)_n$ is non-decreasing and converges point wisely to h.

It remains to prove the convergence of $F_K[h_n]$ to $F_K[h]$. We simply observe that for $x \in \mathbb{R}^d$ we have

$$\begin{aligned}
F_K[h](x) &= \sup_{u \in \tilde{K}} h(x + u) - \delta_K(u) \\
&= \sup_{u \in \tilde{K}} \sup_n h_n(x + u) - \delta_K(u) \\
&= \sup_n \sup_{u \in \tilde{K}} h_n(x + u) - \delta_K(u) \\
&= \lim_{n \to \infty} \uparrow F_K[h_n](x) .
\end{aligned}$$

3. a. Let $\bar{x} \in \mathbb{R}^d$ and ϕ be a smooth test function such that

$$0 = F_K[h](\bar{x}) - \phi(\bar{x}) = \min_{x \in \mathbb{R}^d}(F_K[h] - \phi)(x) \text{ (strict)}.$$

Then,

$$F_K[h](\bar{x}) \geq F_K[h](\bar{x} + y) - \delta_K(y), \ \forall y \in \mathbb{R}^d.$$

Using the test function ϕ,

$$\phi(\bar{x}) \geq \phi(\bar{x} + y) - \delta_K(y), \ \forall y \in \mathbb{R}^d.$$

In particular for $y = \epsilon\zeta$, we obtain

$$\frac{\phi(\bar{x}) - \phi(\bar{x} + \epsilon\zeta)}{\epsilon} \geq -\delta_K(\zeta).$$

Letting ϵ go to 0 yields $\delta_K(\zeta) - \partial\phi(\bar{x})\zeta \geq 0$.
   b. If v is a differentiable super-solution of (5.19) then, for all $x \in \mathbb{R}^d$, $y \in \tilde{K}$, $t \in [0,1]$,

$$\delta_K(y) - \partial_x v(x + ty)y \geq 0$$

leading to

$$\int_0^1 \big(\delta_K(y) - \partial_t v(x + ty)\big)dt \geq 0$$

and then

$$v(x) \geq v(x + y) - \delta_K(y) .$$

Taking the supremum over y, we obtain $v \geq F_K[v]$.
   c. If $[\partial h]^{\top} \in K$, it satisfies (5.19) (recall Lemma 2.4). Since we already know that $F_K[h] \geq h$, we conclude $F_K[h] = h$.

5.3

1. p^g being a solution to (5.14), it is a super-solution of the PDE satisfied by $p_{BS}^g(\sigma)$, for all $\underline{\sigma} \leq \sigma \leq \bar{\sigma}$. Since they satisfy the same boundary condition, the inequality follows from the comparison theorem.
2. It follows from the linearity of $x \mapsto S_T^{t,x}$, when σ is constant, and the convexity of g.

3. p_{BS}^g being convex it is now a solution to (5.14). The comparison theorem implies $p_{BS}^g(\bar{\sigma}) = p^g$.
4. If g is concave, $p_{BS}^g(\underline{\sigma}) = p^g$.

5.4

1. a. $\partial_t p + \frac{1}{2}[\bar{\sigma}(x)]^2 \partial_{xx}^2 p = 0$ and $p(T, \cdot) = g$.
   b. Quantity of risky asset: $\partial_x g$. Cash amount: $p - x\partial_x p$.
2. a.

$$Z_{t_{i+1}}^{0,x,N} := Z_{t_i}^{0,x,N} + \partial\bar{\sigma}(S_{t_i}^{0,x,N})Z_{t_i}^{0,x,N}(W_{t_{i+1}} - W_{t_i}) , \; i = 0,\ldots,N-1,$$

   b. $(S^{0,x,N}, Z^{0,x,N})$ is a discrete-time approximation (Euler scheme) of (S^{0,x_0}, Y^{0,x_0}).
3. a. Adapt the proof of Exercise 4.6, Question 2.b.
   b. One checks that $Y^{t,x}$ is a positive martingale with expectation equal to 1.
   c. We observe that $I[h](t,x) = \mathbb{E}^{\mathbb{Q}^{t,x}}[\partial h(S_T^{t,x})]$ and then $m \le I[h] \le M$ as soon as $m \le \partial h \le M$.
   d. Observe that

$$p_h(0,x_0) + \int_0^T \partial p_h(t, S_t^{0,x_0})dW_t = h(S_T^{0,x_0}) \ge g(S_T^{0,x_0})$$

   as $h \in H$. We thus have $p_h(0,x_0) \in \Gamma$ for all $h \in H_K$.
   e. A dynamic programming principle would tell us that

$$p_K(0,x_0) = \inf\{v | \exists\, \phi \in \mathcal{A}_K \text{ s.t. } V_T^{v,\phi} := v + \int_0^T \phi_s dS_s^{0,x_0} \ge p_K(T, S_T^{0,x_0}) \; \mathbb{P}\text{--a.s.}\}.$$

   But also we should have $p_K \in H_K$ (derivative taking its values in K – the constraint) and thus $p_K(0,x_0) \ge \mathbb{E}\left[p_K(T, S_T^{0,x_0})\right] \ge \inf_{h \in H_K} p_h(0,x_0)$.
4. $\hat{h}(x) = [x - (A-1)]_+ - [x - A]_+$. We define $h_n(x) := \int_{A-3}^x \phi_n(y)dy$ where ϕ_n is a mollification of $\mathbf{1}_{[A-1,A]}$, e.g.

$$\phi_n(x) = n\{x-A+1+\frac{1}{n}\}\mathbf{1}_{[A-1-\frac{1}{n},A-1)}(x)+\mathbf{1}_{[A-1,A)}(x)+n\{A+\frac{1}{n}-x\}\mathbf{1}_{[A,A+\frac{1}{n})}(x).$$

One easily checks that $h_n \in H_K$, $h_n \to \hat{h}$ and $p_{\hat{h}}(0,x_0) = \lim_{n\to\infty} p_{h_n}(0,x_0)$.

Moreover, we observe that there is no $h \in H_K$ s.t. $h(x) < \hat{h}(x)$ for some $x \in \mathbb{R}$. Indeed, if this is the case then necessarily $x \in (A-1,A)$ and therefore

$$h(A) = h(x) + \int_x^A \partial h(y)dy < \hat{h}(x) + (A-x) < 1 = g(A),$$

which contradicts the fact that $h \in H_K$.

5. $p_K = p_{\hat{h}}$ and it suffices to use the (un-constrained) delta-hedging strategy associated to $\hat{h}$.

5.5

1. We observe that $\psi(x) := -\epsilon x + \varphi(x)$ is a smooth test function of h at x_0. Hence, $\partial \psi(x_0) \geq 0$, which implies $\partial \varphi(x_0) \geq \epsilon$. This shows that φ is increasing on an interval (x_1, x_2) containing x_0.

2. For all $x \in [x_1, x_2]$, set $w_1(x) = h^\epsilon(x)$ and $w_2(x) = h^\epsilon(x_2)$. The comparison theorem implies that $w_2(x_0) \leq w_1(x_0)$. φ being increasing, we get the contradiction $h^\epsilon(x_2) - \varphi(x_2) < h^\epsilon(x_0) - \varphi(x_0)$.

3. The previous question proves that $h^\epsilon(x_2) \geq h^\epsilon(x_1)$. The result is obtained by letting ϵ go to zero.

Chapter 6
Hedging Under Loss Constraints

In the previous chapters, we have always obtained the pricing equations by "duality". In incomplete markets, we have explained how to price an option using an indifference utility criterion, which requires to solve first two control problems.

We present in this section a direct approach to obtain the hedging price of a contingent claim, in the almost sure sense of super-replication or in the sense of a risk criterion (*quantile hedging, expected shortfall, utility indifference*). This approach, based on the notion of stochastic target, was initiated by Soner and Touzi [55] for the super-replication criterion, and then extended by Bouchard, Elie and Touzi [10] for the hedging under risk control, see also [8, 13] and [14].

The main advantage of this approach is that it is not based on duality formulae. We are going to use a dynamic programming principle directly written on the initial hedging problem. It is then possible to use this approach in very general models for which it is not possible to establish a duality formula (large trader model, high frequency trading model with price impact, etc.).

6.1 Super-Replication: A Direct Approach

6.1.1 Framework

In this section, we first consider a general market model. The price of the d liquid assets $S^{t,x,y,\phi}$ can be impacted by the financial strategy ϕ and n auxiliary factors $Y^{t,x,y,\phi}$. The Brownian motion has here dimension $d+n$. The prices and factors dynamics are given by

$$Z_s^{t,z,\phi} = z + \int_t^s b_Z(u, Z_u^{t,z,\phi}, \phi_u) du + \int_t^s \sigma_Z(u, Z_u^{t,z,\phi}, \phi_u) dW_u \qquad (6.1)$$

© Springer International Publishing Switzerland 2016
B. Bouchard, J.-F. Chassagneux, *Fundamentals and Advanced Techniques in Derivatives Hedging*, Universitext, DOI 10.1007/978-3-319-38990-5_6

where $Z^{t,z,\phi} = (S^{t,x,y,\phi}, Y^{t,x,y,\phi})$ for $z = (x,y)$, i.e. $S_t^{t,x,y,\phi} = x$ and $Y^{t,x,y,\phi} = y$. Here, b_Z and σ_Z take their values in $\mathbb{R}^{d+n}$ and $\mathbb{M}^{d+n,d+n}$. We assume in the sequel that the component $S^{t,x,y,\phi}$ takes its values in $(0,\infty)^d$ for all initial condition $(t,x,y) \in [0,T] \times (0,\infty)^d \times \mathbb{R}^n$, and we denote $\mathcal{Z} := (0,\infty)^d \times \mathbb{R}^n$. This arbitrary choice is done just to clarify the presentation. Obviously, other cases can be studied using the same approach.

We assume also that b_Z and σ_Z are continuous, Lipschitz in space, uniformly in time and with respect to the control value.

The wealth dynamics $V^{t,v,z,\phi}$ is written as follows

$$V_s^{t,v,z,\phi} = v + \int_t^s b_V(u, V_u^{t,v,z,\phi}, Z_u^{t,z,\phi}, \phi_u)du$$

$$+ \int_t^s \sigma_V(u, V_u^{t,v,z,\phi}, Z_u^{t,z,\phi}, \phi_u)dW_u. \tag{6.2}$$

We consider financial strategies ϕ that are predictable and valued in a closed convex set $K \subset \mathbb{R}^{d+n}$, to take into account possible portfolio constraints. We denote $\mathcal{A}_K$ the set of such strategies. The functions b_V and σ_V are assumed to be continuous, Lipschitz in the V variable, uniformly with respect to the other variables.

Given a measurable function g with polynomial growth, we study the super-replication problem

$$p(t,z) := \inf \left\{ v \in \mathbb{R} : \exists \, \phi \in \mathcal{A}_K \text{ s.t. } V_T^{t,v,z,\phi} \geq g(Z_T^{t,z,\phi}) \right\}. \tag{6.3}$$

Example 6.1 (Large trader model) By setting $b_V(\cdot, \phi) = \sum_{i=1}^d \phi^i b_Z^i$ and $\sigma_V(\cdot, \phi) = \sum_{i=1}^d \phi^i \sigma_Z^i$, we obtain $dV_s^{\cdot,\phi} = \phi_s' dS^{\cdot,\phi}$. We thus model the underlying liquid asset whose dynamics are impacted by factors Y, say a stochastic volatility factor. The main difference with classical stochastic volatility models is that the number of assets held in the portfolio has also an impact on the price dynamics (and factors). This kind of model is called large trader models: If a trader holds a large share of the market, she can influence the price evolution.

In such models, duality approaches presented in Chap. 2 do not allow to obtain the price characterisation in terms of the set of martingale measures. Actually, there is no clear notion of martingale measure in this case. Indeed, the price dynamics depending on the trader's strategy, one would have to consider different martingale measures for each financial strategy.

Example 6.2 (Portfolio liquidation models) We consider the following one dimensional dynamics

$$S_s^{t,z,\phi} = x + \int_t^s b(S_u^{t,z,\phi})du + \int_t^s \sigma(S_u^{t,z,\phi})dW_u - \int_t^s \beta(S_u^{t,z,\phi}, Y_u^{t,z}, \phi_u)du,$$

$$V_s^{t,v,z,\phi} = v + \int_t^s \phi_u S_u^{t,z,\phi}du.$$

This type of models is used for portfolio liquidation problem in a high frequency setting with large portfolio. In this case, ϕ stands for the rate at which S is sold on the market. Selling the asset has an impact on its value which is taken into account by the function $\beta \geq 0$ (impact function), the larger ϕ is, the stronger the impact is, i.e. the quicker we sell the asset, the more the asset price will be pushed down. The factor Y typically stands for the market volume, i.e. the activity rate of the other market participants. In principle, the highest the activity rate is, the easiest it is to sell quickly the asset without pushing down the price.

In this model, the notion of martingale measure does not make sense. A similar model will be studied in Example 6.16 below, for the pricing of guaranteed portfolio liquidation price.

6.1.2 Dynamic Programming Principle

In Chap. 5, the pricing equations were obtained by using the dual formulation that allows to work with a standard control problem, and for which one can apply the classical techniques of stochastic control.

Since here we cannot use a priori such a dual formula, we need a dynamic programming principle directly written on the super-replication problem (6.3). It appeared for the first time in [55], see also [56]. It is commonly called the Geometric Dynamic Programming Principle (GDPP).

Theorem 6.3 (GDPP) *Let $(t, z) \in [0, T] \times \mathcal{Z}$.*

(GDPP1)
If $v > p(t, z)$ then there exists $\phi \in \mathcal{A}_K$ such that for all $\theta \in \mathcal{T}_t$

$$V_\theta^{t,v,z,\phi} \geq p(\theta, Z_\theta^{t,z,\phi}) \; \mathbb{P} - a.s.$$

(GDPP2) If there exists $\phi \in \mathcal{A}_K$ and $\theta^\phi \in \mathcal{T}_t$ such that

$$V_{\theta^\phi}^{t,v,z,\phi} > p(\theta^\phi, Z_{\theta^\phi}^{t,z,\phi}) \; \mathbb{P} - a.s.$$

then $v \geq p(t, z)$.

The very technical proof of this result requires the use of measurable selection arguments, and is largely outside the scope of this book. Nevertheless, we can roughly explain the main ideas of the proof, which are, in fact, quite natural.

If $v > p(t, z)$ then there exists $\phi \in \mathcal{A}_K$ such that $V_T^{t,v,z,\phi} \geq g(Z_T^{t,z,\phi})$. From the flow property, this amounts to say that starting from the initial condition $(V_\theta^{t,v,z,\phi}, Z_\theta^{t,z,\phi})$ at θ, it is possible to match the super-replication criterion following the strategy ϕ. By definition of p, this implies that $V_\theta^{t,v,z,\phi} \geq p(\theta, Z_\theta^{t,z,\phi})$. Reciprocally, if for ϕ and θ^ϕ given, $V_{\theta^\phi}^{t,v,z,\phi} > p(\theta^\phi, Z_{\theta^\phi}^{t,z,\phi})$, then it is possible to find a control $\bar{\phi}$ which allows to verify the super-replication criterion between θ^ϕ and T. This means that the control $\phi\mathbf{1}_{[t,\theta^\phi)} + \bar{\phi}\mathbf{1}_{[\theta^\phi,T]}$ allows to super-replicate the payoff starting from the initial conditions (v, z) at t. We thus necessarily have $v \geq p(t, z)$.

Remark 6.4 An equivalent form of (GDPP2) is: For all $v < p(t, z)$ and $\phi \in \mathcal{A}_K$,

$$\mathbb{P}\left[V_{\theta^\phi}^{t,v,z,\phi} > p(\theta^\phi, Z_{\theta^\phi}^{t,z,\phi})\right] < 1.$$

6.1.3 Pricing Equation

To understand the pricing equation which is satisfied by the price function p, we first follow an informal reasoning starting from the assertion (GDPP1). We assume that p is smooth and that (GDPP1) is true for $v = p(t, z)$, i.e. the infimum is reached in definition (6.3). Let then $\phi \in \mathcal{A}_K$ be such that $V_t^{t,v,z,\phi} = p(t, Z_t^{t,z,\phi})$ and $V_\cdot^{t,v,z,\phi} \geq p(\cdot, Z_\cdot^{t,z,\phi})$ after this date.

This implies necessarily that

$$0 \leq d\left(V_t^{t,v,z,\phi} - p(t, Z_t^{t,z,\phi})\right)$$

$$= \left(b_V(\cdot, \phi_t) - \mathcal{L}_Z^{\phi_t} p\right)(t, V_t^{t,v,z,\phi}, Z_t^{t,z,\phi})dt$$

$$+ \Delta\sigma^{\phi_t}[p](t, V_t^{t,v,z,\phi}, Z_t^{t,z,\phi})dW_t$$

$$= \left(b_V(\cdot, \phi_t) - \mathcal{L}_Z^{\phi_t} p\right)(t, v, z)dt + \Delta\sigma^{\phi_t}[p](t, v, z)dW_t$$

where, for $u \in K$,

$$\mathcal{L}_Z^u p(t, z) := \partial_t p + \partial p b_Z(t, z, u)$$

$$+ \frac{1}{2}\text{Tr}[\sigma_Z(t, z, u)\sigma_Z(t, z, u)'\partial^2 p]$$

$$\Delta\sigma^u[p](t, v, z) := \sigma_V(t, v, z, u) - \partial p \sigma_Z(t, z, u) \in \mathbb{R}^{d+n}.$$

This is possible only if the term in dW_t is zero, i.e.

$$\phi_t \in \mathcal{N}[p](t, v, z) := \{u \in K : \Delta\sigma^u[p](t, v, z) = 0\}.$$

The previous inequality, implies then that

$$\left(b_V(\cdot, \phi_t) - \mathcal{L}_Z^{\phi_t} p\right)(t, v, z) \geq 0.$$

In other words, since $v = p(t, z)$, the function p must satisfy

$$\sup_{u \in \mathcal{N}[p](t, p(t,z), z)} \left(b_V(\cdot, p, \cdot, u) - \mathcal{L}_Z^u p\right)(t, z) \geq 0.$$

We are going to show in the proof below that (GDPP2) implies the converse inequality when the elements u are restricted to belong to the following notion of "interior" of K:

$$\overset{\circ}{\mathcal{N}}[p] := \mathcal{N}[p] \cap \operatorname{int}(K).$$

In the sequel, we use the convention $\sup \emptyset = -\infty$ and make the following assumptions:

(H)$_K$ $\left(|b_V(\cdot, u)| + \|b_Z(\cdot, u)\| + \|\sigma_Z(\cdot, u)\|^2\right) / |\sigma_V(\cdot, u)|$ is locally bounded, uniformly with respect to $u \in K$.

(H)$_\mathcal{N}$ Let $(t, v, z) \in [0, T) \times \mathbb{R} \times \mathcal{Z}$ and φ be a $C^{1,2}$ function such that there exists $u \in \overset{\circ}{\mathcal{N}}[\varphi](t, v, z)$, then there exists an open neighbourhood (open to the right only for the time variable) B of (t, v, z) and a locally Lipschitz function $\hat{u}$ such that $\hat{u} \in \mathcal{N}[\varphi]$ on B and $\hat{u}(t, v, z) = u$.

Theorem 6.5 *Assume that p is continuous[1] on $[0, T) \times \mathcal{Z}$. If **(H)$_K$** holds true, then p is a viscosity super-solution of*

$$\sup_{u \in \mathcal{N}[\varphi](\cdot, \varphi(\cdot), \cdot)} \left(b_V(\cdot, \varphi, \cdot, u) - \mathcal{L}_Z^u \varphi\right) \geq 0 \quad on \ [0, T) \times \mathcal{Z}. \tag{6.4}$$

*Moreover, if **(H)$_\mathcal{N}$** is satisfied, then p is a viscosity sub-solution of*

$$\sup_{u \in \overset{\circ}{\mathcal{N}}[\varphi](\cdot, \varphi(\cdot), \cdot)} \left(b_V(\cdot, \varphi, \cdot, u) - \mathcal{L}_Z^u \varphi\right) \leq 0 \quad on \ [0, T) \times \mathcal{Z}. \tag{6.5}$$

The proof of this result is postponed to the end of the section.

We discuss below simple examples, already studied in the previous chapters.

Example 6.6 (Black and Scholes Model) We consider the Black and Scholes model where $n = 0$, $d = 1$, $K = \mathbb{R}$, $dS = \mu S dt + \sigma S dW$ with $\sigma > 0$ and $\mu \in \mathbb{R}$, and $dV^\phi = \phi dS + (V^\phi - \phi S) r dt$ with the interest rate $r > 0$. It is clear that **(H)$_K$** and **(H)$_\mathcal{N}$** are satisfied. Here, $u \in \mathcal{N}[\varphi] = \overset{\circ}{\mathcal{N}}[\varphi]$ is equivalent to $ux\sigma = x\sigma \partial_x \varphi$ and we obtain

$$0 = \sup_{ux\sigma = x\sigma \partial_x \varphi} \left(ux\mu + (\varphi - ux)r - \partial_t \varphi - \mu x \partial_x \varphi - \frac{1}{2} \sigma^2 x^2 \partial_{xx}^2 \varphi\right)$$

$$= r\varphi - \partial_t \varphi - rx \partial_x \varphi - \frac{1}{2} \sigma^2 x^2 \partial_{xx}^2 \varphi.$$

This is the linear PDE associated to the Black and Scholes model.

---

[1] To simplify the presentation.

Example 6.7 (Black and Scholes Model with stochastic volatility) We consider
again the Black and Scholes model but this time with a stochastic volatility:

$$dS = \mu S dt + \sigma(Y) S dW^1,$$

$$dY = b_Y dt + \sigma_Y dW^2,$$

$$dV^\phi = \phi^1 dS + (V^\phi - \phi^1 S) r dt,$$

where $d = n = 1$ and $W = (W^1, W^2)$ is the Brownian motion, whose components
are independent, and $K = \mathbb{R} \times \{0\}$ (it is not possible to invest in the volatility).
Since $\sigma_Z(t, z, u) = diag\,[\sigma(y)x, \sigma_Y]$, $u \in \mathcal{N}[\varphi]$ is equivalent to $u^1 x \sigma = x \sigma \partial_x \varphi$ and
$\partial_y \varphi = 0$, we find again in particular the fact that p does not depend on y, and that

$$0 \le r\varphi - \partial_t \varphi - rx\partial_x \varphi - \frac{1}{2}\sigma^2(y)x^2\partial_{xx}^2 \varphi \;, \quad \text{for all } y \in \mathbb{R}.$$

This implies that

$$0 \le r\varphi - \partial_t \varphi - rx\partial_x \varphi - \inf_{y\in\mathbb{R}} \frac{1}{2}\sigma^2(y)x^2\partial_{xx}^2 \varphi \;.$$

In this case, $\mathring{\mathcal{N}}[\varphi] = \emptyset$ ($int(K) = \emptyset$), but since p does not depend on y, it is possible
to go back to the case where $\mathcal{N}[\varphi] = \{u \in \mathbb{R} : u\sigma = x\sigma\partial_x\varphi\}$. Once this is done,
we obtain that the previous inequality is in fact an equality, in the viscosity sense.

Example 6.8 (Black and Scholes Model with constraints) We work in the setting of
Example 6.6 but we consider now the case where K is a closed interval $[\underline{u}, \bar{u}]$ with
0 in its interior, $(\mathbf{H})_K$ and $(\mathbf{H})_\mathcal{N}$ are then satisfied. Then $u \in \mathcal{N}[\varphi]$ is equivalent to
$\partial_x\varphi \in K$, i.e. $\min\{\partial_x\varphi - \underline{u}, \bar{u} - \partial_x\varphi\} \ge 0$ and $u \in \mathring{\mathcal{N}}[\varphi]$ is equivalent to $\min\{\partial_x\varphi - \underline{u}, \bar{u} - \partial_x\varphi\} > 0$. The maximum in the equation for the sub-solution reaches $-\infty$
when the inequality is an equality. We then obtain the equation

$$\min\{r\varphi - \partial_t \varphi - rx\partial_x \varphi - \frac{1}{2}\sigma^2 x^2\partial_{xx}^2 \varphi, \min\{\partial_x\varphi - \underline{u}, \bar{u} - \partial_x\varphi\}\} = 0.$$

In other words, the Black-Scholes equation is satisfied as long as the constraint
on the delta is not saturated, and then the discounted price process behaves as a
martingale under the risk neutral measure. When the constraint is saturated, the
discounted price process is not necessarily a martingale, but only a supermartingale.

If there exists a smooth solution w to (6.4), it is possible to obtain an hedging
strategy. The proof of this result is a simple consequence of Itô's lemma.

Theorem 6.9 (Verification) *Assume that $w \in C^{1,2}([0, T) \times \mathcal{Z})$ satisfies the boundary condition $\liminf_{t' \downarrow t, z' \to z} w(t', z') \ge g(z)$ for all $z \in \mathcal{Z}$, and that there exists a*

measurable function $\hat{u}$ such that

$$\hat{u} \in \mathcal{N}[w] \text{ and } b_V(\cdot, w(\cdot), \cdot, \hat{u}(\cdot)) - \mathcal{L}_Z^{\hat{u}(\cdot)} w \geq 0 \quad on \ [0, T) \times \mathcal{Z}. \qquad (6.6)$$

Assume moreover that there exists a strong solution on $[t, T]$ to

$$\hat{Z}_s = z_o + \int_{t_o}^s b_Z(u, \hat{Z}_u, \hat{u}(u, \hat{V}_u, \hat{Z}_u)) du$$

$$+ \int_{t_o}^s \sigma_Z(u, \hat{Z}_u, \hat{u}(u, \hat{V}_u, \hat{Z}_u)) dW_u$$

$$\hat{V}_s = v_o + \int_{t_o}^s b_V(u, \hat{V}_u, \hat{Z}_u, \hat{u}(u, \hat{V}_u, \hat{Z}_u)) du,$$

$$+ \int_{t_o}^s \sigma_V(u, \hat{V}_u, \hat{Z}_u, \hat{u}(u, \hat{V}_u, \hat{Z}_u)) dW_u,$$

where $(t_o, z_o) \in [0, T) \times \mathcal{Z}$ is fixed and $v_o := w(t_o, z_o)$.
  Then,

$$V_T^{t_o, v_o, z_o, \hat{\phi}} \geq g(Z_T^{t_o, z_o, \hat{\phi}})$$

with

$$\hat{\phi} := \hat{u}(\cdot, \hat{V}, \hat{Z}) \in \mathcal{A}_K.$$

Proof of the super-solution property   To simplify the proof, we only consider the case where K is bounded. If this is not the case, one needs to use assumption $(\mathbf{H})_K$. Let φ be a smooth test function and $(t, z) \in [0, T) \times \mathcal{Z}$ be such that $\min_{[0,T) \times \mathcal{Z}}(p - \varphi) = (p - \varphi)(t, z) = 0$, where the minimum is considered to be strict. Suppose that (6.4) is not satisfied at this point. Then, by continuity (K is a compact set), there exist $\varepsilon, \eta > 0$ and an open neighbourhood B of (t, z) such that

$$b_V(t', v', z', u) - \mathcal{L}_Z^u \varphi(t', z') \leq 0$$
$$\forall \ u \in K \text{ s.t. } \|\Delta\sigma^u[\varphi](t', v', z')\| \leq \varepsilon \qquad (6.7)$$
$$\forall \ (t', z') \in B \text{ and } v' \in \mathbb{R} \text{ s.t. } |v' - \varphi(t', z')| \leq \eta.$$

Let $v > \varphi(t, z) = p(t, z)$ and $\phi \in \mathcal{A}_K$ be the associated control given by (GDPP1). We consider θ_1 the first exit time of B by $(s, Z_s^{t,z,\phi})_{s \geq t}$, observe that $\theta_1 \leq T$ by construction for $B \subset [0, T) \times \mathcal{Z}$. We introduce θ_2 the first time when $V^{t,v,z,\phi} - \varphi(\cdot, Z^{t,z,\phi}) \geq \eta$. Since (t, z) is a point of strict minimum of $(p - \varphi)$, with value 0 there, we have $(p - \varphi) \geq \zeta$ on ∂B, for $\zeta > 0$. By definition of $\theta := \theta_1 \wedge \theta_2$, it follows that

$$V_\theta^{t,v,z,\phi} - \varphi(\theta, Z_\theta^{t,z,\phi}) \geq \eta \mathbf{1}_{\theta=\theta_2}$$
$$+ \mathbf{1}_{\theta<\theta_2} \left(V_\theta^{t,v,z,\phi} - p(\theta, Z_\theta^{t,z,\phi}) + \zeta \right).$$

From (GDPP1), $V_\theta^{t,v,z,\phi} - p(\theta, Z_\theta^{t,z,\phi}) \geq 0$ and the previous inequality imply

$$V_\theta^{t,v,z,\phi} - \varphi(\theta, Z_\theta^{t,z,\phi}) \geq \eta \mathbf{1}_{\theta=\theta_2} + \mathbf{1}_{\theta<\theta_2} \zeta \geq \eta \wedge \zeta > 0. \tag{6.8}$$

A simple application of Itô's lemma leads to

$$V_\theta^{t,v,z,\phi} - \varphi(\theta, Z_\theta^{t,z,\phi}) = v - \varphi(t,z) + \int_t^\theta \delta b_s ds + \int_t^\theta \delta \sigma_s dW_s,$$

where

$$\delta b := \left(b_V(\cdot, \phi) - \mathcal{L}_Z^\phi \varphi \right) (\cdot, V^{t,v,z,\phi}, Z^{t,z,\phi}),$$

$$\delta \sigma := \Delta \sigma^\phi [\varphi](\cdot, V^{t,v,z,\phi}, Z^{t,z,\phi}).$$

Using (6.7), we then deduce

$$V_\theta^{t,v,z,\phi} - \varphi(\theta, Z_\theta^{t,z,\phi}) \leq v - \varphi(t,z) + \int_t^\theta \delta b_s \mathbf{1}_A(s) ds + \int_t^\theta \delta \sigma_s dW_s \tag{6.9}$$

with

$$A := \{s \in [t, T] : \|\delta \sigma_s\| \geq \varepsilon\}.$$

Now, let $\tilde{\mathbb{P}} \sim \mathbb{P}$ be given by

$$\frac{d\tilde{\mathbb{P}}}{d\mathbb{P}} = e^{-\frac{1}{2} \int_t^\theta \|\chi_s\|^2 ds - \int_t^\theta \chi_s' dW_s}$$

with

$$\chi := \delta b \mathbf{1}_A \|\delta \sigma\|^{-2} \delta \sigma'.$$

Note that χ is bounded by definition of A, the continuity of the coefficients, the definition of θ and the fact that K is a compact set. From Girsanov Theorem,

$$\tilde{W} := W + \int_t^{\cdot \wedge \theta} \chi_s ds$$

is a Brownian motion under $\tilde{\mathbb{P}}$. Moreover, (6.9) can be re-written as

$$V_\theta^{t,v,z,\phi} - \varphi(\theta, Z_\theta^{t,z,\phi}) \le v - \varphi(t,z) + \int_t^\theta \delta \sigma_s d\tilde{W}_s.$$

Since $\delta \sigma \mathbf{1}_{[t,\theta]}$ is bounded, we obtain $\mathbb{E}^{\tilde{\mathbb{P}}}[V_\theta^{t,v,z,\phi} - \varphi(\theta, Z_\theta^{t,z,\phi})] \le v - \varphi(t,z) = v - p(t,z)$, since $\varphi = p$ at (t,z). For $0 < v - \varphi(t,z) < \eta \wedge \zeta$, we obtain a contradiction to (6.8). $\qquad\square$

Proof of the sub-solution property Let φ be a test function and $(t,z) \in [0,T) \times \mathcal{Z}$ be such that $\max_{[0,T) \times \mathcal{Z}}(p - \varphi) = (p - \varphi)(t,z) = 0$, where the maximum is strict. Assume that (6.5) is not satisfied at (t,z). In particular, this implies that $\overset{\circ}{\mathcal{N}}[\varphi](t, \varphi(t,z), z) \ne \emptyset$. From assumption $(\mathbf{H})_\mathcal{N}$, and the continuity of the coefficients, there exist a locally Lipschitz application $\hat{u}$, an open neighbourhood B of (t,z), and $\eta > 0$ such that

$$\begin{aligned}
b_V(t', v', z', \hat{u}(t', v', z')) - \mathcal{L}_z^{\hat{u}(t',v',z')}\varphi(t', z') &\ge 0, \\
\hat{u}(t', v', z') &\in \mathcal{N}[\varphi](t', v', z'), \\
\forall\, (t', z') \in B \text{ and } v' \in \mathbb{R} \text{ s.t. } |v' - \varphi(t', z')| &\le \eta.
\end{aligned} \tag{6.10}$$

Let $v < p(t,z) = \varphi(t,z)$. We note $(\hat{V}, \hat{Z})$ the solution of (6.1) and (6.2) with $\phi = \hat{\phi} := \hat{u}(\cdot, \hat{V}, \hat{Z}) \in \mathcal{A}_K$, for the initial condition (t,z,v). We just have to solve the SDE. The function $\hat{u}$ being locally Lipschitz, the solution is well defined up to the stopping time θ defined as the minimum between the first time θ_1 where $(s, \hat{Z}_s)_{s \ge t}$ leaves B and the first time θ_2 when $|\hat{V} - \varphi(\cdot, \hat{Z})| \ge \eta$. From (6.10) and the definition of $\mathcal{N}[\varphi]$, we have $d(\hat{V} - \varphi(\cdot, \hat{Z})) \ge 0$ on $[t, \theta]$. In other words, the difference is almost surely increasing. In particular, it cannot be equal to $-\eta$ at θ_2 if $v > p(t,z) - \eta = \varphi(t,z) - \eta$, which is assumed in the following. We then obtain

$$\hat{V}_\theta - p(\theta, \hat{Z}_\theta) \ge \varphi(\theta, \hat{Z}_\theta) - p(\theta, \hat{Z}_\theta) + \eta \mathbf{1}_{\theta=\theta_2} + (v - p(t,z)) \mathbf{1}_{\theta < \theta_2}.$$

Since (t,z) reaches a strict maximum, equal to zero, of $p - \varphi$, there exists $\zeta > 0$ such that $p - \varphi \le -\zeta$ on ∂B, and $p \le \varphi$ everywhere. The previous inequality then implies

$$\hat{V}_\theta - p(\theta, \hat{Z}_\theta) \ge \eta \mathbf{1}_{\theta=\theta_2} + \mathbf{1}_{\theta < \theta_2}(\zeta + v - p(t,z)).$$

Setting $v < p(t,z)$ such that $\zeta + v - p(t,z) > 0$, it follows finally that $\hat{V}_\theta - p(\theta, \hat{Z}_\theta) > 0\ \mathbb{P} - $a.s. Since $v < p(t,z)$, this contradicts (GDPP2). $\qquad\square$

6.1.4 Terminal Condition of the Pricing Equation

It remains to characterise the boundary condition at T. By construction, $p(T, \cdot) = g$. However, the condition $\mathcal{N}[p] \neq \emptyset$, implicit in equation (6.4), recall the convention $\sup \emptyset = -\infty$, leads to a constraint on p and ∂p in the interior of the domain: we must find $u \in K$ such that $\sigma_V(\cdot, p(\cdot), \cdot, u) = \partial p \sigma_Z$, in the viscosity sense. We have obtained in the Example 6.8 a constraint on the gradient $\partial_x p \in [\underline{u}, \bar{u}]$. Such constraint propagates necessarily up to the boundary and have to be taken into account. In Example 6.8, the constraint $\partial p \in [\underline{u}, \bar{u}]$ implies that $p(t, x') - \bar{u}(x' - x) \leq p(t, x) \leq p(t, x') - \underline{u}(x' - x)$ if $x' \geq x$. If g does not satisfy the same constraint, we cannot have $p(T-, \cdot) = g$. This is the same phenomenon as the one observed in Sect. 5.2.2.

Theorem 6.10 *Assume that p is continuous with polynomial growth on $[0, T) \times \mathcal{Z}$, that*

$$p_T(z) := \lim_{t' \uparrow T, z' \to z} p(t', z')$$

is well defined for all $z \in \mathcal{Z}$, and that g is continuous on $\mathcal{Z}$ with polynomial growth. If $(\mathbf{H})_K$ is satisfied, then p_T is a viscosity super-solution of

$$(\varphi - g) \geq 0 \text{ and } \mathcal{N}[\varphi] \neq \emptyset \quad \text{on } \mathcal{Z}.$$

If moreover $(\mathbf{H})_{\mathcal{N}}$ is satisfied, then p_T is a viscosity sub-solution to

$$(\varphi - g)\mathbf{1}_{\overset{\circ}{\mathcal{N}}[\varphi] \neq \emptyset} \leq 0 \quad \text{on } \mathcal{Z}.$$

Let us consider first the examples of the previous section to obtain more explicit statements.

Example 6.11 (Black and Scholes Model with stochastic volatility) In the framework of the Black and Scholes model in Example 6.6, we have $K = \mathbb{R}$ and $\mathcal{N}[\varphi] = \{\partial_x \varphi\}$, so that $\mathcal{N}[\varphi] = \overset{\circ}{\mathcal{N}}[\varphi] \neq \emptyset$. This implies $p_T = g$, which is the natural boundary condition for the Black and Scholes model. In the setting of the stochastic volatility model in Example 6.7, we know already that p does not depend on y, which implies in particular that $\mathcal{N}[\varphi] \neq \emptyset$. We thus have, $p_T(s) \geq g(s, y)$ for all $y \in \mathbb{R}$, i.e. $p_T(s) \geq \sup_y g(s, y)$.

Example 6.12 (Black and Scholes Model with portfolio constraints) We now go back to Example 6.8. Arguing as in this example, it follows that the equation on the boundary is given by

$$\min \{\varphi - g, \min\{\partial_x \varphi - \underline{u}, \bar{u} - \partial_x \varphi\}\} = 0.$$

The minimal solution to this equation[2] is the smallest function $\hat{g}$ above g which satisfies the growth condition $\hat{g}(x') - \bar{u}(x' - x) \leq \hat{g}(x) \leq \hat{g}(x') - \underline{u}(x' - x)$ for all $x' \geq x$. We encounter here again the *face-lift* phenomenon, see Chap. 5.

Proof of the super-solution property at the boundary To simplify the proof, we only consider the case where K is a compact set. Let φ be a test function and $z \in \mathcal{Z}$ be such that

$$\min_{\mathcal{Z}}(p_T - \varphi) = (p_T - \varphi)(z) = 0.$$

Since p, g have polynomial growth, it is always possible to chose φ with polynomial growth of order $\bar{q} \geq 1$ as p and g, see Remark 4.9. Let a sequence $(t_n^o, z_n^o) \to (T, z)$ be such that $p(t_n^o, z_n^o) \to p_T(z)$, with $t_n^o < T$. We consider, for $n \geq 1$,

$$\varphi_n(t', z') := \varphi(z') - \frac{T - t_n^o}{\sqrt{T - t'}} - \|z - z'\|^{4\bar{q}}.$$

We obtain that

$$\min_{[t_n^o, T] \times \mathcal{Z}} (p - \varphi_n) = (p - \varphi_n)(t_n, z_n) \text{ with } (t_n, z_n) \in [t_n^o, T) \times \mathcal{Z} .$$

Indeed, the minimum exists because the penalising term in space dominates the growth of $|p - \varphi|$. Moreover, $t_n < T$ because the penalising term in time explodes for $t' \to T$. From the definition of p and the super-solution property, we thus have

$$V_T^{t_n, v_n, z_n, \phi_n} \geq g(Z_T^{t_n, z_n, \phi_n}) \text{ and } \Delta\sigma^{u_n}[\varphi_n](t_n, \varphi_n(t_n, z_n), z_n) = 0$$

for a given $\phi_n \in \mathcal{A}_K$, $v_n := p(t_n, z_n) + n^{-1}$, and $u_n \in K$.

We are going to show later that, up to a subsequence,

$$(t_n, z_n) \to (T, z) , \ (p, \varphi_n, \partial\varphi_n)(t_n, z_n) \to (p_T, \varphi, \partial\varphi)(z),$$

$$u_n \to u \in K. \tag{6.11}$$

The domain K being a compact set, the function g being continuous, usual estimates allow us to obtain that, up to a subsequence,

$$p_T(z) = \limsup_n V_T^{t_n, v_n, z_n, \phi_n} \geq \liminf_n g(Z_T^{t_n, z_n, \phi_n}) = g(z).$$

Moreover, the convergence of the terms in the right-hand side of (6.11) implies that the volatility term remains equal to zero at the limit: $\Delta\sigma^u[\varphi](T, \varphi(z), z) = 0$, which amounts to say that $\mathcal{N}[\varphi] \neq \emptyset$.

---

[2]See Exercise 5.2 for a rigorous formulation under specific assumptions.

We thus have to prove (6.11). First, recall that $t_n < T$ and moreover that $t_n \to T$ since $t_n \in [t_n^o, T] \to \{T\}$. We observe then by the minimum property,

$$(p - \varphi_n)(t_n, z_n) \le (p - \varphi_n)(t_n^o, z) \le C,$$

where $C > 0$ is a constant independent of n. Then, since p and φ have polynomial growth of order $\bar{q}$, we have

$$C \ge (p - \varphi_n)(t_n, z_n) \ge -C(1 + \|z_n\|^{\bar{q}}) + \|z - z_n\|^{4\bar{q}}$$

possibly by increasing the value of C. We thus have that $(z_n)_{n\ge1}$ is bounded. For a subsequence, still denoted $(z_n)_{n\ge1}$, we have then $z_n \to \bar{z} \in \mathcal{Z}$. By the minimum property of z and of (t_n, z_n)

$$(p_T - \varphi)(z) \le (p_T - \varphi)(\bar{z}) + \liminf_n \left(\frac{T - t_n^o}{\sqrt{T - t_n}} + \|z - z_n\|^{4\bar{q}} \right)$$

$$= \liminf_n (p - \varphi_n)(t_n, z_n).$$

Moreover,

$$\liminf_n (p - \varphi_n)(t_n, z_n) \le \liminf_n (p - \varphi_n)(t_n^o, z_n^o)$$

$$= (p_T - \varphi)(z).$$

This implies that

$$\frac{T - t_n^o}{\sqrt{T - t_n}} + \|z - z_n\|^{4\bar{q}} \to 0$$

and that $(p - \varphi_n)(t_n, z_n) \to (p_T - \varphi)(z)$. In particular, $z_n \to z$. The convergence of the gradient term is a straightforward consequence. □

Proof of the sub-solution property at the boundary To simplify the proof, we work with K a compact set. Let φ be a test function and $z \in \mathcal{Z}$ be such that

$$\max_{\mathcal{Z}} (p_T - \varphi) = (p_T - \varphi)(z) = 0.$$

We assume that $\overset{\circ}{\mathcal{N}}[\varphi](z) \ne \emptyset$ and that

$$p_T \ge g + \eta, \text{ in a neighbourhood of } z, \text{ with } \eta > 0. \tag{6.12}$$

Let us introduce

$$\varphi(t', z') := \varphi(z') + \sqrt{T - t'} + \|z - z'\|^{4\bar{q}},$$

where $\bar{q}$ is the polynomial growth power of p. Let (t_n^o, z_n^o) be a sequence converging to (T, z) and such that $p(t_n^o, z_n^o) \to p_T(z)$, with $t_n^o < T$. The maximum on $[t_n^o, T] \times \mathcal{Z}$ of $p - \varphi$ is reached at a point (t_n, z_n) and

$$(t_n, z_n) \to (T, z) \,, \quad (p, \varphi, \partial\varphi)(t_n, z_n) \to (p_T, \varphi, \partial\varphi)(z). \tag{6.13}$$

This is proved by the same reasoning as in the previous proof.

Assume that, first, the number of n such that $t_n = T$ is finite. Then, after possibly passing to a sub-sequence, it is possible to reduce to the case where

$$t_n < T \quad \text{for all } n \geq 1. \tag{6.14}$$

Moreover, we have

$$\partial_t \varphi(t_n, z_n) = -\frac{1}{2\sqrt{T - t_n}} \to -\infty.$$

For n fixed but large enough, combining this with the condition $\mathring{\mathcal{N}}[\varphi](z) \neq \emptyset$ implies that

$$\sup_{u \in \mathring{\mathcal{N}}[\varphi_n](\cdot, \varphi(\cdot), \cdot)} \left(b_V(\cdot, \varphi, \cdot, u) - \mathcal{L}_Z^u \varphi \right)(t_n, z_n) > 0.$$

We can the follow line by line the proof of the sub-solution property working with an initial condition $v < p(t_n, z_n)$, recall that $t_n < T$. The only difference is that we have to work with $\theta := \theta_1 \wedge \theta_2 \wedge T$. When $\theta = T$, we obtain by using (6.12) that $\hat{V}_\theta \geq \varphi_n(\theta, \hat{Z}_\theta) \geq p_T(\theta, \hat{Z}_\theta) \geq g(\hat{Z}_\theta) + \eta = p(T, \hat{Z}_\theta) + \eta$. The other cases, $\theta \in \{\theta_1, \theta_2\} \setminus \{T\}$, are dealt with as in the sub-solution proof. In every case, we obtain a contradiction to (GDPP2).

To conclude the proof, we have to study the case where the number of n such that $t_n = T$ is infinite. Then, up to a subsequence, we can consider that $t_n = T$ for all n. In this case, we use the fact that $p(T, \cdot) = g$ by definition, so that $(p - \varphi)(t_n, z_n) = (g - \varphi)(z_n) - \|z - z_n\|^{4\bar{q}} \to (g - \varphi)(z) = (g - p_T)(z) < 0$ from (6.12), while (6.13) implies $(p - \varphi)(t_n, z_n) \to (p_T - \varphi)(z) = 0$. $\qquad\square$

6.2 Hedging Under Loss Control

As seen in Chap. 5, the super-replication criterion may lead to degenerate results that do not correspond to a realistic pricing.

In this case, it seems reasonable to consider a price and a hedging strategy associated to an indifference utility criterion or other risk criteria. We have already studied this kind of problem in Chap. 3 in the setting of classical financial markets, by duality arguments. These are based on the link between the set of reachable wealth and martingale measures. In incomplete markets, these characterisations are not always useful. Moreover, we have already mentioned some interesting cases for which the notion of martingale measure does not exist.

Using the arguments introduced in the previous section, we can nevertheless characterise the hedging price defined by using a given risk criterion as a solution of a (non-linear) PDE. This is the topic of this section.

Given a "loss function" ℓ and an expected loss level m, we consider the problem:

$$p(t, z, m) := \inf \left\{ v \in \mathbb{R} : \exists \, \phi \in \mathcal{A}_K \text{ s.t. } \mathbb{E}\left[L_T^{t,v,z,\phi} \right] \geq m \right\}, \qquad (6.15)$$

where

$$L_T^{t,v,z,\phi} := \ell \left(V_T^{t,v,z,\phi} - g(Z_T^{t,z,\phi}) \right).$$

The function ℓ is assumed to be non-decreasing and to have polynomial growth.

Example 6.13 (Quantile hedging) Considering $\ell(r) := \mathbf{1}_{r \geq 0}$ and $m \in [0, 1]$, the problem rewrites

$$p(t, z, m) := \inf\{v \in \mathbb{R} : \exists \, \phi \in \mathcal{A}_K \text{ s.t. } \mathbb{P}[V_T^{t,v,z,\phi} \geq g(Z_T^{t,z,\phi})] \geq m\}.$$

This is the minimum price at which the option must be sold so that it can be super-replicated with probability at least m. The case $m = 1$ corresponds to the usual super-replication criterion in the almost-sure sense.

Example 6.14 (Quadratic loss function) When $\ell(r) := -(r^-)^2$ and $-m \in \mathbb{R}_-$, then the problem consists in finding the minimum initial wealth v such that there exists $\phi \in \mathcal{A}_K$ satisfying

$$\mathbb{E}\left[\left(\left[V_T^{t,v,z,\phi} - g(Z_T^{t,z,\phi}) \right]^- \right)^2 \right] \leq m.$$

The loss coming from the hedging is $\left[V_T^{t,v,z,\phi} - g(Z_T^{t,z,\phi}) \right]^-$. We now want that the expectation of the loss squared is controlled by m.

Example 6.15 (Indifference pricing) Let U be a utility function and $m \in \mathbb{R}$. We want to find the smallest initial wealth $p(t, z, m)$ such that there exists $\phi \in \mathcal{A}_K$ satisfying

$$\mathbb{E}\left[U \left(V_T^{t,v,z,\phi} - g(Z_T^{t,z,\phi}) \right) \right] \geq m.$$

We want the expected utility to be at least m when the option is sold. If

$$m := \sup_{\phi \in \mathcal{A}_K} \mathbb{E}\left[U \left(V_T^{t,v_o,z,\phi} \right) \right]$$

then $p(t, z, m) - v_o$ is the indifference price introduced in Chap. 3. Obviously we need to be able to compute this last quantity. This is possible if the initial market is complete, i.e. the incompleteness appears only through the option which is sold. This is the case when we want to hedge an option whose payoff depends on a very liquid underlying and some other randomness: By example, when the option is written on the value of the wheat production of a given producer which depends on the wheat market price and the production level is linked to the weather conditions.

Example 6.16 (Option on liquidation price) When a large investor wants to rebalance an important part of her portfolio, she cannot do it instantaneously as she will move the market price. She can then rely on a broker. The broker will send market orders in an optimal way for his own account, seeking to find a compromise between the risk of moving (unfavorably) the price by sending the orders too rapidly and the market risk if he sends the orders too slowly. A very popular contract is the guaranteed VWAP (volume weighted average price): The broker guarantees to his client at maturity an average price to buy or sell the asset corresponding to a given percentage of the average market price observed during the rebalancing period. The main issue is then to find the price of this guarantee.

We give here a simple modelling example. The goal is to sell N_o shares of an underlying S. We assume that the instantaneous volume of market transactions is deterministic, given by ϑ_s at time s. The sell orders are made at a rate ϕ, a predictable process with values in a compact set $K = [0, \hat{u}]$. The set of such processes is denoted $\mathcal{A}_K$. The *broker*'s trade has an impact on the dynamics of the underlying S, given by

$$S^{t,x,\phi}_\cdot = x + \int_t^\cdot \mu(s, S^{t,x,\phi}_s)dr + \int_t^\cdot \sigma(s, S^{t,x,\phi}_s)dW_s$$
$$- \int_t^\cdot \beta(s, S^{t,x,\phi}_s, \vartheta_s, \phi_s)ds$$

where β is a positive continuous function which represents the sell order impact. Besides, selling at a rate ϕ_s at time s gives only $\gamma(s, S^{t,x,\phi}_s, \phi_s)\phi_s ds$ where γ is another positive continuous function satisfying typically $\gamma(\cdot, x, \cdot) \leq x$. This property means that selling is not made at the observed price but at a smaller price. The wealth dynamics are thus given by

$$V^{t,v,x,\phi}_\cdot = v + \int_t^\cdot \gamma(s, S^{t,x,\phi}_s, \vartheta_s, \phi_s)\phi_s ds.$$

If n shares have already been sold at t, the dynamics of the number of liquidates shares are

$$N^{t,n}_\cdot = n + \int_t^\cdot \phi_s ds.$$

We denote

$$W^{t,w,x,\phi} = w + \int_t^{\cdot} S_s^{t,x,\phi} \vartheta_s ds$$

the cumulated value of all the orders sent if this one has already reached the level $w \geq 0$ at t, and $C^{t,c} := c + \int_t^{\cdot} \vartheta_s ds$ the cumulated volume of the orders sent to the market by the other agents. Here, we only take into account the trades of the other agents when computing W and C.

If the value of the underlying is x at time 0, the best action for the *broker* is to set a selling price v at 0 for which he can find a strategy ϕ satisfying $V_T^{0,v,x,\phi}/N_o \geq \kappa W_T^{0,0,x,\phi}/C_T^{0,0}$ and $N_T^{0,0} = N_o$, with $\kappa \in (0,1)$ representing the percentage of the average observed price $W_T^{0,w,x,\phi}/C_T^{0,0}$ which has been guaranteed. It is not difficult to see that in general this is not possible (it is even worse if the market transactions volume is not deterministic). It is thus essential to weaken the super-replication criterion by considering a loss function ℓ, i.e. by defining a price $p(0,x,0,0,0,m)$, where $p(t,x,n,w,c,m)$ is given by the smallest v such that there exists $\phi \in \mathcal{A}_K$ satisfying

$$\mathbb{E}\left[\ell\left(V_T^{t,v,x,\phi}/K_o - \kappa W_T^{t,w,x,\phi}/C_T^{t,c}\right)\right] \geq m \text{ and } N_T^{t,n} = N_o.$$

6.2.1 Problem Reduction

The first step to solve (6.15) consists in rewriting the problem as in (6.3) where the constraint is set in the $\mathbb{P}-$a.s. sense. In this section, the set K will be a compact set.

To rewrite the problem, we introduce the set $\mathcal{U}$ of processes α which are square integrable. If K is a compact set, the growth condition on ℓ, g and on the coefficients appearing in equations (6.1) and (6.2) imply that $\ell\left(V_T^{t,v,z,\phi} - g(Z_T^{t,z,\phi})\right)$ is square integrable. From the martingale representation theorem see Theorem 2.22, it is possible to find $\alpha \in \mathcal{U}$ such that

$$\ell\left(V_T^{t,v,z,\phi} - g(Z_T^{t,z,\phi})\right) = M^{t,m_o,\alpha}(T)$$

with

$$M^{t,m_o,\alpha} := m_o + \int_t^{\cdot} \alpha_s dW_s$$

and

$$m_o := \mathbb{E}\left[\ell\left(V_T^{t,v,z,\phi} - g(Z_T^{t,z,\phi})\right)\right].$$

In particular, if $\phi \in \mathcal{A}_K$ is such that $m_o \geq m$, then

$$\ell\left(V_T^{t,v,z,\phi} - g(Z_T^{t,z,\phi})\right) = M^{t,m_o,\alpha}(T)$$

$$= m_o - m + M^{t,m,\alpha}(T)$$

$$\geq M^{t,m,\alpha}(T).$$

Reciprocally, if $(\phi, \alpha) \in \mathcal{A}_K \times \mathcal{U}$ are such that the previous inequality is satisfied, then

$$\mathbb{E}\left[\ell\left(V_T^{t,v,z,\phi} - g(Z_T^{t,z,\phi})\right)\right] \geq \mathbb{E}\left[M^{t,m,\alpha}(T)\right] = m,$$

since $M^{t,m,\alpha}$ is a martingale.

In other words, by introducing the new controlled process $M^{t,m,\alpha}$, we end up with a criterion similar to (6.3).

Proposition 6.17 *The quantity* $p(t, z, m)$ *is equal to*

$$\inf\left\{v \in \mathbb{R} : \exists\, (\phi, \alpha) \in \mathcal{A}_K \times \mathcal{U} \text{ s.t. } V_T^{t,v,z,\phi} \geq \Psi\left(Z_T^{t,z,\phi}, M_T^{t,m,\alpha}\right)\right\}, \quad (6.16)$$

where

$$\Psi(z, m) := g(z) + \ell^{-1}(m) \quad (6.17)$$

with ℓ^{-1} *the right inverse of* ℓ.

Remark 6.18

(i) The process $M^{t,m_o,\alpha}$ is obtained as the conditional expectation of the random variable $\ell\left(V_T^{t,v,z,\phi} - g(Z_T^{t,z,\phi})\right)$. Its terminal value, $M_T^{t,m_o,\alpha}$, takes values in the range I_ℓ of the function ℓ. When I_ℓ is convex, $m_o \in I_\ell$ and it is possible to consider only processes α such that $M^{t,m_o,\alpha} \in I_\ell$ on $[t, T]$.

(ii) If there exists a control ϕ such that $\mathbb{E}\left[\ell\left(V_T^{t,v,z,\phi} - g(Z_T^{t,z,\phi})\right)\right] = m$ for $v = p(t, z)$, then $M_s^{t,m,\alpha} = \mathbb{E}\left[\ell\left(V_T^{t,v,z,\phi} - g(Z_T^{t,z,\phi})\right)|\mathcal{F}_s\right]$ for all $s \in [t, T]$. In other words, $M^{t,m,\alpha}$ gives the dynamics of the conditional expectation along the optimal trajectory.

6.2.2 Pricing Equation

Given the problem reformulation obtained in Proposition 6.17, it is sufficient a priori to apply the results of Theorem 6.5 to the system augmented with the process $M^{t,m,\alpha}$ to obtain the equation satisfied by p. The main difficulty comes from the fact that the assumptions $(\mathbf{H})_K$ and $(\mathbf{H})_\mathcal{N}$ are in general not satisfied anymore for the augmented system. Nevertheless, it is possible to adapt the proof of Theorem 6.5 to take this into account, see [11, Section 3].

We thus adapt to this new setting the operators introduced in the previous section. We then define

$$\mathcal{L}^{u,a}_{Z,M}p := \partial_t p + \partial_z p b_Z(\cdot, u) + \frac{1}{2}\mathrm{Tr}[\sigma_Z(\cdot, u)\sigma_Z(\cdot, u)'\partial^2_{zz}p]$$

$$+ \frac{1}{2}\left[\|a\|^2 \partial^2_{mm}v + 2\partial^2_{mz}v\sigma_Z(\cdot, u)a\right],$$

$$\Delta_Z\sigma^u[p] := \sigma_V(\cdot, u) - \partial_z p\sigma_Z(\cdot, u),$$

and

$$\mathcal{N}[p](t, v, z, m) := \{(u, a) \in K \times \mathbb{R}^n : \Delta_Z\sigma^u[p](t, v, z) = a'\partial_m p\}.$$

From now on, we assume that

the range I_ℓ of ℓ is convex,

i.e. it is an interval, and we denote $\mathrm{int}(I_\ell)$ its interior.

Theorem 6.19 *Assume that p is continuous on $[0, T) \times \mathcal{Z} \times \mathrm{int}(I_\ell)$. If $(\mathbf{H})_K$ is satisfied, then p is a viscosity super-solution of*

$$\max\left\{\sup_{(u,a)\in\mathcal{N}[\varphi](\cdot,\varphi(\cdot),\cdot)} \left(b_V(\cdot, \varphi, \cdot, u) - \mathcal{L}^{u,a}_{Z,M}\varphi\right), -\partial_m\varphi\right\} \geq 0$$

on $[0, T) \times \mathcal{Z} \times \mathrm{int}(I_\ell)$. Moreover, p is a viscosity sub-solution of

$$\min\left\{\sup_{(u,a)\in\mathcal{N}[\varphi](\cdot,\varphi(\cdot),\cdot)} \left(b_V(\cdot, \varphi, \cdot, u) - \mathcal{L}^{u,a}_{Z,M}\varphi\right), \partial_m\varphi\right\} \leq 0 \qquad (6.18)$$

on $[0, T) \times \mathcal{Z} \times \mathrm{int}(I_\ell)$.

Note that new terms $-\partial_m\varphi$ and $\partial_m\varphi$ appear in the super and sub-solution properties. Without these new terms, the statement is not correct. They allow to go back to the situation where the assumption $(\mathbf{H})_K$ is satisfied for the augmented system. However, as the function ℓ is non-decreasing, the function p is also non-decreasing at m. If this function is strictly increasing at m, i.e. $\partial_m\varphi > 0$, then we obtain the simpler formulation of Theorem 6.5.

Note also that we do not assume that $(\mathbf{H})_\mathcal{N}$ is satisfied to obtain the sub-solution property and that we have replaced $\overset{\circ}{\mathcal{N}}[\varphi]$ by $\mathcal{N}[\varphi]$ in the sub-solution property. This is due to the new term $\partial_m\varphi$. Indeed, if (6.18) is not satisfied, then $\partial_m\varphi > 0$. For any

$u \in K$, it is possible to find a such that $(u, a) \in \mathcal{N}[\varphi](\cdot, \varphi(\cdot), \cdot)$. Indeed, it is sufficient to set $a' = \Delta_Z \sigma^u[p]/\partial_m \varphi$. The proof of the sub-solution part of Theorem 6.19 can thus be easily modified.

Example 6.20 Let us go back to Example 6.13 with the dynamics corresponding to the hedging problem in Black and Scholes model:

$$dV^{t,v,x,\phi} = \phi V^{t,v,x,\phi} dS^{t,x}/S^{t,x} \quad \text{and} \quad dS^{t,x} = S^{t,x}(\mu dt + \sigma dW)$$

where $\mu \in \mathbb{R}$ and $\sigma > 0$. The function ℓ is given by $\ell(y) = \mathbf{1}_{y \geq 0}$, which implies that $\mathrm{int}(I_\ell) = (0, 1)$. If p is smooth and satisfies $\partial_m p > 0$, and if $K = \mathbb{R}$, i.e. there is no portfolio constraint, then the equation obtained in the previous theorem can be written

$$0 = \sup_{(u,a) \in \mathcal{N}[\varphi](\cdot, \varphi(\cdot), \cdot)} \left(up\mu - \partial_t p - \mu x \partial_x p - \frac{1}{2} \left(x^2 \sigma^2 \partial^2_{xx} p + 2x\sigma a \partial^2_{xm} p + a^2 \partial^2_{mm} p \right) \right)$$

with $\mathcal{N}[p](t, x, y, m)$ defined by

$$\{(u, a) \in \mathbb{R}^2 : u\sigma p(t, x, m) = \sigma x \partial_x p(t, x, m) + a \partial_m p(t, x, m)\},$$

or equivalently

$$\sup_{a \in \mathbb{R}} \left(a \frac{\mu}{\sigma} \partial_m p - \partial_t p - \frac{1}{2} \left(x^2 \sigma^2 \partial^2_{xx} p + 2x\sigma a \partial^2_{xm} p + a^2 \partial^2_{mm} p \right) \right) = 0.$$

This implies in particular that $\partial^2_{mm} p \geq 0$, so p is convex in m. If p is strictly convex, we obtain, by straightforward computations,

$$- \partial_t p - \frac{1}{2} x^2 \sigma^2 \partial^2_{xx} p + \frac{1}{2} \frac{((\mu/\sigma)\partial_m p - x\sigma \partial^2_{xm} p)^2}{\partial^2_{mm} p} = 0. \tag{6.19}$$

Example 6.21 Let us now go back to Example 6.16. From the dynamics and its coefficients, we have that

$$\mathcal{N}[p] = \{(u, a) \in K \times \mathbb{R} : \sigma \partial_x p + a \partial_m p = 0\}.$$

If p is smooth and strictly increasing at m, then it should satisfy, for $m \in \mathrm{Int}(I_\ell)$,

$$\sup_{u \in K} (u\gamma(\cdot, u) + \beta(\cdot, u)\partial_x p - u\partial_n p) - \mathcal{L}_{SWC} p - \mathcal{H} p = 0 \tag{6.20}$$

with

$$\mathcal{L}_{SWC}p := \partial_t p + \mu \partial_x p + \frac{1}{2}\sigma^2 \partial^2_{xx}p + x\vartheta\, \partial_w p + \vartheta\, \partial_c p,$$

$$\mathcal{H}p := \frac{1}{2}\left((\sigma\partial_x p/\partial_m p)^2 \partial^2_{mm}p - (\sigma^2\partial_x p/\partial_m p)\partial^2_{xm}p\right).$$

But this equation can be satisfied only if n belongs to $[\inf_{u\in K} N_o/(u(T-t)), N_o]$. If n does not belong to this interval then the constraint $N_T^{t,n,\phi} = N_o$ cannot be satisfied. Indeed, p is solution to *(6.20)* on the domain D defined by

$$\{(t,x,n) \in [0,T) \times (0,\infty) \times \mathbb{R}_+ : n \in [\frac{N_o}{\hat{u}(T-t)}, N_o]\} \times \mathbb{R}^2_+ \times \text{int}(I_\ell).$$

When $n = \underline{n}_t := N_o/(\hat{u}(T-t))$, the only possibility to satisfy the terminal liquidation constraint is to chose the maximal control $\hat{u}$ and the equation becomes

$$(\hat{u}\gamma(\cdot,\hat{u}) + \beta(\cdot,\hat{u})\partial_x p - \hat{u}\partial_n p) - \mathcal{L}_{SWC}p - \mathcal{H}p = 0.$$

But if $n = N_o$ then the only possibility is to follow the zero control and the equation becomes

$$\beta(\cdot,0)\partial_x p - \mathcal{L}_{SWC}p - \mathcal{H}p = 0.$$

6.2.3 Time Boundary Condition

The boundary condition in time, when $t \to T$, must be carefully studied. The natural condition is

$$p(T,z,m) = \Psi(z,m)$$

where the function Ψ is defined in (6.17). However, in Example 6.20, we have shown that the function p is convex in m. If this is not the case for Ψ, then the natural condition cannot be the correct one and Ψ must be replaced by its convex envelop in m.

Let us study our two main examples.

Example 6.22 Let us consider Example 6.20. We assume that p is smooth, that $g \geq 0$ and that $K = \mathbb{R}$, i.e. there is no portfolio constraint. The convex envelop at $m \in [0,1] = I_\ell$ of $\Psi(x,m) = g(x)\mathbf{1}_{m>0}$ is then given by $\hat{\Psi}(x,m) := mg(x)$. Assume that $v > p(t,x,m)$. Then, following the arguments of Proposition 6.17 and of Remark 6.18, there exist $\phi \in \mathcal{A}_K$ and $\alpha \in \mathcal{U}$ such that

$$\mathbf{1}_{V_T^{t,v,x,\phi} \geq g(S_T^{t,x})} \geq M_T^{t,m,\alpha} \quad \text{and } M_T^{t,m,\alpha} \in [0,1],$$

i.e., since $g \geq 0$,

$$V_T^{t,v,x,\phi} \geq g(S_T^{t,x})\mathbf{1}_{M_T^{t,m,\alpha}>0} \geq M_T^{t,m,\alpha} g(S_T^{t,x}) \quad \text{and } M_T^{t,m,\alpha} \in [0,1].$$

Let $\mathbb{Q}_t \sim \mathbb{P}$ be the probability measure for which $S^{t,x}$ is a martingale, then

$$v \geq \mathbb{E}^{\mathbb{Q}_t}\left[M_T^{t,m,\alpha}g(S_T^{t,x})\right].$$

Denote by H_t the density associated to $\mathbb{Q}_t$

$$H_t := e^{-\frac{\mu^2}{2\sigma^2}(T-t)-\frac{\mu}{\sigma}(W_T-W_t)}.$$

Assume moreover that g is bounded by L, to simplify. Then, since $M_T^{t,m,\alpha}, m \in [0,1]$ and $\mathbb{E}[M_T^{t,m,\alpha}] = m$,

$$
\begin{aligned}
v &\geq \mathbb{E}\left[H_t M_T^{t,m,\alpha}g(S_T^{t,x})\right] \\
&\geq \mathbb{E}\left[H_t m g(S_T^{t,x})\right] + \mathbb{E}\left[(M_T^{t,m,\alpha} - m)g(x)\right] \\
&\quad + \mathbb{E}\left[(M_T^{t,m,\alpha} - m)(g(S_t^{t,x}) - g(x))\right] \\
&\quad + \mathbb{E}\left[(H^t - 1)(M_T^{t,m,\alpha} - m)g(S_T^{t,x})\right] \\
&\geq \mathbb{E}\left[H_t m g(S_T^{t,x})\right] - 2\mathbb{E}\left[|g(S_t^{t,x}) - g(x)|\right] - 2L\mathbb{E}\left[|H^t - 1|\right].
\end{aligned}
$$

This implies that

$$p(t,x,m) \geq \mathbb{E}\left[H_t m g(S_T^{t,x})\right] - 2\mathbb{E}\left[|g(S_t^{t,x}) - g(x)|\right] - 2L\mathbb{E}\left[|H^t - 1|\right].$$

Since $H^t \to 1$ and $g(S_T^{t,x}) \to g(x_o)$ if $t \to T$ and $x \to x_o$, we deduce by using the dominated convergence theorem that

$$\liminf_{(t,x,m)\to(T,x_o,m_o)} p(t,x,m) \geq m_o g(x_o).$$

We now use the fact that p is convex in m and that $m = m \times 1 + (1-m) \times 0$:

$$p(t,x,m) \leq m p(t,x,1) + (1-m)p(t,x,0).$$

But $p(t,x,0) = 0$ since doing nothing and starting with an initial wealth equal to zero are enough to hedge with a zero probability... Moreover, we have $p(t,x,1) =$

$\mathbb{E}^{Q_t}\left[g(S_T^{t,x})\right]$ since $m = 1$ corresponds to the case where the hedging is almost sure. Since $p(t, x, 1) \to g(x)$ when $t \to T$, this implies that

$$\limsup_{(t,x,m)\to(T,x_o,m_o)} p(t, x, m) \le m_o g(x_o).$$

We thus have shown that the time boundary condition is given by:

$$\lim_{(t,x,m)\to(T,x_o,m_o)} p(t, x, m) = m_o g(x_o) = \hat{\Psi}(x_o, m_o). \tag{6.21}$$

Example 6.23 Let us consider now Example 6.21 and assume that ℓ is continuous with linear growth. Since $K = [0, \hat{u}]$ is a compact set, it is easy to show that

$$\sup_{\phi\in\mathcal{A}_K} \mathbb{E}\left[|V_T^{t,v,x,\phi} - v_o|\right] \to 0 \text{ if } (t, v, x) \to (T, v_o, x_o).$$

We then easily deduce that

$$p(t, x, n, w, c, m) \to \ell^{-1}(m_o) + \kappa w_o/c_o$$

if

$$(t, x, n, w, c, m) \in \bar{D} \to (T, x_o, N_o, w_o, c_o, m_o)$$

with $m_o \in \text{Int}(I_\ell)$. In particular, if $I_\ell = \mathbb{R}$, then this boundary condition and the PDE obtained in Example 6.21 yield a full characterisation of the price dynamics for the guaranteed VWAP contract.

6.3 Comments

This direct approach has three advantages. First, it permits to work on hedging problems directly without using dual formulations established in Part I. This allows in particular not to forget what super-replication really means: super-replicating (or replicating) does not always boil down to computing a price as an expectation. Besides, it yields pricing equations in very general models. Finally, it allows to obtain pricing equations even when the criterion is not the super-replication criterion but only a loss control criterion.

We refer to [8, 10, 13] and [14] for a precise study of the hedging and pricing problem under loss control. These articles study respectively the *quantile hedging* criterion in a complete market model, hedging problems with constraint on the set of trajectories between 0 and T, an evaluation criterion for the guaranteed VWAP contract and an hedging criterion with constraints on the P&L distribution.

Let us note finally that similar techniques can be used in the problem of portfolio management under risk constraint [9], or in models with parameters uncertainty [11].

6.4 Problems

6.1 (Hedging price via the direct approach) We consider a model of Black Scholes type where the risky asset's volatility depends on the trader's strategy. Let $\phi \in \mathcal{A}$ be a financial strategy (number of risky assets), where $\mathcal{A}$ is the set of predictable processes which are square integrable (i.e. $\mathbb{E}\left[\int_0^T |\phi_s|^2 ds\right] < \infty$). The price dynamics are given by

$$S_s^{t,x,\phi} = x + \int_t^s \sigma(\phi_r)dW_r, \quad t \le s \le T,$$

where $x \in \mathbb{R}$ is the initial value of the underlying at date t. We assume that σ is continuous and that

$$a \in \mathbb{R} \mapsto (a\sigma(a), \sigma(a)) \text{ is bounded.} \tag{6.22}$$

We work with a risk-free interest rate set to 0, so that the wealth process V with initial value y at date t is written

$$V_s^{t,x,v,\phi} = v + \int_t^s \phi_r dS_r^{t,x,\phi}, \quad t \le s \le T. \tag{6.23}$$

The goal of this exercise is to study the super-replication price of an European option with payoff $g(S_T^{t,x,\phi})$ at maturity T,

$$p(t,x) := \inf\{v \in \mathbb{R} : \exists \phi \in \mathcal{A} \text{ s.t. } V_T^{t,x,v,\phi} \ge g(S_T^{t,x,\phi})\}.$$

We assume that g is bounded.

1. Using (6.22), show that $V^{t,x,v,\phi}$ is a martingale on $[t, T]$ for all $\phi \in \mathcal{A}$.
2. Assuming that the infimum is reached in the definition of p, show that $p(t,x) \ge \mathbb{E}\left[g(S_T^{t,x,\hat{\phi}})\right] =: \bar{p}(t,x)$ for a strategy $\hat{\phi} \in \mathcal{A}$.
3. Show that there exists $\psi \in \mathcal{A}$ such that $\bar{p}(t,x) + \int_t^T \psi_s dW_s = g(S_T^{t,x,\hat{\phi}})$.
4. Can $\hat{\phi}$ and ψ be such that $\psi = \hat{\phi}\sigma(\hat{\phi})$, i.e. such that $V_T^{t,x,\bar{p}(t,x),\hat{\phi}} \ge g(S_T^{t,x,\hat{\phi}})$?

From now on, we assume that p is a bounded $C^{1,2}([0, T] \times \mathbb{R})$ function. We want to show that p is solution of

$$- F\varphi(t, x) := -\mathcal{L}^{\psi(\partial\varphi(t,x))}\varphi(t, x) = 0 \text{ on } [0, T) \times \mathbb{R}, \qquad (6.24)$$

with, for $a \in \mathbb{R}$, $\mathcal{L}^a \varphi(t, x) = \partial_t \varphi(t, x) + \frac{1}{2}\sigma(a)^2\partial^2\varphi(t, x)$, and $\psi(p)$ defined as the unique solution a to

$$a\sigma(a) = \sigma(a)p ,$$

i.e. $\psi(p)\sigma(\psi(p)) = \sigma(\psi(p))p$. In the sequel, we assume that ψ is Lipschitz.

Part I: We start with the sub-solution property. Let $(t_0, x_0) \in [0, T) \times \mathbb{R}$. Assume that

$$- Fp(t_0, x_0) > 0 . \qquad (6.25)$$

5. Define φ by $\varphi(t, x) := p(t, x) + |t - t_0|^2 + |x - x_0|^4$. Show that (6.25) implies $-F\varphi > 0$ on the ball $B_\varepsilon(t_0, x_0)$ with center (t_o, x_o) and radius ε, for some $\varepsilon > 0$.

Let $v_0 = p(t_0, x_0) - (\varepsilon \wedge \zeta)/2$ with $-\zeta := \max_{\partial B_\varepsilon(t_0, x_0)} v - \varphi < 0$. Let (S^0, V^0) be the solution on $[t_o, \theta]$ of

$$S_t^0 = x_0 + \int_{t_0}^t \sigma\left(\psi(\partial\varphi(s, S_s^0))\right) dW_s ,$$

$$V_t^0 = v_0 + \int_{t_0}^t \psi(\partial\varphi(s, S_s^0))\sigma\left(\psi(\partial\varphi(s, S_s^0))\right) dW_s ,$$

with $\theta := \inf\{s \geq t_0 : (s, S_s^0) \notin B_\varepsilon(t_0, x_0) \text{ or } |V_s^0 - \varphi(s, S_s^0)| \geq \varepsilon\}$.

6. Show that $V_\theta^0 - v(\theta, S_\theta^0) \geq V_\theta^0 - \varphi(\theta, S_\theta^0) \geq -(\varepsilon \wedge \zeta)/2 > -\varepsilon$.
7. Deduce that $V_\theta^0 - \varphi(\theta, S_\theta^0) \geq \varepsilon > 0$ if $|V_\theta^0 - \varphi(\theta, S_\theta^0)| \geq \varepsilon$.
8. Deduce also that $V_\theta^0 - v(\theta, S_\theta^0) \geq V_\theta^0 - \varphi(\theta, S_\theta^0) + \zeta \geq \zeta/2 > 0$ if $(\theta, S_\theta^0) \in \partial B_\varepsilon(t_0, x_0)$.
9. Conclude that $V_\theta^0 - v(\theta, S_\theta^0) > 0$.
10. Deduce from the previous point that (6.25) cannot be satisfied.

Part II: We now prove the super-solution property.
Let $(t_0, x_0) \in [0, T) \times \mathbb{R}$ and assume that

$$- Fp(t_0, x_0) < 0 . \qquad (6.26)$$

We define $\varphi(t, x) = p(t, x) - |t - t_0|^2 - |x - x_0|^4$ and admit that the previous point implies

$$-\mathcal{L}^a \varphi < -\eta \qquad (6.27)$$

for $(t, x, a) \in B_\varepsilon(t_0, x_0) \times \mathbb{R}$ s.t. $|a\sigma(a) - \partial\varphi(t, x)\sigma(a)| \leq \varepsilon$,

for some $\varepsilon, \eta > 0$. Let $\phi \in \mathcal{A}$ and $(S^0, V^0) := (S^{t_0, x_0, \phi}, V^{t_0, x_0, v_0, \phi})$ for $v_0 := p(t_0, x_0) + (\zeta \wedge \varepsilon)/2$ with

$$\zeta := \min_{\partial B_\varepsilon(t_0, x_0)} v - \varphi > 0 .$$

Moreover, let $\theta := \inf\{s \geq t_0 : (s, S_s^0) \notin B_\varepsilon(t_0, x_0) \text{ or } |V_s^0 - \varphi(s, S_s^0)| \geq \varepsilon\}$. For a bounded predictable process λ, we define the local martingale L,

$$L_t := 1 - \int_{t_0}^{t \wedge \theta} L_s \lambda_s \delta_s dW_s \quad \text{with } \delta := \phi\sigma(\phi) - \partial\varphi(\cdot, S^0)\sigma(\phi) .$$

11. Show that, for $t \in [t_0, \theta]$,

$$d\left(L_t[V_t^0 - \varphi(t, S_t^0)]\right) = L_t\left(-\mathcal{L}^{\phi_t}\varphi(t, S_t^0) - \lambda_t|\delta_t|^2\right) dt + \gamma_t dW_t$$

with $\gamma := L\delta(1 - \lambda[V^0 - \varphi(\cdot, S^0)])$.
12. Deduce from (6.27) and (6.22) that there exists λ such that, on $[t_0, \theta]$,

$$d\left(L_t[V_t^0 - \varphi(t, S_t^0)]\right) \leq \gamma_t dW_t.$$

13. Deduce from the previous point and (6.22) that $L(V^0 - \varphi(\cdot, S^0))$ is a super-martingale on $[t_0, \theta]$.
14. Applying the geometric dynamic programming principle, we can find $\phi \in \mathcal{A}$ such that

$$V_\theta^0 - p(\theta, S_\theta^0) \geq 0 .$$

Show that this implies $V_\theta^0 - \varphi(\theta, S_\theta^0) \geq (\varepsilon \wedge \zeta)$.
15. Deduce that $L_\theta\left(V_\theta^0 - \varphi(\theta, S_\theta^0)\right) \geq L_\theta(\varepsilon \wedge \zeta)$.
16. Deduce $(\zeta \wedge \varepsilon)/2 \geq \mathbb{E}\left[L_\theta\left(V_\theta^0 - \varphi(\theta, S_\theta^0)\right)\right] \geq \zeta \wedge \varepsilon$.
17. Conclude using the previous step that (6.26) cannot be satisfied.

6.2 (Pricing with different interest rates) We consider a financial market with one risky asset, whose price process satisfies

$$S_t = S_0 + \int_0^t \mu(S_s)S_s ds + \int_0^t \sigma(S_s)S_s dW_s$$

where W is a one-dimensional Brownian motion defined on a probability space $(\Omega, \mathcal{A}, \mathbb{P})$. The function $x \mapsto \mu(x)x$, $\Sigma : x \mapsto \sigma(x)x$ are Lipschitz-continuous and μ, σ^{-1} are bounded. In the market, there are two different interest rates: $r > 0$ is the rate for lending money, and $R > r$ is the rate for borrowing money. The goal of this exercise is to characterise the hedging price of an European option paying $g(S_T)$ at maturity $T > 0$. Here, g is a Lipschitz function.

1. Dynamics of the wealth process. We denote by ϕ the quantity of risky asset held in the portfolio, by α the amount of money held on the cash account and V the value of the wealth process.

   a. Explicit the infinitesimal change of value of the cash account between the date t and $t + dt$ using r and R.

   b. Recall the definition of a self-financing strategy (α, ϕ) and show that

   $$V_t = v + \int_0^t (rV_s + (\mu(S_s) - r)S_s\phi_s - (R - r)[V_s - \phi_s S_s]^-) \, ds + \int_0^t \phi_s \Sigma(S_s) dW_s$$

   where v is the initial wealth. We work here with strategies ϕ belonging to the set $\mathcal{U}$ of predictable processes satisfying $\mathbb{E}\left[\int_0^T |\phi_s \Sigma(S_s)|^2 ds\right] < \infty$.

   c. Compare with the classical framework.

2. In the sequel, we denote $f(x, y, z) = -ry - (\mu(x) - r)z/\sigma(x) + (R - r)[y - z/\sigma(x)]^-$. We observe in particular that the dynamics of the wealth process starting from v and following the strategy ϕ reads

   $$V_t^{v,\phi} = v - \int_0^t f(S_s, V_s^{v,\phi}, \Sigma(S_s)\phi_s) ds + \int_0^t \phi_s \Sigma(S_s) dW_s .$$

   We want to characterise the minimal super-hedging price given by

   $$p := \inf \mathcal{G} \quad \text{with} \quad \mathcal{G} = \{v \in \mathbb{R} \mid \exists \phi \in \mathcal{U}, \ V_T^{v,\phi} \geq g(S_T)\} .$$

   We assume that there exists a unique smooth solution u to the following semi-linear PDE

   $$\partial_t u + \mu(x)x\partial_x u + \frac{1}{2}\Sigma(x)^2 \partial_{xx}^2 u + f(x, u, \Sigma(x)\partial_x u) = 0 ,$$

   with $u(T, x) = g(x)$. In the following, we suppose moreover that u and its derivatives have polynomial growth.

   a. Let $Y_t := u(t, S_t)$ and $Z_t := \Sigma(S_t)\partial_x u(t, S_t)$. Compute the dynamics of Y and give the value of Y_T.

   b. By determining the hedging strategy, deduce that $u(0, S_0) \geq p$.

3. We are now going to prove the converse inequality.

a. Let $v \in \mathcal{G}$ and $V^{v,\phi}$ be the associated wealth process such that $V_T^{v,\phi} \geq g(S_T)$. We set $\delta := V^{v,\phi} - Y$ and $\beta := \phi\Sigma(S) - Z$. Show that

$$\delta_t = \delta_T + \int_t^T (a_s\delta_s + b_s\beta_s)ds - \int_t^T \beta_s dW_s ,$$

for some bounded processes a and b to determine.

b. Let Γ be the process given by

$$\Gamma_t = e^{\int_0^t (a_s - \frac{1}{2}b_s^2)ds + \int_0^t b_s dW_s} .$$

Compute the dynamics of $(\Gamma_t\delta_t)_{t\in[0,T]}$.

c. Deduce from the previous question the sign of δ_0 and conclude.

6.3 (Explicit resolution of the quantile hedging problem) We work in the setting of Example 6.20. In this exercise, we are going to show how to obtain an explicit solution to the quantile hedging problem using a Fenchel transform approach. We assume that $g \geq 0$, and that the function p is continuous, C^∞ and strictly convex with respect to m on $[0, T) \times (0, \infty) \times (0, 1)$.

We define the Fenchel transform of the price p with respect to m:

$$\tilde{p}(t, x, q) := \sup_{m \in (-\infty, 1]} (mq - p(t, x, m)) \tag{6.28}$$

for $(t, x, q) \in [0, T] \times (0, \infty) \times \mathbb{R}$. Note that $p(t, x, m) = \infty$ if $m > 1$, which explains why we restrict the supremum to m lower than 1 in the previous definition.

1. We start by studying the Fenchel transform to obtain a PDE characterisation. We set $(t, x, q) \in [0, T) \times (0, \infty) \times \mathbb{R}$.

a. Assume that the maximum in (6.28) is reached at $m \in (0, 1)$.

   i. Show that $m = J(t, x, q)$ where J is the inverse with respect to m of $\partial_m p$.
   ii. Deduce that

$$\tilde{p}(t, x, q) = qJ(t, x, q) - p(t, x, J(t, x, q)).$$

   iii. Compute the partial derivatives of order 1 and 2 of p using the ones of $\tilde{p}$.
   iv. Using (6.19), show that at (t, x, q) we have

$$-\partial_t\tilde{p} - \frac{1}{2}\sigma^2 x^2 \partial_{xx}^2\tilde{p} - \frac{\mu}{\sigma}qx\sigma\partial_{xq}^2\tilde{p} - \frac{\mu^2}{2\sigma^2}q^2\partial_{qq}^2\tilde{p} = 0. \tag{6.29}$$

b. Assume now that the maximum is reached at $m \in \{0, 1\}$ and that the maximum is reached at the same point in a neighbourhood of (t, x, q). Show that $\tilde{p}$ still satisfies the equation (6.29) (use the fact that $p(t, x, 0) = 0$ and that $p(t, x, 1)$ is the hedging price in the Black and Scholes model).

c. Show that $\tilde{p}$ is equal to ∞ for $q < 0$ and 0 for $q = 0$.

2. From now on, we assume that the function $\tilde{p}$ is well defined as soon as $q \geq 0$ and that it satisfies (6.29) if $(t, x, q) \in [0, T) \times (0, \infty)^2$.

a. Using (6.21), show that $\tilde{p}(T, x, q) = [q - g(x)]^+$.
b. Deduce that

$$\tilde{p}(t, x, q) = \mathbb{E}^{\mathbb{Q}_t}\left[[Q_T^{t,q} - g(S_T^{t,x})]^+\right]$$

where $\mathbb{Q}_t$ is the risk neutral measure in the Black and Scholes model starting at t, and $Q^{t,q}$ is the solution to

$$Q^{t,q} = q + \int_t^{\cdot} \frac{\mu}{\sigma} Q_s^{t,q} dW_s^{\mathbb{Q}_t},$$

with

$$W_s^{\mathbb{Q}_t} = W_s + \frac{\mu}{\sigma} s, \quad s \leq T.$$

c. Write $d\mathbb{Q}_t/d\mathbb{P}$ in terms of $Q^{t,q}$.

3. We fix $(t, x, m) \in [0, T) \times (0, \infty)^2$ and recall that

$$p(t, x, m) = \sup_{q>0} (mq - \tilde{p}(t, x, q)) \tag{6.30}$$

since p is convex in m (and $\tilde{p}(t, x, q) = \infty$ if $q < 0$).

a. Show that the supremum is reached in (6.30) at a point $\hat{q} > 0$ (recall that $m \in (0, 1)$).
b. From now on, we assume that $\mathbb{P}\left[Q_T^{t,q} = g(S_T^{t,x})\right] = 0$ for all $q > 0$. Prove that the maximum point $\hat{q}$ satisfies

$$m = \mathbb{E}^{\mathbb{Q}_t}\left[Q_T^{t,1} 1_{\{Q_T^{t,\hat{q}} \geq g(S_T^{t,x})\}}\right].$$

c. Deduce that

$$m = \mathbb{P}[A] \quad \text{where } A := \{Q_T^{t,\hat{q}} \geq g(S_T^{t,x})\}.$$

d. How to compute $\hat{q}$?

4. We are now going to identify the quantile hedging strategy. In the following, we set $v = p(t, x, m)$.

a. Show that

$$v = m\hat{q} - \tilde{p}(t, x, \hat{q}) = \mathbb{E}^{\mathbb{Q}_t}\left[g(S_T^{t,x})\mathbf{1}_A\right].$$

b. Deduce that there exists $\phi \in \mathcal{A}$ such that

$$V_T^{t,v,\phi} = g(S_T^{t,x})\mathbf{1}_A.$$

c. Show that ϕ satisfies

$$\mathbb{P}\left[V_T^{t,v,\phi} = g(S_T^{t,x})\right] = m.$$

d. Conclude.

Corrections

6.1

1. Observe that $\mathbb{E}\left[\int_0^T |\phi_r \sigma(\phi_r)|^2 dr\right] < +\infty$.
2. $Y_T^{t,x,p(t,x),\hat{\phi}} \geq g(S_T^{t,x,\hat{\phi}})$ and the result follows since Y is a martingale.
3. Follows from the martingale representation theorem.
4. No in general. But if $Y_T^{t,x,p(t,x),\hat{\phi}} = g(S_T^{t,x,\hat{\phi}})$, then equality has to hold by the martingale representation theorem.
5. By (6.25) and the definition of φ, we have $-F\varphi(t_0, x_0) > 0$. The result then follows by continuity of F, ψ, φ and its derivatives.
6. From the definition of φ, $Y_\theta^0 - p(\theta, S_\theta^0) \geq Y_\theta^0 - \varphi(\theta, S_\theta^0)$ and obviously $-\frac{\epsilon \wedge \zeta}{2} > -\epsilon$. To prove the middle inequality, apply Itô's formula to get

$$Y_\theta^0 - \varphi(\theta, S_\theta^0) = y_0 - \varphi(t_0, x_0) - \int_{t_0}^\theta F\varphi(s, S_s^0)ds > -\frac{\epsilon \wedge \zeta}{2} \qquad (6.31)$$

from 5.
7. Straightforward from 6: $Y_\theta^0 - \varphi(\theta, S_\theta^0) > -\epsilon$.
8. $Y_\theta^0 - p(\theta, S_\theta^0) = Y_\theta^0 - \varphi(\theta, S_\theta^0) + \varphi(\theta, S_\theta^0) - p(\theta, S_\theta^0) \geq Y_\theta^0 - \varphi(\theta, S_\theta^0) + \zeta$ and the result follows from 6.
9. Combine 7. and 8.
10. 9. contradicts the GDPP.
11. Apply Itô's formula.
12. Set $\lambda_t = -\mathcal{L}^{\phi_t}\varphi(t, S_t^0)\mathbf{1}_{|\delta_t|>\epsilon}/|\delta_t|^2$ for $t \in [t_0, \theta]$.
13. The drift is always non-positive and the integrability requirement comes from (6.22).

14. Observe that $\theta = \theta^1 \wedge \theta^2$ with θ^1 the exit time of $B_\epsilon(t_0, x_0)$ and

$$\theta^2 = \inf\{s \geq t_0 \mid |Y_s^0 - \varphi(s, S_s^0)| \geq \epsilon\} \wedge T.$$

We compute

$$Y_\theta^0 - \varphi(\theta, S_\theta^0) = Y_\theta^0 - p(\theta, S_\theta^0) + p(\theta, S_\theta^0) - \varphi(\theta, S_\theta^0) \geq p(\theta, S_\theta^0) - \varphi(\theta, S_\theta^0) \geq 0,$$

the last inequality coming from the definition of φ.

This implies: $Y_\theta^0 - \varphi(\theta, S_\theta^0) = \epsilon$ if $\theta = \theta^2$, and $p(\theta, S_\theta^0) - \varphi(\theta, S_\theta^0) \geq \zeta$ otherwise.

15. Observe that L is an exponential martingale.
16. Using the supermartingale property, we obtain

$$\frac{\zeta \wedge \epsilon}{2} = y_0 - \varphi(t_0, x_0) \geq \mathbb{E}\left[L_\theta \left(Y_\theta^0 - \varphi(\theta, S_\theta^0)\right)\right].$$

From the previous question and the fact that $\mathbb{E}[L_\theta] = 1$, we get

$$\mathbb{E}\left[L_\theta \left(Y_\theta^0 - \varphi(\theta, S_\theta^0)\right)\right] \geq \zeta \wedge \varepsilon.$$

17. $\zeta > 0$ and $\epsilon > 0$, the previous question yields then an obvious contradiction.

6.2

1. a. $d\alpha_t = (\alpha_t r \mathbf{1}_{\{\alpha_t > 0\}} + \alpha_t R \mathbf{1}_{\{\alpha_t < 0\}})dt = r\alpha_t dt - (R - r)[\alpha_t]^- dt$
   b. (α, ϕ) self-financing means $dV_t = d\alpha_t + \phi_t dS_t$. Using the previous question, we have

$$dV_t = r\alpha_t dt - (R - r)[\alpha_t]^- dt + \phi_t dS_t$$

   then observing that $\alpha_t = V_t - \phi_t S_t$ yields the result.
   c. Dynamics are non linear in terms of V and ϕ, but the wealth at time t still only depends on v and ϕ, the quantity of risky asset.
2. a. Apply Itô's formula and use the PDE satisfied by u to get

$$Y_t = g(S_T) + \int_t^T f(S_s, Y_s, Z_s)ds - \int_t^T Z_s dW_s.$$

   b. Set $\phi_s = Z_s / \Sigma(S_s)$ and observe that $u(0, S_0)$ is indeed a (super-)hedging price. Thus $u(0, S_0) \geq p$.

3. a. For $t \in [0, T]$, set

$$a_t = \frac{f(S_t, V_t^{v,\phi}, \Sigma(S_t)\phi_t) - f(S_t, Y_t, \Sigma(S_t)\phi_t)}{\delta_t} \mathbf{1}_{\{\delta_t \neq 0\}},$$

and $\quad b_t = \frac{f(S_t, Y_t, \Sigma(S_t)\phi_t) - f(S_t, Y_t, Z_t)}{\beta_t} \mathbf{1}_{\{\beta_t \neq 0\}}.$

These processes are bounded as f is Lipschitz in y and z.

b. Applying Ito's formula, we compute $d(\Gamma_t \delta_t) = (\beta_t + \delta_t b_t)\Gamma_t dW_t$.

c. Observe that $\delta_0 = \mathbb{E}[\Gamma_T \delta_T] \geq 0$ as $\delta_T \geq 0$ by assumption. Thus $v \geq u(0, S_0)$. This being true for all $v \in \mathcal{G}$, we obtain $p \geq u(0, S_0)$. From 2.b, we thus conclude $p = u(0, S_0)$.

6.3

1. a.   i. From the first order condition in the optimisation problem, we get $q = \partial_m p(t, x, m)$.
      ii. At the optimum, we have $\tilde{p}(t, x, q) = mq - p(t, x, m)$ and the result follows from the previous question.
     iii. Direct computations.
      iv. Insert the expressions obtained in the previous question into (6.19).

   b. Say the maximum is reached at $m = 0$, in a neighbourhood $\mathcal{V}$ of (t, x, q). On $\mathcal{V}$, we thus have $\tilde{p}(t, x, q) = 0$ and it satisfies obviously (6.29).
   Let us denote $\bar{p}(t, x) = p(t, x, 1)$ the hedging price in the Black-Scholes model. We then have, if the maximum is reached at $m = 1$, in a neighbourhood $\mathcal{V}$ of (t, x, q), $\tilde{p}(t, x, q) = q - \bar{p}(t, x)$. From this, we compute $\partial_t \tilde{p} = -\partial_t \bar{p}$, $\partial_{xx}^2 \tilde{p} = -\partial_{xx}^2 \bar{p}$, $\partial_{xq}^2 \tilde{p} = \partial_{qq}^2 \tilde{p} = 0$. We thus have

$$-\partial_t \bar{p} - \frac{1}{2}\sigma^2 x^2 \partial_{xx}^2 \bar{p} - \frac{\mu}{\sigma} qx\sigma \partial_{xq}^2 \bar{p} - \frac{\mu^2}{2\sigma^2} q^2 \partial_{qq}^2 \bar{p} = -\partial_t \bar{p} - \frac{1}{2}\sigma^2 x^2 \partial_{xx}^2 \bar{p}.$$

   But the right hand side is equal to 0 as $\bar{p}$ is the hedging price in the Black Scholes model.

   c. Since $p(t, x, m) \geq p(t, x, 0)$, $\tilde{p}(t, x, 0) = 0$ if $q = 0$. Since $p(t, x, m) = 0$ for $m \leq 0$, $\tilde{p}(t, x, q) = +\infty$ for $q < 0$.

2. a. From (6.21), we have $p(T, x, m) = mg(x)$. The result follows from direct computations.

   b. This follows from Feynman-Kac theorem, applied to (6.29) and the previous question.

   c. $d\mathbb{Q}_t/d\mathbb{P} = (Q^{t,1})^{-1}$.

3. a. For $m \in (0, 1)$, $\tilde{p}(t, x, q) - mq = \mathbb{E}^{\mathbb{Q}}\left[[qQ^{t,1} - g(S_T^{t,x})]^+ - mqQ_T^{t,1}\right] \to \infty$
   as $q \to \infty$ (by Fatou's Lemma), and $mq - \tilde{p}(t, x, q) \to 0$ as $q \to 0$ (by
   dominated convergence theorem). Since p is strictly convex (by assumption)
   and non-decreasing in m, and since $p(t, x, 0) = 0$, we must have $p(t, x, m) > 0$
   and therefore $\hat{q} \in (0, 1)$.

  b. At the maximum, $m = \partial_q \tilde{p}(t, x, \hat{q}) = \mathbb{E}^{\mathbb{Q}_t}\left[Q_T^{t,1} \mathbf{1}_{\{Q_T^{t,\hat{q}} \geq g(S_T^{t,x})\}}\right]$ (this last
   equality is obtained by differentiating under the expectation $q \mapsto \tilde{p}(t, x, q) = \mathbb{E}^{\mathbb{Q}_t}\left[[Q_T^{t,q} - g(S_T^{t,x})]^+\right]$).

  c. This follows from 2.c.

  d. Solve the equation given in 3.c.

4. a. From 3.a,

$$p(t, x, m) = m\hat{q} - \tilde{p}(t, x, \hat{q}) = \mathbb{E}^{\mathbb{Q}_t}\left[Q_T^{t,\hat{q}} \mathbf{1}_{\{Q_T^{t,\hat{q}} \geq g(S_T^{t,x})\}} - [Q_T^{t,\hat{q}} - g(S_T^{t,x})]^+\right],$$

   the second equality coming from 3.b. Simple algebra concludes the proof.

  b. Use the representation theorem applied to the option whose payoff is $g(S_T^{t,x})\mathbf{1}_A$.

  c. From 4.b, $\{V_T^{t,v,\phi} = g(S_T^{t,x})\} = A$.

  d. ϕ allows indeed to hedge with probability m.

Part III
Practical Implementation in Local and Stochastic Volatility Models

Chapter 7
Local Volatility Models

We present here the main characteristics of local volatility models in which the volatility of the risky assets is a function of time and of the *spot* value of the underlying. It is a standard in the industry. They are flexible enough to fit the vanilla option prices of all maturities, while preserving the completeness of the market. This permits a clear identification of the hedging strategy, see Chap. 4.

7.1 Black and Scholes Model and Implicit Volatility

The *Black and Scholes model* corresponds, in its multidimensional version, to the model presented in the Exercise 2.4. We simply recall that, when $d = n = 1$, the price of the risky asset solves

$$S_t = S_0 + \int_0^t S_s \mu ds + \int_0^t S_s \sigma dW_s \ , \ t \in \mathbb{T}, \tag{7.1}$$

in which μ and σ are constants, and that the risk-free interest rate is a constant r.

The Black and Scholes model is very popular because it allows to compute the price of a large class of European option explicitly. In theory, it should be easily calibrated[1] since the only important parameters is the volatility (the interest rate is given by the market).

However, the assumption of a constant volatility is not consistent with market data. More precisely, assume that the risk-free interest rate r is fixed and that we observe the prices C_i's of European calls of maturity T_i and of strike k_i, $i = 1, \ldots, I$,

---

[1] Find the values of the parameters which better fit the prices of the options listed in the market.

© Springer International Publishing Switzerland 2016
B. Bouchard, J.-F. Chassagneux, *Fundamentals and Advanced Techniques in Derivatives Hedging*, Universitext, DOI 10.1007/978-3-319-38990-5_7

written on the same underlying S of spot price S_0 today. It should hold that

$$C_i = BS(0, S_0, k_i, T_i, \sigma) \quad \text{for all } i = 1, \ldots, I$$

where σ is the volatility and $BS(t, x, k, T, \sigma)$ is the price at time t in the Black and Scholes model of a European call option of maturity T and of strike k when $S_t = x$.

In particular, it is possible to infer from the above the volatility parameter σ by inverting the Black and Scholes formula, see Exercise 2.4, i.e. find σ such that

$$C_i = BS(0, S_0, k_i, T_i, \sigma).$$

The solution $\sigma_{\text{imp}}(T_i, k_i)$ of this equation is called *implicit volatility*. In the Black and Scholes model, the implicit volatility should not depend on i. But, this is not true in practice. In place, for a same maturity, the implicit volatility is not constant in terms of the strike. It is often U shaped, which refers to the presence of a *volatility smile*.

One way to solve this issue is to consider a *local volatility model*, namely to allow σ to be a function of t and S_t:

$$dS_t/S_t = rdt + \sigma(t, S_t)dW_t$$

where W is a Brownian motion under the risk-neutral measure $\mathbb{Q}$.

In this case, it is necessary to calibrate the volatility, i.e. to look for a local volatility function σ such that the theoretical prices match the one listed on the market, typically call option prices.

7.2 Local Volatility Surface

In this section, we briefly describe common ways of computing the local volatility map $(t, x) \mapsto \sigma(t, x)$.

7.2.1 Dupire's Approach

The first approach is due to [26]. We assume from now on that the law of S_τ given $S_t = x > 0$ admits a smooth density $f(\tau, y; t, x)$ under $\mathbb{Q}$ for $t < \tau \leq T$. We first provide a partial differential equation satisfied by f.

7.2.1.1 Fokker-Planck Equation

Proposition 7.1 *Fix $(t,x) \in [0,\infty)\times(0,\infty)$. If $(\sigma, f(\cdot;t,x)) \in C^{1,2}((t,\infty)\times(0,\infty))$ then $f(\cdot;t,x)$ satisfies*

$$\partial_\tau f(T,y;t,x) = -\partial_y \left[ryf(T,y;t,x)\right] + \frac{1}{2}\partial_y^2 \left[\sigma^2(T,y)y^2 f(T,y;t,x)\right] \tag{7.2}$$

on $(t,\infty)\times(0,\infty)$.

The above equation is called *Fokker-Planck equation* or *backward Kolmogorov's equation*.

Proof Fix $g \in C_b^\infty$ with compact support contained in $(0,\infty)$. Then $(v(s,S_s) := \mathbb{E}^Q[g(S_T) \mid \mathcal{F}_s])_{s\leq T}$ is a martingale. This implies that, for all $t \leq \tau < \tau+\varepsilon \leq T$,

$$\mathbb{E}^Q\left[v(\tau+\varepsilon, S_{\tau+\varepsilon}) - v(\tau, S_\tau) \mid S_t = x\right] = 0 .$$

Since f is smooth and $g \in C_b^\infty$, v is smooth, and by Itô's lemma

$$\int_\tau^{\tau+\varepsilon} \int \mathcal{L}v(s,y)f(s,y;t,x)dyds = 0 ,$$

where

$$\mathcal{L}v(t,x) := \partial_t v(t,x) + rx\partial v(t,x) + \frac{1}{2}\sigma(t,x)^2 x^2 \partial^2 v(t,x) .$$

By assuming that σ is bounded,[2] one have

$$\lim_{y\to 0} v(s,y) = \lim_{x\to 0} g(x) = 0 \text{ and } \lim_{y\to\infty} v(s,y) = \lim_{x\to\infty} g(x) = 0$$

since the support of g is compact and contained in $(0,\infty)$. By integrating by parts, we then obtain

$$0 = \int \left([v(s,y)f(s,y;t,x)]_\tau^{\tau+\varepsilon} - \int_\tau^{\tau+\varepsilon} v(s,y)\partial_\tau f(s,y;t,x)ds\right) dy$$
$$- \int_\tau^{\tau+\varepsilon}\int \left(v(s,y)\partial_y(ryf(s,y;t,x))\right) dyds$$
$$+ \int_\tau^{\tau+\varepsilon}\int \left(\frac{1}{2}v(s,y)\partial_y^2(\sigma^2(s,y)y^2 f(s,y;t,x))\right) dyds .$$

---

[2]Otherwise, we use a localisation argument.

We now observe that

$$\int [v(s,y)f(s,y;t,x)]_\tau^{\tau+\varepsilon} dy$$

$$= \int \left(v(\tau+\varepsilon,y)f(\tau+\varepsilon,y;t,x) - v(\tau,y)f(\tau,y;t,x) \right) dy$$

$$= v(t,x) - v(t,x) = 0 .$$

By dividing by ε the above equation and by sending this parameter to 0, we get

$$0 = \int v(\tau,y) \left(\partial_y f(\partial_k^2 \tau, y; t, x) + \partial_y(ryf(\tau,y;t,x)) - \frac{1}{2}\partial_y^2(\sigma^2(\tau,y)y^2 f(\tau,y;t,x)) \right) dy .$$

By sending $\tau \to T$, so that $v(\tau,y) \to g(y)$, we obtain

$$0 = \int g(y) \left(\partial_\tau f(T,y;t,x) + \partial_y(ryf(T,y;t,x)) - \frac{1}{2}\partial_y^2(\sigma^2(T,y)y^2 f(T,y;t,x)) \right) dy .$$

Since g is arbitrary, this implies that f solves (7.2). $\square$

7.2.1.2   Dupire's Formula and Equation

We now use the above result to obtain the *Dupire's equation*. If g is the payoff of a European call of maturity τ, strike k and of price $\Pi_C(0,x;\tau,k)$, then

$$\Pi_C(0,x;\tau,k) = e^{-r\tau} \int_k^\infty (y-k)f(\tau,y;0,x)dy .$$

If Π_C is smooth in its two last variables, we then obtain, by differentiating and then using (7.2), that

$$\partial_\tau \Pi_C(0,x;\tau,k) = -r\Pi_C(0,x;\tau,k) + e^{-r\tau}\int_0^\infty (y-k)^+ \partial_\tau f(\tau,y;0,x)dy$$

$$= -r\Pi_C(0,x;\tau,k)$$

$$+ e^{-r\tau} \int_0^\infty (y-k)^+ \left(-\partial_y(ryf(\tau,y;0,x)) \right) dy$$

$$+ e^{-r\tau}\int_0^\infty (y-k)^+ \left(\frac{1}{2}\partial_y^2(\sigma^2(\tau,y)y^2 f(\tau,y;0,x)) \right) dy .$$

Integrating by parts and using the notation d for the Dirac mass at 0 leads to

$$\partial_\tau \Pi_C(0, x; \tau, k) = -re^{-r\tau} \int_0^\infty (y - k) \mathbf{1}_{\{y \geq k\}} f(\tau, y; 0, x) dy$$

$$+ e^{-r\tau} \int_0^\infty ry \mathbf{1}_{\{y \geq k\}} f(\tau, y; 0, x) dy$$

$$+ e^{-r\tau} \left(\int_0^\infty d(y - k) \frac{1}{2} \sigma^2(\tau, y) y^2 f(\tau, y; 0, x) \right) dy$$

$$= -rk \partial_k \Pi_C(0, x; \tau, k) + \frac{1}{2} \sigma^2(\tau, k) k^2 \partial_k^2 \Pi_C(0, x; \tau, k)$$

because

$$\partial_k \Pi_C(0, x; \tau, k) = \int \mathbf{1}_{\{y \geq k\}} f(\tau, y; 0, x) dy \,,$$

$$\partial_k^2 \Pi_C(0, x; \tau, k) = \int d(y - k) f(\tau, y; 0, x)) dy \,.$$

Hence, if the price $\Pi_C(0, x; \tau, k)$ is smooth with respect to (τ, k), it satisfies *Dupire's equation* (7.3) below.

Proposition 7.2 *Under the conditions of Proposition 7.1, the price $\Pi_C(0, x; \cdot)$ satisfies*

$$\partial_\tau \Pi_C(0, x; \tau, k) = -rk \partial_k \Pi_C(0, x; \tau, k) + \frac{1}{2} \sigma^2(\tau, k) k^2 \partial_k^2 \Pi_C(0, x; \tau, k) \qquad (7.3)$$

for $(\tau, k) \in (0, T] \times (0, \infty)$, with initial condition $\Pi_C(0, x; 0, k) = [x - k]^+$ at $\tau = 0$ and boundary condition $\Pi_C(0, x; \tau, 0) = x$ for all $\tau \in [0, T]$.

This equation has two important consequences:

1. By solving (7.3), one can compute call prices for all maturities τ and strike k at the same time. This is very important from the calibration point of view!
2. The local volatility σ must satisfies *Dupire's formula*:

$$\sigma^2(\tau, k) = 2 \frac{\partial_\tau \Pi_C(0, x; \tau, k) + rk \partial_k \Pi_C(0, x; \tau, k)}{k^2 \partial_k^2 \Pi_C(0, x; \tau, k)} \,. \qquad (7.4)$$

If we knew all the call prices for all (τ, k) in $(0, T] \times (0, \infty)$, one would be able to compute the corresponding local volatility by the above. This local volatility model would perfectly match the vanilla option prices of all maturities. In practice, only few prices are available but we can interpolate the observed price to *estimate* what the other values should be. Since (7.4) is based on the derivatives of this price, this procedure can be rather unstable, and very sensitive to the choice of the interpolation procedure.

Remark 7.3 (Possible extensions) A pure probabilistic proof of the above can be found in [38]. It is based on a time reversion technique for stochastic differential equation. It can also be applied to jump diffusion models. In the same paper, one can find extensions to digital option or barrier options, see also [50].

7.2.2 Calibration of the Volatility Curve on a Finite Number of Calls

Another technique consists in choosing a parametrised family $\{\sigma(a), a \in A \subset \mathbb{R}^M\}$ and to try to approximate a finite number of calls, e.g. by minimising

$$\min_a \sum_{i=1}^{I} \omega_i |\Pi_i(\sigma(a)) - C_i|^2 \,,$$

where the ω_i are positive weights and $\Pi_i(\sigma) := e^{-rT_i}\mathbb{E}^{\mathbb{Q}}\left[[S_{T_i} - k_i]^+\right]$.
There are many natural choices for the weights:

1. They can put more weights on the more liquid options.
2. They can also be used to give as much importance to far of the money options for which C_i is much less than the values of deep in the money options. One way to proceed is to take

$$\omega_i = 1/C_i^2,$$

which amounts to looking at the relative error.
3. They can also depend on the level of precision of *bid/ask* intervals of each maturity/*strike*. Let us denote by C_i^b and C_i^a the bid and ask prices. The calibration error should not exceed $C_i^a - C_i^b$. It is therefore natural to use the weight

$$\omega_i = 1/(C_i^a - C_i^b)^2.$$

4. Finally, one can take

$$\omega_i = 1/(\mathcal{V}ega(k_i, T_i, \sigma_{\mathrm{imp}}(T_i, k_i)))^2$$

where $\mathcal{V}ega$ is the derivative of the Black and Scholes price with respect to the volatility. A first order Taylor expansion shows that this essentially amounts to looking at the error in terms of implicit volatility.

However, this alone can lead to unstable results and very irregular curves. In order to correct this problem, it is often necessary to add a penalty term to force the solution to be more regular.

7.2.2.1 Tikhonov's Regularisation

In [21], see also the references therein, it is suggested to use a Tikhonov's regularisation and to solve

$$\sum_{i=1}^{I} \omega_i |\Pi_i(\sigma(a)) - C_i|^2 + \alpha_1 \|\sigma(a) - \sigma(a_0)\|_H^2 + \alpha_2 \|\partial_a \sigma(a) - \partial_a \sigma(a_0)\|_H^2$$

where ω_i are positive weights, $\alpha_1, \alpha_2 > 0$ are given, a_0 is a fixed parameter which serves as a reference, and

$$\|h\|_H^2 := \int_0^T \int_0^\infty |h(t,x)|^2 dxdt .$$

The first penalty term aims at stabilising the solution, the second at ensuring more regularity.

In order to minimise this quantity, it is natural to use the gradient method. The computation of the gradient of the penalty term depends on the parametrisation but is in general not difficult. The problem is to estimate the gradient of $\sum_{i=1}^{I} \omega_i |\Pi_i(\sigma(a)) - C_i|^2$, i.e.

$$2 \sum_{i=1}^{I} \omega_i \partial_a \Pi_i(\sigma(a)) (\Pi_i(\sigma(a)) - C_i) . \tag{7.5}$$

For this, one can:

- Use finite difference methods: Given $a \in A$, all the $\Pi_i(\sigma(a))$, $i = 1, \ldots, I$, are computed by solving (7.3). This provides estimates of $\partial_a \Pi_i(\sigma(a))$ used to estimate the gradient of (7.5).
- Use a tree approximation: In [21], the author proposes to estimate $(\Pi_i(\sigma(a)))_{i \leq I}$ by using a trinomial tree discretisation of (7.3) and to compute $(\partial_a \Pi_i(\sigma(a)))_{i \leq I}$ by deriving the numerical estimate. In this approach, a is the local volatility at each node of the tree. One can then use a gradient approach to minimise

$$\sum_{i=1}^{I} \omega_i |\tilde{\Pi}_i(\sigma(a)) - C_i|^2 + \alpha_1 \|\sigma(a) - \sigma(a_0)\|_H^2 + \alpha_2 \|\partial_a \sigma(a) - \partial_a \sigma(a_0)\|_H^2 .$$

Then, the local volatility is interpolated between the nodes to provide a whole curve.

7.2.2.2 Entropy Regularisation

In [3], the author suggests to look for the value of the parameter a satisfying

$$\Pi_i(\sigma(a)) = C_i, \quad \forall\, i \le I,$$

which minimises

$$\mathbb{E}^a\left[\int_0^T \eta(\sigma(a)_s^2 - \sigma_0^2)ds\right]$$

where $\mathbb{E}^a$ is the expectation under the law of S when its volatility is given by $\sigma(a)$, σ_0 is a constant, and η is a positive function, that is coercive, strictly convex, and reaches its minimum at 0. Typically, $\eta(y) = y^2$. This leads to optimising over $\lambda \in \mathbb{R}^I$ and $a \in A$ the Lagrangian

$$V(0, S_0; \lambda, \sigma(a)) := \mathbb{E}^a\left[-\int_0^T \eta(\sigma(a)_s^2 - \sigma_0^2)ds\right] + \sum_{i=1}^I \lambda^i(\Pi_i(\sigma(a)) - C_i).$$

We look for a solution to

$$\inf_{\lambda \in \mathbb{R}^I} \sup_{a \in A} V(0, S_0; \lambda, \sigma(a)). \tag{7.6}$$

When $\{\sigma(a), a \in A\}$ is the set of processes of the form $\sigma(t, S_t)$ with values in a set of the form $[\sigma_{min}, \sigma_{max}]$, one can use a dynamic programming argument to obtain a characterisation of the corresponding value function in terms of a Hamilton-Jacobi-Bellman equation, from which the local volatility curve $\hat\sigma$ can be inferred, see [3].

This approach gives good results in terms of calibration. Unfortunately, no criteria imposes to $\hat\sigma$ to be smooth and the curves obtained by this procedure are typically quite irregular.

7.2.3 The Recalibration Issue

In practice, a model is calibrated every day or week. Each time, a new calibration is performed, the parameters used for the hedging are changed as well, and so is the hedging strategy. This can lead to losses (or gains). In any case, one should be careful to choose a model that is stable enough so as to minimise this effect.

Otherwise, one needs to take the recalibration into account in the hedging procedure, as explained in Exercise 7.5 below.

7.3 Impact of the Gamma on the Hedging

The *gamma* is the second derivative of the price with respect to the spot price. Hence, it measures the sensitivity of the *delta*, which provides the hedging strategy, to the underlying price moves. In order to limit the number of times the portfolio needs to be re-adjusted, which has a cost in practice, one should try to keep it small. It also governs the quality of the hedge in case of a misspecification of the model (which is always the case in practice).

7.3.1 Impact of a Volatility Misspecification

We assume here that the true volatility σ is a square integrable predictable process and that

$$S_t = S_0 + \int_0^t S_s r ds + \int_0^t S_s \sigma_s dW_s \ , \ t \in \mathbb{T} \ . \tag{7.7}$$

Suppose that the trader evaluates the price p of an option of payoff g and maturity T by using the local volatility model $\tilde{\sigma}(t, S_t)$. In view of the results of Chap. 4, if p is smooth, it satisfies

$$\partial_t p(t, x) + rx \partial p(t, x) + \frac{1}{2} \tilde{\sigma}(t, x)^2 x^2 \partial^2 p(t, x) = rp(t, x) \ , \tag{7.8}$$

$$p(T, x) = g(x) \ .$$

Moreover, the hedging strategy consists in having a number $\phi_t = \partial p(t, S_t)$ of shares at t. Let us set $V = V^{p(0,S_0),\phi}$. In view of Itô's lemma and the above equation, the dynamics of $Z := V - p(\cdot, S)$ is given by

$$dZ_t = rV_t dt - \left(\partial_t p(t, S_t) + rS_t \partial p(t, S_t) + \frac{1}{2} \sigma_t^2 S_t^2 \partial^2 p(t, S_t) \right) dt$$

$$= rV_t dt - \left(rp(t, S_t) + \frac{1}{2}(\sigma_t^2 - \tilde{\sigma}(t, S_t)^2) S_t^2 \partial^2 p(t, S_t) \right) dt$$

$$= \left(rZ_t + \frac{1}{2}(\tilde{\sigma}(t, S_t)^2 - \sigma_t^2) S_t^2 \partial^2 p(t, S_t) \right) dt \ .$$

Since $p(T, S_T) = g(S_T)$, the hedging error $V_T - g(S_T)$ is

$$Z_T = V_T - p(T, S_T) = \frac{1}{2} \int_0^T e^{r(T-t)} (\tilde{\sigma}(t, S_t)^2 - \sigma_t^2) S_t^2 \partial^2 p(t, S_t) dt \ .$$

We thus observe that:

1. If $\partial^2 p \geq 0$, an over-estimation of the volatility leads to a gain, an under-estimation leads to a loss.
2. It is the opposite if $\partial^2 p \leq 0$.
3. If $\partial^2 p \approx 0$, the quality of the hedge is not very sensitive to the realised volatility. In this case, we say that the hedge is gamma-neutral.

In particular, if we have an uncertainty on the way the volatility should be modelled, one can look for a function $(t, x) \mapsto \tilde{\sigma}(t, x)$ such that the term $(\tilde{\sigma}(t, S_t)^2 - \sigma_t^2)\partial^2 p(t, S_t)$ remains non-negative. If we only know that $(\sigma_t)_t$ takes values in $[\underline{\sigma}, \bar{\sigma}]$, this amounts to take $\tilde{\sigma} = \bar{\sigma}$ when p is convex, and $\tilde{\sigma} = \underline{\sigma}$ when p is concave. Since $\tilde{\sigma}$ and p are linked by (7.8), this means that p should solve

$$rp - \partial_t p - rx\partial p - \frac{1}{2}\left[\bar{\sigma}^2 \mathbf{1}_{\{\partial^2 p\} \geq 0\}} + \underline{\sigma}^2 \mathbf{1}_{\{\partial^2 p\} < 0\}}\right] x^2 \partial^2 p = 0$$

with boundary condition $p(T, \cdot) = g$. We retrieve the *Black-Scholes-Barenblatt equation* of Sect. 5.2.1.
If g is convex, it is easy to see that the solution of the above equation is in fact convex, i.e. p solves

$$rp - \partial_t p - rx\partial p - \frac{1}{2}\bar{\sigma}^2 x^2 \partial^2 p = 0.$$

The hedge is obtained by considering the upper-bound on the volatility. If g is concave, p is concave and the hedge should be based on the lower-bound.

7.3.2   *Impact of Discrete Rebalancing*

In practice, it is obviously impossible to change the composition of the portfolio (to *rebalance* the portfolio) in continuous time, which leads to a hedging error. In this section, we analyse the $\mathbb{L}^2$-error generated when rebalancing only at times $t_i := iT/n$, $i = 0, \ldots, N$. For ease of notations, we restrict to the one-dimensional Black and Scholes model.

We assume that the trader wants to hedge an option of payoff g, assumed to be Lipschitz (hence a.e. differentiable by the Rademacher's theorem), by following the discrete time delta-hedging strategy

$$\hat{\phi}_t = \phi_t \quad \text{for } t \in [t_i, t_{i+1}) \text{ where } \phi := \partial p(\cdot, S)'.$$

The value of his portfolio at T is

$$V_T^{p,\hat{\phi}} = e^{rT}\left(p + \int_0^T \hat{\phi}_s' \text{diag}\left[\tilde{S}_s\right]\sigma dW_s^{\mathbb{Q}}\right).$$

In view of the previous section and the Itô's isometry, the $\mathbb{L}^2(\mathbb{Q})$-hedging error is

$$\text{Err}_N := e^{rT}\left\|e^{-rT}g(S_T) - \tilde{V}_T^{p,\hat{\phi}}\right\|_{\mathbb{L}^2(\mathbb{Q})}$$

$$= e^{rT}\left\|\int_0^T (\phi_s - \hat{\phi}_s)' \text{diag}\left[\tilde{S}_s\right]\sigma dW_s^{\mathbb{Q}}\right\|_{\mathbb{L}^2(\mathbb{Q})}$$

$$= e^{rT}\mathbb{E}^{\mathbb{Q}}\left[\int_0^T \left\|(\phi_s - \hat{\phi}_s)' \text{diag}\left[\tilde{S}_s\right]\sigma\right\|^2 ds\right]^{\frac{1}{2}}.$$

Hence,

$$e^{-2rT}\text{Err}_N^2 \le 2\sum_{i=0}^{N-1}\mathbb{E}^{\mathbb{Q}}\left[\int_{t_i}^{t_{i+1}} \left\|(\phi_s'\text{diag}\left[\tilde{S}_s\right] - \phi_{t_i}'\text{diag}\left[\tilde{S}_{t_i}\right])\sigma\right\|^2 ds\right] + 2B_N$$

$$\le 2\sum_{i=0}^{N-1}\mathbb{E}^{\mathbb{Q}}\left[\int_{t_i}^{t_{i+1}} \left\|\psi_s - \psi_{t_i}\right\|^2 ds\right] + 2B_N$$

where $\psi_s := \phi_s'\text{diag}\left[\tilde{S}_s\right]\sigma$ and

$$B_N := \sum_{i=0}^{N-1}\mathbb{E}^{\mathbb{Q}}\left[\int_{t_i}^{t_{i+1}} \left\|\phi_{t_i}'(\text{diag}\left[\tilde{S}_{t_i}\right] - \text{diag}\left[\tilde{S}_s\right])\sigma\right\|^2 ds\right].$$

We now observe that

$$p(t, S_t) = \mathbb{E}^{\mathbb{Q}}\left[e^{-r(T-t)}g(S_T) \mid \mathcal{F}_t\right]$$

implies (by differentiating with respect to S_t) that

$$e^{-rt}\partial p(t, S_t)\text{diag}\left[S_t\right] = \mathbb{E}^{\mathbb{Q}}\left[e^{-rT}\partial g(S_T)\text{diag}\left[S_T\right] \mid \mathcal{F}_t\right].$$

We deduce that the process given by the right-hand side term is a square-integrable martingale. Otherwise stated $(\psi_t)_{t \le T}$ is a square-integrable $\mathbb{Q}$-martingale. In particular, there exists a square-integrable process η such that

$$\psi_t = \psi_0 + \int_0^t \eta_s' dW_s^{\mathbb{Q}} \ , t \le T,$$

which implies that

$$\mathbb{E}^{\mathbb{Q}}\left[\int_{t_i}^{t_{i+1}} \|\psi_s - \psi_{t_i}\|^2 \, ds\right] = \mathbb{E}^{\mathbb{Q}}\left[\int_{t_i}^{t_{i+1}} \int_{t_i}^{s} \|\eta_v\|^2 \, dv ds\right] \leq \frac{T}{N}\mathbb{E}^{\mathbb{Q}}\left[\int_{t_i}^{t_{i+1}} \|\eta_v\|^2 \, dv\right] .$$

We deduce that

$$e^{-2rT}\mathrm{Err}_N^2 \leq \frac{2T}{N}\mathbb{E}^{\mathbb{Q}}\left[\int_0^T \|\eta_v\|^2 \, dv\right] + 2B_N .$$

It remains to study the term B_N. We have

$$B_N \leq C \max_{i \leq N-1} \mathbb{E}^{\mathbb{Q}}\left[\|\phi_{t_i}\|^4\right]^{\frac{1}{2}} \mathbb{E}^{\mathbb{Q}}\left[\sup_{s \in [t_i, t_{i+1}]} \left\|\tilde{S}_{t_i} - \tilde{S}_s\right\|^4\right]^{\frac{1}{2}}$$

where $C > 0$ is a constant independent of N. We then obtain by the arguments already used above that

$$\max_{i \leq N-1} \mathbb{E}^{\mathbb{Q}}\left[\|\phi_{t_i}\|^4\right]^{\frac{1}{2}} < \infty.$$

It is moreover standard to show that

$$\max_{i \leq N-1} \mathbb{E}^{\mathbb{Q}}\left[\sup_{s \in [t_i, t_{i+1}]} \left\|\tilde{S}_{t_i} - \tilde{S}_s\right\|^4\right]^{\frac{1}{2}} \leq \bar{C}/N$$

where $\bar{C} > 0$ is a constant independent of N.

We finally get

$$\mathrm{Err}_N \leq \hat{C}/\sqrt{N} ,$$

where $\hat{C} > 0$ is a constant independent of N.

Otherwise stated, the error Err_N is of order of $\frac{1}{\sqrt{N}}$. However, it depends on the regularity of ψ through $\mathbb{E}^{\mathbb{Q}}\left[\int_0^T \|\eta_v\|^2 \, dv\right]^{\frac{1}{2}}$. This last term is intimately related to the gamma $\partial^2 p$ of the option price (apply Itô's lemma to ∂p). It can be reduced by using a gamma-hedging strategy, see Exercise 7.3 below.

Looking for the $\mathbb{L}^2$-error under $\mathbb{Q}$ is not a very good criteria since $\mathbb{Q}$ is a pure mathematical construction and does not reflect the trader's view of the future (which is modelled by $\mathbb{P}$). However, one can obtain a similar bound for the error computed under $\mathbb{P}$, but it requires more technical developments. This can also be extended to quite general diffusion models by appealing to the notion of tangent process, see Sect. 4.2.3.

7.4 Example: The CEV Model

The CEV model (or Cox's model) is a local volatility model of the form $\sigma_t = \sigma S_t^\rho$, i.e.

$$dS_t = \mu S_t dt + \sigma S_t^\rho dW_t \tag{7.9}$$

where $\sigma > 0$ and $0 \le \rho < 1$ are given parameters. The term CEV means *Constant Elasticity Variance*:

$$(\partial(\sigma S_t^\rho)/\partial S_t)S_t/(\sigma S_t^\rho) = \rho .$$

Otherwise stated, the elasticity of the volatility σS_t^ρ in terms of S is constant, equal to ρ/σ. In this model, $\sigma_t/S_t = \sigma S_t^{\rho-1}$ is non-increasing as a function of S_t: the conditional variance of the rate or return is non-increasing with respect to the underlying, which matches several empirical studies.

The existence of such a process has been studied, among others, by [25]. Like in the CIR model, see Exercise 7.4 below, this is essentially a squared Bessel process. A squared Bessel process of dimension $d \in \mathbb{N}$ corresponds to the square of the Euclidean norm of a d-dimensional Brownian motion. Here, we will need to consider squared Bessel processes X^δ of dimension $\delta \in \mathbb{R}$, and so δ is possibly not an integer.

Definition 7.4 Fix $\delta \ge 0$ and $x \ge 0$. We call *squared Bessel process* of dimension δ starting from x, the unique strong solution $X^{\delta,x}$ to the equation

$$X_t = x + 2 \int_0^t \sqrt{|X_s|} d\hat{W}_s + \delta t , \; t \ge 0 , \tag{7.10}$$

where $\hat{W}$ is a Brownian motion of dimension 1.

As mentioned above, if W is Brownian motion of dimension $\delta \in \mathbb{N}$ then $\rho := \|W\|$ satisfies

$$\rho_t^2 = 2 \sum_{i=1}^\delta \int_0^t W_s^i dW_s^i + \delta t = 2 \int_0^t \rho_s d\hat{W}_s + \delta t$$

where

$$\hat{W} := \sum_{i=1}^\delta \int_0^{\cdot} (W_s^i/\rho_s) dW_s^i$$

is a Brownian motion, so that $\|W\|^2$ is a squared Bessel process of dimension δ.

The uniqueness of a solution is guaranteed by the following (see [52, Theorem 3.5, Chapter IX]).

Theorem 7.5 *Fix b and a two Borel functions from $\mathbb{R}$ to $\mathbb{R}$ such that b is Lipschitz and*

$$|a(x) - a(y)|^2 \leq \varphi(|x - y|)$$

where φ satisfies $\int_{0+}^{\infty} dz/\varphi(z) = \infty$. Fix W a Brownian motion of dimension 1. Then, the equation

$$X_t = X_0 + \int_0^t b(X_s)ds + \int_0^t a(X_s)dW_s$$

has at most one strong solution.

To conclude to the existence of a strong solution to (7.10) when $\delta \in \mathbb{N}$, we use the standard argument:

1. We have shown the existence of a weak solution, i.e. existence of a couple $(X, \hat{W})$ satisfying (7.10).
2. We have shown that given W, this equation admits at most one solution: We say that pathwise uniqueness holds.
3. We use the fact that the existence of a weak solution combined with the pathwise uniqueness implies existence of a unique strong solution, see [52].

The case $\delta \in \mathbb{R} \setminus \mathbb{N}$ follows from deeper weak existence results that can be found in [52].

Remark 7.6 It is clear that if $\delta = 0$ then the point 0 is absorbing. For $0 < \delta < 2$, one can show that it is reflecting, and that 0 is never reached if and only if $\delta \geq 2$, see [52].

We can now construct the process S. First, we consider the process $X := X^{\delta,x}$ with $x := S_0^{\frac{2}{2-\delta}}$ and call ζ the first hitting time of 0 by X. Given $\nu > 0$ and $\delta < 2$, we consider the time change

$$\tau_t^{\nu,\delta} := \frac{\sigma^2}{2\nu(2-\delta)}(1 - e^{-2\nu t/(2-\delta)}) \,.$$

We then define $Y^{\nu,\delta}$ by

$$Y_t := e^{\nu t}\left(X_{\tau_t^{\nu,\delta} \wedge \zeta}\right)^{1-\delta/2} . \tag{7.11}$$

To go on, we now need a *time change formula*, see [52].

Theorem 7.7 *Let W be a Brownian motion of dimension 1. Fix α a positive adapted increasing and absolutely continuous process such that $\alpha_0 = 0$ and $\mathbb{E}[\alpha_t] < \infty$ for all $t \geq 0$. Then, for all predictable process ξ, $\mathbb{P} - a.s.$ square-integrable,*

$$\int_0^{\alpha_t} \xi_s dW_s = \int_0^t \xi_{\alpha_s} \sqrt{\alpha'_s} d\tilde{W}_s$$

where α' is the density of α, $\tilde{W}$ is defined by

$$\tilde{W}_t = \int_0^{\alpha_t} \sqrt{c_s} dW_s$$

and c is defined by $c_{\alpha_t} := (\alpha'_t)^{-1}$.

By using Itô's formula and Theorem 7.7, we now obtain that $Y^{\nu,\delta}$ satisfies

$$dY_t^{\nu,\delta} = \begin{cases} \nu Y_t^{\nu,\delta} dt + \sigma \left(Y_t^{\nu,\delta} \right)^{\frac{1-\delta}{2-\delta}} dW_t^{(\nu,\delta)} & \text{on } \tau_t^{\nu,\delta} \leq \zeta, \\ 0 & \text{on } \tau_t^{\nu,\delta} > \zeta, \end{cases}$$

where $W^{(\nu,\delta)}$ is the Brownian motion defined by

$$W_t^{(\nu,\delta)} = \int_0^{\tau_t^{\nu,\delta}} \frac{2-\delta}{\sqrt{\sigma^2 - 2\nu(2-\delta)s}} dW_s.$$

If we now take $\delta = \delta_\rho := (1 - 2\rho)/(1 - \rho)$, one can check that Y^{μ,δ_ρ} is a weak solution of (7.9). In view of Theorem 7.5 and the comments just after, there exists a strong solution to (7.9).

Let us observe that $\delta_\rho < 1$, which implies that the probability that S touches 0 is not zero, see Remark 7.6. This is one drawback of this model.

The representation (7.11) provides the law of the process as a transformation of the law of a non central χ^2, whose Fourier transform is known. This permits to obtain quasi-explicit formulae for calls and puts, see [25], which is very useful in terms of calibration, see Sect. 8.3.3. It can also be used to simulate its path exactly on a discrete time grid, see Exercise 7.4 below.

7.5 Problems

7.1 (Forward start option) We consider a one dimensional model 1, i.e. $d = n = 1$. We consider a financial market in which the risk-free interest rate is set to zero, $r = 0$. We assume that there exists a measure $\mathbb{Q} \sim \mathbb{P}$ and a $\mathbb{Q}$-Brownian motion $W^{\mathbb{Q}}$

such that

$$S_t = S_0 + \int_0^t S_s \sigma_s dW_s^{\mathbb{Q}}, \quad t \leq T.$$

The function $t \in [0, T] \mapsto \sigma_t \geq 0$ is deterministic and continuous. We want to evaluate at 0 and hedge an option with payoff $[S_T - \kappa S_1]^+$ where $\kappa > 0$ (we assume $T > 1$). This is a *forward-start* option. We admit that the option can be hedged perfectly and we note

$$p(t, x) := \mathbb{E}^{\mathbb{Q}}[[S_T - \kappa S_1]^+ | S_t = x] \text{ for } (t, x) \in [0, 1] \times (0, \infty).$$

1. By using the fact that $[S_T - \kappa S_1]^+ = S_1[S_T/S_1 - \kappa]^+$, show that

$$p(t, x) = xF(1, \kappa, \int_1^T \sigma_s^2 ds) \text{ if } t \in [0, 1]$$

where, for $y, K, \gamma^2 > 0$,

$$F(y, K, \gamma^2) = \mathbb{E}\left[[ye^Y - K]^+\right] \text{ with } Y \sim \mathcal{N}(-\gamma^2/2, \gamma^2) \text{ under } \mathbb{P}.$$

Here $\mathcal{N}(-\gamma^2/2, \gamma^2)$ denotes the Gaussian law of mean $-\gamma^2/2$ and of variance γ^2.

2. What is the hedging strategy of the option between the dates 0 and 1? (Express the number of units of shares to hold at each date in terms of S and F.)

3. What is the gamma of the option on $[0, 1]$? Are we exposed to a volatility misspecification on this period? If yes, are we exposed to an increase or a decrease of the volatility?

4. Express the hedging price of the option in terms of S_t, S_1 and of the function F for the dates t between 1 and T. Provide the hedging strategy on this period in terms the derivatives of F and S.

5. Are we exposed to a volatility misspecification on the period $[1, T]$? If yes, are we exposed to an increase or a decrease of the volatility?

6. We now assume that the model is given by

$$S_t = S_0 + \int_0^t S_s \sigma(S_s) dW_s^{\mathbb{Q}}, \quad t \leq T,$$

where $x \in (0, \infty) \mapsto x\sigma(x) > 0$ is Lipschitz and bounded, and $W^{\mathbb{Q}}$ is the Brownian motion under the unique risk neutral measure $\mathbb{Q}$.

How is the analysis of question 3 modified on the period $[0, 1]$?

7.2 (Hedging error in a model with stochastic correlation) We consider a financial market made of two risky asset S^1 and S^2 with dynamics

$$S_t^i = S_0^i + \int_0^t S_s^i \sigma_i dZ_s^i \ , \quad i = 1, 2 \ ,$$

where

$$Z_t^1 := W_t^1 \text{ and } Z_t^2 := \int_0^t \rho_s dW_s^1 + \int_0^t (1 - \rho_s^2)^{\frac{1}{2}} dW_s^2 \ , \ \rho_t = f(W_t^3) \ ,$$

for $t \leq T$.

Here $W = (W^1, W^2, W^3)$ is a 3 dimensional $\mathbb{P}$-Brownian motion. The function f is twice differentiable with values in $(0, 1)$ and with derivatives of order 1 and 2 bounded, and $S_0^i, \sigma_i > 0$ for $i = 1, 2$. In the following, we assume that the risk-free interest rate is equal to zero and we denote by $\mathcal{M}$ the set of equivalent probability measures which turn $S = (S^1, S^2)$ into martingales on $[0, T]$.

1. Completeness/incompleteness of the market:

a. Show that if $\mathbb{Q} \in \mathcal{M}$ then

$$\frac{d\mathbb{Q}}{d\mathbb{P}} = \frac{d\mathbb{Q}^\nu}{d\mathbb{P}} := e^{-\frac{1}{2}\int_0^T |\nu_s|^2 ds + \int_0^T \nu_s' dW_s}$$

where $\nu = (\nu^1, \nu^2, \nu^3)$ is a predictable process with values in $\mathbb{R}^3$ satisfying $\int_0^T |\nu_s|^2 ds < \infty$ $\mathbb{P}$-a.s. In the following, we let $\mathcal{U}$ denote the set of predictable processes ν such that $\mathbb{Q}^\nu$ defined above is an equivalent martingale measure.

b. Show that there exists an infinite number of martingale measures.

c. Show that $\nu \in \mathcal{U}$ implies that $\nu^1 = \nu^2 = 0$ $dt \times d\mathbb{P}$-a.e. on $[0, T]$.

d. Deduce that Z^i is a Brownian motion under each $\mathbb{Q} \in \mathcal{M}$, for $i = 1, 2$.

e. Show that $\mathbb{E}^\mathbb{Q}[|S_T^i|^p] < \infty$ for all $p \geq 1$ and that, if $g : \mathbb{R} \mapsto \mathbb{R}$ has polynomial growth,[3] then $\mathbb{E}^\mathbb{Q}[g(S_T^i)]$ does not depend on $\mathbb{Q} \in \mathcal{M}$, $i = 1, 2$.

f. We fix $i = 1, 2$. Show that for all $g : \mathbb{R} \mapsto \mathbb{R}$ with polynomial growth, there exists a real predictable process ϕ^i such that

$$g(S_T^i) = \mathbb{E}\left[g(S_T^i)\right] + \int_0^T \phi_s^i dS_s^i \text{ and } \mathbb{E}\left[\int_0^T |\phi_s^i S_s^i|^2 ds\right] < \infty \ .$$

g. Find a $\mathcal{F}_T$-measurable bounded random variable G such that there do not exist $p \in \mathbb{R}$ and a predictable process with values in $\mathbb{R}^2$, $\mathbb{P}$-a.s. square integrable,

---

[3]i.e. there exist $C > 0$ and $p \geq 1$ such that $|g(x)| \leq C(1 + |x|^p)$ for all $x \in \mathbb{R}$.

$\phi = (\phi^1, \phi^2)$ such that

$$G = p + \int_0^T \phi_s^1 dS_s^1 + \int_0^T \phi_s^2 dS_s^2 .$$

h. Comment the results of questions 1f and 1g.

2. From now on, we focus on the hedging of the option of *payoff* $\mathcal{S} := [S_T^1 - S_T^2]^+$ paid at T.

   a. We first assume that the pricing and the hedging are done under the assumption that $f = \bar{\rho} \in (-1, 1)$ is a constant. We denote

$$p(t, s^1, s^2) := \mathbb{E}\left[\mathcal{S} \mid S_t^1 = s^1, s_t^2 = s^2\right]$$

   and we assume that this function is $C^{1,2}([0, T) \times (0, \infty)^2)$ and continuous on $[0, T] \times (0, \infty)^2$.

     i. Show that, under the above conditions, the function p is solution on $[0, T) \times (0, \infty)^2$ of

$$0 = -\mathcal{L}^{\bar{\rho}} p$$

$$:= -\partial_t p - \frac{1}{2}\left[(s^1\sigma_1)^2 \partial^2_{s^1 s^1} p + (s^2\sigma_2)^2 \partial^2_{s^2 s^2} p\right] - (s^1 s^2 \bar{\rho}\sigma_1\sigma_2)\partial^2_{s^1 s^2} p$$

     with

$$p(T, s^1, s^2) = [s^1 - s^2]^+ \text{ on } (0, \infty)^2 .$$

     ii. What is the hedging strategy of $\mathcal{S}$ if one believes that $f = \bar{\rho}$ is constant? We will denote by $\bar{\phi}^1$ and $\bar{\phi}^2$ the number of units of the assets S^1 and S^2 held when following this strategy.

   b. We now assume that we follow the hedging strategy given by $\bar{\phi}^1$ and $\bar{\phi}^2$ starting from $p(0, S_0^1, S_0^2)$, see 2(a)ii, but that f is not constant. We let V_T denote the value at T of the hedging portfolio.

     i. Show that the hedging error is given by

$$\text{Err} := \mathcal{S} - V_T$$

$$= \int_0^T (\rho_t - \bar{\rho}) S_t^1 S_t^2 \sigma_1 \sigma_2 \partial^2_{s^1 s^2} p(t, S_t^1, S_t^2)dt .$$

     Comment.

ii. We admit that $\mathbb{P}\left[S_T^1 = S_T^2 \mid (S_t^1, S_t^2) = (s^1, s^2)\right] = 0$ if $f = \bar{\rho}$, given $(t, s^1, s^2) \in [0, T) \times (0, \infty)^2$. Deduce that

$$\partial_{s^1} p(t, s^1, s^2) = \mathbb{E}\left[\frac{S_T^1}{s^1} \mathbf{1}_{\{S_T^1 \geq S_T^2\}}\right],$$

and then that

$$\partial_{s^1 s^2}^2 p(t, s^1, s^2) \leq 0, \quad (t, s^1, s^2) \in [0, T) \times (0, \infty)^2.$$

iii. Show that to hedge the option without taking any risk, one should take $\bar{\rho} = \inf_{\mathbb{R}} f$.

7.3 (Gamma hedging and discrete rebalancing) Let W be a one dimensional Brownian motion. We consider a financial market composed of a non-risky asset, with zero return, and of a risky asset which price $S = (S_t)_{t \geq 0}$ is the unique solution of

$$S_t = S_0 + \int_0^t S_s \sigma(S_s) dW_s, \quad t \geq 0.$$

Here, $a : x \in [0, \infty) \to a(x) := x\sigma(x) \in [0, \infty)$ is assumed to be Lipschitz. In the following, we denote by $\mathcal{L}$ the Dynkin operator associated to this process, i.e.

$$\mathcal{L}\varphi(t, x) := \partial_t \varphi(t, x) + \frac{1}{2} a(x)^2 \partial^2 \varphi(t, x)$$

for all functions $\varphi \in C^{1,2}$.

We denote by $\mathcal{A}$ the set of predictable processes ϕ such that

$$\mathbb{E}\left[\int_0^T |\phi_s a(S_s)|^2 ds\right] < \infty$$

for all $T > 0$, and, we assume that, for all $T > 0$ and all $\mathcal{F}_T$-measurable random variables X such that $\mathbb{E}\left[|X|^2\right] < \infty$, there exists $\phi \in \mathcal{A}$ such that $V_T^{\mathbb{E}[X], \phi} = X \mathbb{P}$−a.s. where $V_t^{x, \phi} := x + \int_0^t \phi_s dS_s, \ t \geq 0, \ (x, \phi) \in \mathbb{R} \times \mathcal{A}$.

1. A priori estimates :

   a. Show that, for all $T > 0$ and $p \geq 1$, there exists a constant $C_{T,p} > 0$ such that $\mathbb{E}\left[\sup_{t \leq T} |S_t|^p\right] \leq C_{T,p}$.

   b. Show that S is a martingale under $\mathbb{P}$.

2. Fix G a Borel function with polynomial growth and $T_2 > 0$.

   a. Justify the existence of a function g of $[0, T_2] \times [0, \infty)$ to $\mathbb{R}$ such that $g(t, S_t) = \mathbb{E}\left[G(S_{T_2}) \mid \mathcal{F}_t\right] \mathbb{P}$ − a.s. if $t \leq T_2$.

b. Which PDE should satisfy the function g if it is smooth?

c. We assume that $\mathbb{E}\left[\int_0^{T_2} |\partial g(t, S_t) a(S_t)|^2 \, dt\right] < \infty$. What is the price of the option of payoff $G(S_{T_2})$ paid at T_2 which is compatible with the absence of arbitrage? Express the hedging strategy of the option in terms the derivatives of g and of a.

3. We consider now another Borel function F with polynomial growth and $0 < T_1 < T_2$. We assume that $g \in C_b^{1,2}([0, T_1] \times [0, \infty))$. Given $x \in \mathbb{R}$, $\phi, \alpha \in \mathcal{A}$, we set

$$V_t^{x,\phi,\alpha} := x + \int_0^t \phi_s dS_s + \int_0^t \alpha_s dg(s, S_s) \quad t \in [0, T_1] .$$

We assume that there exist $\bar{\phi}, \bar{\alpha} \in \mathcal{A}$ such that

$$0 = \bar{\phi}_t + \bar{\alpha}_t \partial g(t, S_t) - \partial f(t, S_t)$$
$$= \bar{\alpha}_t \partial^2 g(t, S_t) - \partial^2 f(t, S_t) \quad \mathbb{P}-\text{a.s. } \forall\, t < T_1 , \tag{7.12}$$

where $f \in C^{1,2}([0, T_1) \times [0, \infty))$ satisfies $f(t, S_t) = \mathbb{E}[F(S_{T_1}) \mid \mathcal{F}_t]$ $\mathbb{P}-$a.s. for all $t \le T_1$.

a. Provide a financial interpretation of $V^{x,\bar{\phi},\bar{\alpha}}$.

b. Find $\bar{x} \in \mathbb{R}$ such that $\bar{V} := V^{\bar{x},\bar{\phi},\bar{\alpha}}$ satisfies $\bar{V}_{T_1} = F(S_{T_1})$.

4. Fix $n \in \mathbb{N} \setminus \{0\}$ and $t_i := iT_1/n$, $i \le n$. We note $\eta_t := \max\{t_i,\, i \le n \text{ s.t. } t_i \le t\}$, i.e. $\eta_t = t_i$ if $t \in [t_i, t_{i+1})$, $t \ge 0$. From now on, we consider the piecewise constant strategy $(\tilde{\phi}, \tilde{\alpha})$ defined by $(\tilde{\phi}_t, \tilde{\alpha}_t) := (\bar{\phi}_{\eta_t}, \bar{\alpha}_{\eta_t})$, $t \le T_1$. We note $\tilde{V} := V^{\bar{x},\tilde{\phi},\tilde{\alpha}}$. For sake of simplicity, we assume in addition that g and f are C^∞ with bounded derivatives and that the process $\bar{\alpha}$ is essentially bounded.[4]

a. By using (7.12), show that

$$\tilde{V}_{T_1} - F(S_{T_1})$$

$$= \int_0^{T_1} \bar{\alpha}_{\eta_t} \left(\partial g(t, S_t) - \partial g(\eta_t, S_{\eta_t})\right) a(S_t) dW_t$$

$$- \int_0^{T_1} \left(\partial f(t, S_t) - \partial f(\eta_t, S_{\eta_t})\right) a(S_t) dW_t$$

$$= \int_0^{T_1} A_t a(S_t) dW_t$$

---

[4]Even if this is not realistic.

where $A_t := \int_{\eta_t}^t B_s a(S_s)dW_s + \int_{\eta_t}^t C_s ds$ with

$$B_s := \bar{\alpha}_{\eta_s} \partial^2 g(s, S_s) - \partial^2 f(s, S_s),$$
$$C_s := \bar{\alpha}_{\eta_s} \mathcal{L}\left[\partial g(s, S_s)\right] - \mathcal{L}\left[\partial f(s, S_s)\right].$$

b. By using (7.12) again show that

$$B_s := \int_{\eta_s}^s \left(\bar{\alpha}_{\eta_s} \frac{\partial^3}{\partial x^3} g(u, S_u) - \frac{\partial^3}{\partial x^3} f(u, S_u)\right) a(S_u)dW_u$$

$$+ \int_{\eta_s}^s \left(\bar{\alpha}_{\eta_s} \mathcal{L}\left[\partial^2 g(u, S_u)\right] - \mathcal{L}\left[\partial^2 f(u, S_u)\right]\right) du.$$

c. Show that there exists $C > 0$ such that $\mathbb{E}\left[|B_s|^2\right] \le C/n$ for all $s \le T_1$.
d. Deduce that there exists $C > 0$ such that $\mathbb{E}\left[|A_t|^2\right] \le C/n^2$ for all $t \le T_1$.
e. Deduce that there exists $C > 0$ such that

$$\mathbb{E}\left[|\tilde{V}_{T_1} - F(S_{T_1})|^2\right]^{\frac{1}{2}} \le C/n.$$

7.4 (Cox-Ingersoll-Ross model (CIR)) We call CIR process,[5] the solution of a stochastic differential equation of the form

$$dX_t^x = a(b - X_t^x)dt + \sigma\sqrt{X_t^x}dW_t$$

with initial condition $X_0^x = x \ge 0$, $a, b, \sigma > 0$.

In the following, we shall need a time grid $\pi := \{0 = t_0 < t_1 < \ldots < t_i < \ldots < t_n = T\}$ of $[0, T]$. We assume that $b > \frac{1}{2a}\gamma > 0$ where $\gamma := \sigma^2$.

1. By using the techniques introduced in Sect. 7.4 above, find Z_0 such that $\bar{X}$ defined by $\bar{X}_t := e^{-at}Z_{\frac{\gamma}{4a}(e^{at}-1)}$ satisfies the above equation for a well-chosen Brownian motion, and such that Z a squared Bessel of dimension $\delta := 4ab/\gamma$.
2. Set

$$F(t, x) := (2\lambda L(t) + 1)^{-2ab/\gamma} e^{-\frac{\lambda L(t)\zeta(t,x)}{2\lambda L(t)+1}}$$

where $L(t) = (\gamma/4a)(1 - e^{-at})$ and $\zeta(t, x) = 4xae^{-at}/(\gamma(1 - e^{-at}))$, and show that $(F(t - s, X_s^x))_{s\le t}$ is a martingale such that $F(0, X_t^x) = e^{-\lambda X_t^x}$. Deduce that the Laplace's transform of X_t^x is given by

$$\phi_{X_t^x}(\lambda) = (2\lambda L(t) + 1)^{-2ab/\gamma} e^{-\frac{\lambda L(t)\zeta(t,x_0)}{2\lambda L(t)+1}}.$$

---

[5]This process is widely used to model the interest rate.

3. What is the Laplace transform $\phi_{Y_t^x}$ of $Y_t^x := X_t^x/L(t)$?
4. We assume for this question that $4ab/\gamma =: k \in \mathbb{N}$.

   a. Let N be a Gaussian random variable with variance 1 and mean m. Compute the Laplace's transform ϕ_{N^2} of N^2.
   b. Deduce a way to simulate the increments $(X_{t_{i+1}}^x - X_{t_i}^x)_{i<n}$ when $4ab/\gamma$ is an integer.

5. We now consider the case $4ab/\gamma \in (0,\infty)$.

   a. Given $\alpha, \beta > 0$, let

   $$f_{\alpha,\beta}(x) = \frac{\beta^\alpha}{\Gamma(\alpha)} x^{\alpha-1} e^{-\beta x} \mathbf{1}_{x>0}$$

   be the density of the Gamma distribution $G(\alpha,\beta)$, where Γ is the Gamma function. How can we simulate a uniform deviate (U,V) on $D := \{(u,v) \in (0,\infty)^2 : 0 \leq u \leq \sqrt{f_{\alpha,\beta}(v/u)}\}$ when $\alpha > 1$?
   b. Assume that $\alpha > 1$. Compute the law of V/U?
   c. Deduce from the above how to simulate in the law $G(\alpha,\beta)$ when $\alpha > 1$.
   d. What can we do if $\alpha = 1$?
   e. Recall that, for $\alpha, \beta > 0$, the Laplace's transform of $G(\alpha,\beta)$ is

   $$\phi_{\alpha,\beta}(y) = (1 + y/\beta)^{-\alpha} \quad y \geq 0 .$$

   Fix $\nu > 0$, let M have a Poisson distribution of parameter $p > 0$ and $(\chi_{\nu+2i})_{i\geq 0}$ be a sequence of independent random deviates (independent of M) such that $\chi_{\nu+2i} \sim G((\nu+2i)/2, 1/2)$ for each $i \geq 0$. Compute the Laplace's transform $\phi_{\nu+2M}$ of $\chi_{\nu+2M} := \sum_{i\geq 0} \chi_{\nu+2i} \mathbf{1}_{i=M}$.
   f. How to simulate in the law $\chi_{\nu+2M}$?
   g. Deduce[6] a way to simulate the increments of $(X_{t_{i+1}}^x - X_{t_i}^x)_{i<n}$.

7.5 (Approximate hedging with recalibration[7]) Let us a consider a model in which the interest r is set to zero, and the stock price is S. Let C_t^i be the price at time t of the European call with maturity T_i and strike K_i, $i = 1, \ldots, I$, as observed on the market. We use a local volatility model to hedge a claim $g(S_T)$ of maturity T, in which g is a continuous and bounded function. However, the local volatility model is re-calibrated at each $t \leq T$ using the listed option prices $C_t := (C_t^i)_{i\leq I}$. Namely, for each t, we compute $(s,x) \mapsto p(s,x;C_t)$ as the solution of

$$-\partial_s p(s,x;C_t) - \frac{1}{2}\hat{\sigma}^2(s,x;C_t)\partial_{xx}^2 p(s,x;C_t) = 0, \quad \text{for } t \leq s < T,$$

---

[6]In practice, exact simulation can be very costly if the cardinality of π is big. In this case, one prefers Euler type schemes as in [2].

[7]This exercise is based on ideas developed in [7].

with $p(T, \cdot; C_t) = g$, and where $(s, x, c) \mapsto \hat{\sigma}(s, x; c)$ is assumed measurable, Lipschitz in x, uniformly in (s, c), and bounded. We assume that p is smooth in all its arguments on $[0, T)$.

However, in reality, (S, C) is a continuous semi-martingale with Lebesgue absolutely continuous quadratic variation:

$$\langle (S, C) \rangle = \int_0^{\cdot} \begin{pmatrix} \sigma_s^2 & A_s' \\ A_s & B_s \end{pmatrix} ds,$$

in which σ, A and B are predictable and bounded processes with values in $\mathbb{R}_+$, $\mathbb{R}^I$ and $\mathbb{M}^{I,I}$ respectively.

In the following, we set $P_t := p(t, S_t; C_t)$, $t \le T$. The options C can be traded dynamically. Moreover, we assume that the calibration is perfect in the sense that each price curve $p^i(\cdot; c)$ associated to the i-th option, if the option prices is c, satisfies $p^i(\cdot; c) = c^i$, for all $c \in \mathbb{R}_+^I$.

1. Based on Dupire's formula, justify the form of $\hat{\sigma}$.
2. What is the dynamics of P?
3. How can one hedge the un-bounded variation part of P? Can you relate it to a notion of *vega*-hedging?
4. Assume that $T_1 \ge T$. How can we obtain a P&L of the form

$$\int_0^T \left(\frac{1}{2} \mathrm{Tr} \left[B_t \partial_{cc}^2 p(t, S_t; C_t) \right] dt + \partial_{xc}^2 p(t, S_t; C_t) A_t \right) dt$$

by trading the option C^1?
5. Can we also get rid of the terms $\partial_{cc}^2 p$ and $\partial_{xc}^2 p$ by trading the other liquid options used for calibration?
6. How can we proceed with other options?
7. Assume that we know that (A, B) takes $\mathbb{P} - $ a.s. values in a set $\Theta \subset \mathbb{R}^I \times \mathbb{M}^{I,I}$. If we use the gamma-hedging strategy of the question 4, what PDE should be used for the pricing to ensure that the P&L is $\mathbb{P} - $ a.s. non-negative?

Corrections

7.1

1. We have $p(t, x) = \mathbb{E}^{\mathbb{Q}}[S_1[S_T/S_1 - \kappa]^+ | S_t = x] = x \mathbb{E}^{\bar{\mathbb{Q}}}[[S_T/S_1 - \kappa]^+ | S_t = x]$ with $d\bar{\mathbb{Q}}/d\mathbb{Q} = S_1/\mathbb{E}^{\mathbb{Q}}[S_1|\mathcal{F}_t] = S_1/S_t$. Namely,

$$\frac{d\bar{\mathbb{Q}}}{d\mathbb{Q}} = e^{-\frac{1}{2}\int_t^1 \sigma_s^2 ds + \int_t^1 \sigma_s dW_s}$$

so that $\bar{W} := W - \int_t^{\cdot \wedge 1} \sigma_s ds$ is a $\bar{\mathbb{Q}}$-Brownian motion. Observing that $dW = d\bar{W}$ on $[1, T]$, we see that S_T/S_1 has a log-normal distribution of the prescribed form under $\bar{\mathbb{Q}}$, since σ is deterministic. Hence, the result.

2. We have $\partial_x p(t, x) = F(1, K, \int_1^T \sigma_s^2 ds)$, so we should hold $F(1, K, \int_1^T \sigma_s^2 ds)$ stocks on the period $[0, 1]$.

3. By the above, the gamma is 0 on $[0, 1]$, so we are not exposed to a volatility misspecification on this period, see Sect. 7.3.1.

4. For $t \in [1, T]$, S_1 is known at t, so we just have to hedge a call option of strike $S_1 K$ from time 1. The price is explicit since σ is constant (use the Black and Scholes formula with volatility $(\int_t^T \sigma_s^2 ds/(T - t))^{\frac{1}{2}}$).

5. Yes, see Sect. 7.3.1. Since the payoff is convex, it is easy to see that the price is convex as well, and we are therefore exposed to an increase of the volatility.

6. Because the volatility is local, the analysis of question 3. does not hold anymore. We might be exposed to a misspecification of the volatility (depending on the convexity of p which itself depend on the local volatility function σ).

7.2

1. a. Use the martingale representation theorem.
   b. Since only S^1 and S^2 have to be martingales, the process v^3 in the above representation can be chosen freely (recall Girsanov's theorem).
   c. Apply Girsanov's theorem and use the fact that S^1 and S^2 have to be martingales to obtain the condition: $\sigma_1 v^1 = \sigma_2(\rho v^1 + (1 - \rho^2)^{\frac{1}{2}} v^2) = 0$. Since $\sigma_1, \sigma_2, 1 - \rho > 0$, this is the announced condition.
   d. Apply Girsanov's theorem and use a. and c.
   e. $\mathbb{E}^{\mathbb{Q}}[g(S_T^i)]$ does not depend on $\mathbb{Q} \in \mathcal{M}$ since the Brownian motions Z^1 and Z^2 are Brownian motions (and therefore have the same law) under each $\mathbb{Q} \in \mathcal{M}$. In particular $\mathbb{E}^{\mathbb{Q}}[|S_T^i|^p] < \infty$ for all $p \geq 1$ (see the Laplace transform of a Gaussian distribution).
   f. In view of the preceding question, one could apply Corollary 2.18. Strictly speaking, it does not fit because g is not bounded in this exercise. However, $g(S_T^i)$ is measurable with respect to the filtration generated by the Brownian motion Z^i, and one can apply the martingale representation theorem.
   g. Take $G = \cos(W_T^3)$. It admits a (unique) martingale representation in terms of W^3 and can therefore not be replicated by only using S^1 and S^2.
   h. Vanilla payoffs on a single stock can be hedged perfectly, but the global market is incomplete.

2. a. i. Use Feynman and Kac's formula.
      ii. It is a Δ-hedging strategy on S^1 and S^2 with quantities $\partial_{s^1} p(\cdot, S^1, S^2)$ and $\partial_{s^2} p(\cdot, S^1, S^2)$ (check with Itô's lemma).
   b. i. Adapt the arguments of Sect. 7.3.1. When the cross derivative $\partial^2_{s^1 s^2} p(t, S_t^1, S_t^2)$ is positive and the realised correlation is bigger than $\bar{\rho}$, then the hedger looses money. The other way round when the cross derivative is negative.

    ii. Under this condition, one can differentiate inside the expectation. Then, the derivative in s^1 is clearly non-increasing with s^2, which provides the sign of the cross derivative.

    iii. From the above, one would like that $\bar{\rho} \leq f(W^3)\,dt \times d\mathbb{P}$-a.e. Since $\mathbb{P}[W^3 \in [a - \varepsilon, a + \varepsilon],\ \text{on } [0, T]] > 0$ for all $a \in \mathbb{R}$ and $\varepsilon > 0$, this implies that $\bar{\rho} \leq \inf_{\mathbb{R}} f$.

7.3

1. a. This is the solution of a SDE with Lipschitz coefficients.
   b. This is a square integrable local martingale.
2. a. S is a Markov process.
   b. Apply Feynman and Kac's formula.
   c. The price is the expectation of the payoff, and one should use a delta hedging strategy.
3. a. This is the wealth process generated by a position $\bar{\phi}$ is S and $\bar{\alpha}$ in the liquid option of payoff $G(S_T)$.
   b. $\bar{x} = \mathbb{E}[F(S_{T_1})]$.
4. a. We have

$$\bar{V}_{T_1} = x + \int_0^{T_1} \bar{\phi}_{\eta_t}\,dS_t + \int_0^t \bar{\alpha}_{\eta_t}\,dg(t, S_t),$$

but

$$\bar{\phi}_{\eta_t} + \bar{\alpha}_{\eta_t}\,\partial g(\eta_t, S_{\eta_t}) - \partial f(\eta_t, S_{\eta_t}),$$

so that

$$\bar{V}_{T_1} = x + \int_0^{T_1} \left(\bar{\alpha}_{\eta_t}\,\partial g(\eta_t, S_{\eta_t}) - \partial f(\eta_t, S_{\eta_t}) \right) a(S_t)\,dW_t$$

$$+ \int_0^t \bar{\alpha}_{\eta_t}\,\partial g(t, S_t)a(S_t)\,dW_t.$$

It remains to use that $F(S_{T_1}) = x + \int_0^{T_1} \partial f(t, S_t)a(S_t)\,dW_t$ to obtain the first equality. For the second, we apply Itô's lemma on $[\eta_t, t]$ to $\partial g(\cdot, S)$ and $\partial f(\cdot, g)$.

    b. Apply the same trick but use the second identity in (7.12).

    c. By Jensen's inequality and the Itô's isometry,

$$\mathbb{E}\left[|B_s|^2\right] \leq 2\mathbb{E}\left[\left[\int_{\eta_s}^s \left(\bar{\alpha}_{\eta_s} \frac{\partial^3}{\partial x^3} g(u, S_u) - \frac{\partial^3}{\partial x^3} f(u, S_u) \right) a(S_u)\,dW_u \right]^2 \right]$$

$$+ 2\mathbb{E}\left[\left[\int_{\eta_s}^s \left(\bar{\alpha}_{\eta_s} \mathcal{L}\left[\partial^2 g(u, S_u) \right] - \mathcal{L}\left[\partial^2 f(u, S_u) \right] \right) du \right]^2 \right]$$

$$= 2\mathbb{E}\left[\int_{\eta_s}^{s}\left(\bar{\alpha}_{\eta_s}\frac{\partial^3}{\partial x^3}g(u,S_u)-\frac{\partial^3}{\partial x^3}f(u,S_u)\right)^2 a(S_u)^2 du\right]$$

$$+ 2\mathbb{E}\left[\left[\int_{\eta_s}^{s}\left(\bar{\alpha}_{\eta_s}\mathcal{L}\left[\partial^2 g(u,S_u)\right]-\mathcal{L}\left[\partial^2 f(u,S_u)\right]\right)du\right]^2\right].$$

The first term is of order $1/n$, the second of order $1/n^2$, since all involved quantities are bounded.

d. Similarly,

$$\mathbb{E}\left[|A_t|^2\right]\leq 2\mathbb{E}\left[\left[\int_{\eta_t}^{t}B_s a(S_s)dW_s\right]^2\right]+2\mathbb{E}\left[\left[\int_{\eta_t}^{t}C_s ds\right]^2\right]$$

$$= 2\mathbb{E}\left[\int_{\eta_t}^{t}[B_s a(S_s)]^2 ds\right]+2\mathbb{E}\left[\left[\int_{\eta_t}^{t}C_s ds\right]^2\right].$$

In view of the preceding question, both terms are of order of $1/n^2$.

e. Combine the above

7.4

1. First apply the time change formula to compute the dynamics of $(Z_{\frac{\gamma}{4a}(e^{at}-1)})_{t\geq 0}$. Then, apply the integration by part formula to get the dynamics of $\bar{X}$.

2. Apply Itô's lemma and check that the dt terms cancel. Then, $\mathbb{E}[F(0,X_t^x)] = F(t,X_0^x)$. Since $F(0,X_t^x) = e^{-\lambda X_t^x}$, this provides the Laplace transform.

3. $\phi_{Y_t^x}(\lambda) = \phi_{X_t^x}(\lambda/L(t)) = (2\lambda+1)^{-2ab/\gamma}e^{-\frac{\lambda\zeta(t,x)}{2\lambda+1}}$ by definition.

4. a. By direct computations, $\phi_{N^2}(\lambda) = (2\lambda+1)^{-\frac{1}{2}}e^{-\lambda m^2/(2\lambda+1)}$.

   b. If $k := 4ab/\gamma$ is an integer, the above shows that the Laplace transform of $\bar{Y}_t$ is the product of the k Laplace transforms of k squared independent Gaussian random variables, with mean m that can be computed explicitly by comparing the formulas. Hence, the sum of this k squared independent Gaussian random variables and $\bar{Y}$ have the same law. From $\bar{Y}_t$, we deduce $\bar{X}_t^x$ by multiplying by $L(t)$. Finally, we use the fact that the law of $\bar{X}_t^x - \bar{X}_s^x$ given $\bar{X}_s^x = x_s$ is the same as the law of $\bar{X}_{t-s}^{x_s}$, $t \geq s$. Hence, if we can simulate independent Gaussian random variables, we can simulate the increments of $\bar{X}^x$ by drawing independently k independent Gaussian random variables, associated to each time step.

5. a. Simulate a uniform (U',V') on a rectangle containing D, keep the result if and only if $(U',V') \in D$ (rejection method).

   b. Use the change of variable formula to verify that it has the law $G(\alpha,\beta)$ (this is the ratio method).

   c. Combine the above.

   d. When $\alpha = 1$, this is an exponential distribution and we can use the inversion of the cumulating function technique.

e. It is given by

$$\sum_{i \geq 0} (1 + 2y)^{-(\frac{v}{2}+i)} \frac{p^i}{i!} e^{-p} = (1 + 2y)^{-\frac{v}{2}} e^{-2yp}.$$

f. Simulate M, and given its value m, draw randomly in the law χ_{v+2m}.

g. We can choose p and v in e. such that one retrieves $\phi_{Y_t^x}$. Then, we argue as in 4.b.

7.5

1. It can be an approximation of Dupire's formula, based on $(C^i)_{i \leq I}$. It means that we recalibrate continuously the model on the options prices.

2. Apply Itô's Lemma and the PDE satisfied by $p(\cdot; t, C_t)$ to find

$$dP_t = \partial_x p(t, S_t; C_t) dS_t + \partial_c p(t, S_t; C_t) dC_t + \frac{1}{2} \left(\sigma_t^2 - \hat{\sigma}^2(t, S_t; C_t) \right) \partial_{xx}^2 p(t, S_t; C_t) dt$$

$$+ \left(\frac{1}{2} \text{Tr} \left[B_t \partial_{cc}^2 p(t, S_t; C_t) \right] + \partial_{xc}^2 p(t, S_t; C_t) A_t \right) dt.$$

3. Use a Δ-hedging strategy on S and C with quantities $\partial_x p(t, S_t; C_t)$ and $\partial_c p(t, S_t; C_t)$ at time t. Since C_t enters p through $\hat{\sigma}$, the term $\partial_c p(t, S_t; C_t)$ corresponds to a hedge on the evolution of the local volatility model.

4. We use a Γ-hedging strategy to reduce to the case $\partial_{xx}^2 p(\cdot, S; C) = 0$. Let us assume that we can find a sufficiently integrable process λ such that

$$-\partial_{xx}^2 p(t, S_t; C_t) = \lambda_t \partial_{xx}^2 p^1(t, S_t; C_t), \quad t \leq T.$$

By assumption, $p^1(\cdot; c) = c^1$ for all c, so that $p^1(t, S_t, C_t) = C_t^1$ and $\partial_{cc}^2 p^1(\cdot; C) = \partial_{xc}^2 p^1(\cdot; C) = 0$. Then, by question 2,

$$\lambda_t dC_t^1 - \lambda_t \partial_x p^1(t, S_t; C_t) dS_t - \lambda_t \partial_c p^1(t, S_t; C_t) dC_t$$

$$= \frac{\lambda_t}{2} \left(\sigma_t^2 - \hat{\sigma}^2(t, S_t; C_t) \right) \partial_{xx}^2 p^1(t, S_t; C_t) dt.$$

By the previous question again, the announced result holds with the hedge given by holding $\partial_x(\lambda p^1 + p)(\cdot, S; C)$ in S and $\partial_c(\lambda p^1 + p)(\cdot, S; C) - \lambda \mathbf{1}_{i=1}$ in C^i, for $i \leq I$.

5. No. Since $p^i(\cdot; c) = c^i$ by assumption, we have $\partial_{cc}^2 p^i(\cdot; C) = \partial_{xc}^2 p^i(\cdot; C) = 0$.

6. One needs to trade options such that the derivatives ∂_{cc}^2 and ∂_{xc}^2 of their prices are non zero. This would typically require two other options, which might not be liquid in practice.

7. Let us assume a smooth solution $(t, x, c) \mapsto p^{\text{rob}}(t, x, c)$ to

$$-\partial_s p^{\text{rob}} - \frac{1}{2}\hat{\sigma}^2 \partial_{xx}^2 p^{\text{rob}} - \sup_{(a,b)\in\Theta} \left(\frac{1}{2}\text{Tr}\left[b\partial_{cc}^2 p^{\text{rob}}\right] + \partial_{xc}^2 p^{\text{rob}} \, a \right) = 0$$

with boundary condition $p^{\text{rob}}(T, \cdot; \cdot) = g$. Then,

$$dP_t \leq \partial_x p^{\text{rob}}(t, S_t; C_t)dS_t + \partial_c p^{\text{rob}}(t, S_t; C_t)dC_t$$

$$+ \frac{1}{2}\left(\sigma_t^2 - \hat{\sigma}^2(t, S_t; C_t)\right) \partial_{xx}^2 p^{\text{rob}}(t, S_t; C_t)dt$$

and one can apply the gamma-hedging strategy of question 4 to cancel the dt term in the above.

Chapter 8
Stochastic Volatility Models

Stochastic volatility models are used when the option price is very sensitive to volatility (smile) moves, and when they cannot be explained by the evolution of the underlying asset itself, see e.g. [34]. This is typically the case for exotic options.

In Chap. 5, we have already discussed some properties of these models. In particular, we have seen how to compute the super-hedging price when only the underlying asset is traded, which is in general too costly. We have also discussed in Chap. 6 alternatives to the super-hedging. In this chapter, we will again discuss the super-hedging problem, but we will now allow the trader to trade liquid options, in order to control the volatility risk either by trading dynamically or by constructing a semi-static hedging strategy. We will also discuss the calibration issue through the example of the Heston's model.

We refer to [7] for the description of the most common stochastic volatility models, such as the SABR model proposed in [34].

Finally, let us mention that one can combine stochastic and local volatility models as in [1]. We then obtain a *stochastic local volatility models* (SLV): the volatility depends on time, the spot price of the underlying asset, and on another stochastic factor. See also Exercise 8.5.

8.1 Hedging with Liquid Options

A natural way to hedge the volatility risk consists in trading dynamically liquid options (typically calls or puts).

To simplify the presentation, let us assume that one wants to hedge a European option of payoff G, maturity T, and such that G is $\sigma(W_s, s \leq T)$-measurable, where W is a Brownian motion of dimension 2. The risk-free interest rate r is equal to zero,

© Springer International Publishing Switzerland 2016
B. Bouchard, J.-F. Chassagneux, *Fundamentals and Advanced Techniques in Derivatives Hedging*, Universitext, DOI 10.1007/978-3-319-38990-5_8

and we trade a single underlying asset S of dynamics

$$dS_t = \sigma(S_t, Y_t)dW_t \; ; \tag{8.1}$$

$$dY_t = \gamma(S_t, Y_t)dt + \beta(S_t, Y_t)dW_t \; ,$$

where the functions σ, γ and β are measurable and such that a strong solution to the above system exists, for all initial condition.

Here, the process Y is not a tradable asset, which renders the market a priori incomplete. It is impossible to perfectly hedge G, in general, see Chap. 5.

Let us now assume that one can also trade a vanilla option of payoff $g(S_{T'})$ where $T' \geq T$. We let $p(t, x, y)$ denote its price at time t if $(S_t, Y_t) = (x, y)$ and we assume that $p \in C^{1,2}$.

It is then possible to build a portfolio $V^{v,\phi}$ where ϕ_t^1 (resp. ϕ_t^2) denotes the number of units of S (resp. of p) held at t. Its dynamics are given by

$$dV_t^{v,\phi} = \phi_t^1 dS_t + \phi_t^2 dp(t, S_t, Y_t) \; .$$

If the dynamics (8.1) correspond to the risk neutral measure $\mathbb{Q}$ under which p is computed (i.e. W is a $\mathbb{Q}$-Brownian motion), we obtain

$$dV_t^{v,\phi} = \phi_t^1 \sigma(S_t, Y_t)dW_t$$

$$+ \left(\phi_t^2 (\partial_x p(t, S_t, Y_t)\sigma(S_t, Y_t) + \partial_y p(t, S_t, Y_t)\beta(S_t, Y_t)) \right) dW_t$$

since $p(t, S_t, Y_t)$ should be a $\mathbb{Q}$-(local) martingale.

On the other hand, if G is $\mathbb{Q}$-integrable, it admits a representation of the form

$$G = \mathbb{E}^{\mathbb{Q}}[G] + \int_0^T \zeta_s dW_s \; .$$

In order to hedge this option, it then suffices to find a predictable (and sufficiently integrable) process $\phi = (\phi^1, \phi^2)$ such that, $dt \times d\mathbb{P}$ – a.e.,

$$\phi_t^1 \sigma(S_t, Y_t) + \phi_t^2 (\partial_x p(t, S_t, Y_t)\sigma(S_t, Y_t) + \partial_y p(t, S_t, Y_t)\beta(S_t, Y_t)) = \zeta_t$$

on $[0, T]$. The main issue is then whether this system can be inverted. It is the case in general, see e.g. [22].

All works as if the process Y that drives the volatility could be traded directly. It indeed can by combining a liquid option and the underlying asset.

The limit of this approach is that the bid-ask interval of the options is generally rather big, which makes strategies based on a dynamic trading of options costly. In practice, it is often better to design a static or semi-static strategy.

8.2 Static and Semi-static Strategies

In this section, we show how one can hedge European options by only using static positions on call and put, possibly by also trading dynamically the underlying asset.

8.2.1 Decomposition of Payoff on a Basis of Calls and Puts

Fix a vanilla option of payoff $g(S_T)$ where S_T is a real non-negative random variable which represents the terminal value of the underlying asset.

Lemma 8.1 *If g admits derivatives in the sense of distributions up to the order 2, then, for all $\kappa \geq 0$ such that g is differentiable at κ, we have*

$$g(S_T) = g(\kappa) + \partial g(\kappa) \left[(S_T - \kappa)^+ - (\kappa - S_T)^+ \right]$$
$$+ \int_0^\kappa \partial^2 g(k)(k - S_T)^+ dk + \int_\kappa^\infty \partial^2 g(k)(S_T - k)^+ dk .$$

Proof We note d the Dirac mass at 0. We have

$$g(S_T) = \int_0^\kappa g(k)\mathrm{d}(S_T - k)dk + \int_\kappa^\infty g(k)\mathrm{d}(S_T - k)dk .$$

By integrating by parts, we obtain then

$$\int_0^\kappa g(k)\mathrm{d}(S_T - k)dk = g(\kappa)\mathbf{1}_{S_T < \kappa} - \int_0^\kappa \partial g(k)\mathbf{1}_{\{S_T < k\}}dk$$

where

$$\int_0^\kappa \partial g(k)\mathbf{1}_{\{S_T < k\}}dk = \partial g(\kappa)(\kappa - S_T)^+ - \int_0^\kappa \partial^2 g(k)(k - S_T)^+ dk .$$

We proceed similarly for the second term in the first equation. $\qquad\square$

This formula can be easily interpreted. The term $g(\kappa)$ corresponds to a position in cash. The second term is a position in the underlying asset and in cash, in virtue of the call-put parity. The other terms are positions of magnitude $\partial^2 g(k)dk$ in calls and puts of strike k.

By absence of arbitrage opportunity (and under suitable integrability conditions), this implies in particular that the value v_0 of the contract with payoff $g(S_T)$ is given by

$$g(\kappa) + \partial g(\kappa) [S_0 - \kappa] + \int_0^\kappa \partial^2 g(k) P_0(k, T) dk$$
$$+ \int_\kappa^\infty \partial^2 g(k) C_0(k, T) dk$$

where $P_0(k, T)$ and $C_0(k, T)$ are the prices of the puts and the calls of strike k and maturity T on S.

In practice, this formula cannot be applied directly since some calls and puts may not be available on the market, however it suggests to approximate the payoff by a suitable combinations of the liquid ones.

Let us assume that the calls and puts of strikes k_i, $i = 1, \dots, I$, are liquid. We can look for weights c, ω_0, $\omega := (\omega_i)_{i \leq I}$ which minimize

$$\mathbb{E}\left[\ell\left(\Delta(c, \omega_0, \omega, S_T)\right)\right]$$

where ℓ is a given loss function (for instance $\ell(x) = x^2$ or $\ell(x) = (x^+)^2$), and

$$\Delta(c, \omega_0, \omega, S_T) := g(S_T) - V(c, \omega_0, \omega, S_T)$$

is the hedging error if

$$V(c, \omega_0, \omega, S_T) := c + \omega_0 S_T + \sum_{i=1}^I \omega_i (S_T - k_i)^+$$

is the terminal value of the hedging portfolio. Obviously, if v_0 is the price of the option $g(S_T)$, the optimisation is performed under the constraint that the cost of building the hedging portfolio is less than v_0.

This approach has two main interests: (1) It does not depend on the dynamics of S, it only depends on the law of S_T; (2) If the dynamics of S is known, it is possible to hedge the residual part $\Delta(c, \omega_0, \omega, S_T)$. Even with simple models, this semi-static approach can be very efficient if the remaining part to be hedged is small with respect to the original risk $g(S_T)$.

8.2.2 Application to Variance Swaps

A variance swap on an underlying S is an exchange between

$$J_T^n := \frac{1}{T} \sum_{k=1}^n \left(\ln \frac{S_{t_k}}{S_{t_{k-1}}} \right)^2$$

and a fixed premium $\kappa \in \mathbb{R}$. Here, the times t_k constitute an increasing sequence in $[0, T]$ and $t_n = T$.

This is a very popular product, which allows to take directly a position on the realised volatility, without appealing to options, see Sect. 8.1.

If the dynamics of S is of the form (under a risk neutral measure $\mathbb{Q}$ and assuming that the risk-free interest rate is zero for simplicity)

$$dS_t = S_t \sigma_t dW_t$$

then Itô's lemma implies that

$$J_T^n = \frac{1}{T} \sum_{k=1}^{n} \left(-\frac{1}{2} \int_{t_{k-1}}^{t_k} \sigma_t^2 dt + \int_{t_{k-1}}^{t_k} \sigma_t dW_t \right)^2$$

$$= \frac{1}{T} \sum_{k=1}^{n} \left(- \int_{t_{k-1}}^{t_k} (Z_t^k \sigma_t^2 - \sigma_t^2) dt + 2 \int_{t_{k-1}}^{t_k} Z_t^k \sigma_t dW_t \right)$$

with

$$Z_t^k := -\frac{1}{2} \int_{t_{k-1}}^{t} \sigma_s^2 ds + \int_{t_{k-1}}^{t} \sigma_s dW_s \quad \text{on } [t_{k-1}, t_k] .$$

By sending the time step to 0, i.e. $n \to \infty$, we obtain

$$J_T^n \sim J_T := \frac{1}{T} \int_0^T \sigma_t^2 dt .$$

Indeed, if by example $\int_0^T \sigma_t^4 dt \in \mathbb{L}^2$ then (denoting by $C > 0$ a generic constant independent of n)

$$\left\| \sum_{k=1}^{n} \int_{t_{k-1}}^{t_k} Z_t^k \sigma_t^2 dt \right\|_{\mathbb{L}^1} \leq C \max_k \sup_{t \in [t_{k-1}, t_k]} \left\| Z_t^k \right\|_{\mathbb{L}^2}$$

where

$$\left\| Z_t^k \right\|_{\mathbb{L}^2}^2 \leq C \left(\sqrt{t_k - t_{k-1}} + t_k - t_{k-1} \right) .$$

By similar argument,

$$\left\| \sum_{k=1}^{n} \int_{t_{k-1}}^{t_k} Z_t^k \sigma_t dW_t \right\|_{\mathbb{L}^2} \to 0 .$$

From now on, we hence assume that the *payoff* of random the leg is J_T in place of J_T^n. In view of the above discussion, the hedging error $J_T - J_T^n$ is small if the time step is small.

We now observe that

$$\ln\left(\frac{S_T}{S_0}\right) = -\frac{1}{2}\int_0^T \sigma_t^2 dt + \int_0^T \sigma_t dW_t,$$

so that

$$J_T = -\frac{2}{T}\ln\left(\frac{S_T}{S_0}\right) + \int_0^T \frac{2}{TS_t} dS_t.$$

Up to the last term, which is a portfolio trading only S, the random leg is essentially a short position short on a "log-contract" of payoff $\ln\left(\frac{S_T}{S_0}\right)$ on S. Since such a contract does not exist on the market, it has to be hedged dynamically or by using the semi-static approach described in Sect. 8.2: By Lemma 8.1 applied to $\kappa = S_0$

$$\ln\left(\frac{S_T}{S_0}\right) = \frac{S_T - S_0}{S_0} - \int_0^{S_0} \frac{1}{k^2}(k - S_T)^+ dk - \int_{S_0}^\infty \frac{1}{k^2}(S_T - k)^+ dk.$$

This moreover implies that the value v_0 of the payoff $\ln\left(\frac{S_T}{S_0}\right)$ is given by

$$v_0 = -\int_0^{S_0} \frac{1}{k^2} P_0(k, T) dk - \int_{S_0}^\infty \frac{1}{k^2} C_0(k, T) dk$$

where $P_0(k, T)$ and $C_0(k, T)$ are the prices of puts and calls with strike k and maturity T on S.

We refer to [49] for a detailed analysis of this type of products.

8.3   Example: The Heston's Model

8.3.1   The Model

The Heston's model of [35] is a stochastic volatility model of the form

$$dS_t = \mu S_t dt + \sigma_t S_t dW_t^1,$$
$$d\sigma_t^2 = a(b - \sigma_t^2) dt + \gamma \sigma_t dZ_t,$$

where $Z = \sqrt{\rho}W^1 + \sqrt{1 - \rho}W^2$ with $\rho \in [0, 1]$, $a, b > 0$. Otherwise stated, σ^2 is a CIR process, see Exercise 7.4. Note that $e^{-abt}\sigma_t^2$ follows dynamics similar to the process S of the CEV model of the preceding section. It is also related to a *squared Bessel process*. Let us also notice that (S, σ^2) solves an equation with non-Lipschitz coefficients and that the moments of S can explode for certain values of the parameters, see for instance [4].

This model is very popular, as it can take into account various types of correlations between the asset and the volatility, depending on ρ and the sign of γ. This last parameter also controls the volatility of the volatility (Vo-Vol).

Here, a is the mean reverting strength, and b the long term volatility. This model produces a symmetric smile when the correlation parameter ρ is zero. The volatility of the volatility controls the slope of the *skew* (bigger when γ increases). A negative correlation produces a negative skew. Finally, let us observe that the probability that the volatility vanishes is positive if $2ab < \gamma^2$, see Chap. 7.

The Heston's model with jumps of [6] adds a compound Poisson process J, independent of W, to the dynamics of S:

$$dS_t = \mu S_t dt + \sigma_t S_t dW_t^1 + S_{t-} dJ_t .$$

If the intensity of the jump is constant and their size follows a Gaussian distribution, then the characteristic function of S_T is known. Hence, the model is more flexible but still tractable, see Sect. 8.3.3 below.

This model corrects a flaw in the Heston's model that can account for the skew of the large maturities but not of the small. The introduction of jumps increases the risk on a small time scale and thus increases the skew of these small maturities. On a long time scale, the jumps are averaged out, and have therefore a low impact on long maturity options. It is thus possible to calibrate separately the short term skew (with the jumps) and the long-term (with the correlation between the two Brownian motions).

8.3.2 Fourier's Transform Computation

The models in which the Fourier's transform of $\ln(S_T)$ is known are very tractable from a numerical point of view as will be explained in Sect. 8.3.3. It is the case in the Heston model, which belongs to the general class of affine models.

Let us explain here how it can be computed. We let $\mu = 0$, which amounts to working directly under the risk neutral measure, assuming for simplicity that the risk-free rate is zero. We make the change of variable $X = \ln(S)$ and compute $\psi(t, x, v; \theta) := \mathbb{E}\left[e^{i\theta X_T} \mid (X_t, \sigma_t^2) = (x, v)\right]$. This can be done by first applying the Feynman-Kac theorem (Theorem 4.2) to obtain a PDE characterisation of ψ:

$$\mathcal{L}\psi = 0,$$

where $\mathcal{L}$ is the Dynkin operator of the diffusion (X, σ^2), with the terminal condition $\psi(T, x, v; \theta) := e^{i\theta x}$. We then look for a particular solution in the form

$$\psi(t, x, v; \theta) = e^{C(t,T;\theta)+D(t,T;\theta)v+i\theta x}$$

for some functions with complex values to be determined. Inserting this in the above PDE, we obtain the explicit form of C and D. A verification argument, see Chap. 4, allows us to conclude. We refer to [35] for the explicit form of C and D.

8.3.3 FFT Techniques for the Calibration on Call Prices

We have shown in the preceding section that the Fourier transform of the law of X_T is known. We now use the idea suggested in [16] to construct an algorithm that allows to compute very quickly the price of calls with different strikes at the same time. It can then be used to calibrate the model's parameters to the prices observed on the market.[1]

8.3.3.1 First Approach

Let $C_T(k)$ be the price of the call with maturity T and strike e^k. We denote by q_T the density of the law of $\ln(S_T)$ and by ϕ_T its Fourier's transform. Then

$$\phi_T(\theta) = \int e^{i\theta x} q_T(x) dx \text{ and } C_T(k) = \int_k (e^x - e^k) q_T(x) dx .$$

Since C_T is not integrable (as it goes to a constant for k very negative), we instead work with $c_T(k) := e^{\alpha k} C_T(k)$ for a fixed $\alpha > 0$.

Then the transform of c_T satisfies

$$\psi_T(\theta) := \int e^{i\theta k} c_T(k) dk$$

$$= \int e^{i\theta k} \int_k e^{\alpha k} (e^x - e^k) q_T(x) dx dk$$

$$= \int q_T(x) \int_{-\infty}^{x} e^{i\theta k} (e^{x+\alpha k} - e^{(1+\alpha)k}) dk dx$$

$$= \frac{\phi_T(\theta - (\alpha + 1)i)}{\alpha^2 + \alpha - \theta^2 + i(2\alpha + 1)\theta} .$$

---

[1]One can also use asymptotic results, see e.g. [7].

In order to insure that ψ_T is well-defined, c_T should be integrable on the negative orthant. This amounts to assuming that $\psi_T(0) < \infty$, i.e. $\phi_T(-(\alpha+1)i) < \infty$ or equivalently $\mathbb{E}\left[S_T^{1+\alpha}\right] < \infty$. This imposes a restriction on α in terms of the parameters of model, but it can be tuned explicitly since ϕ_T is known.

To obtain call prices, its remains to invert the transformation:

$$C_T(k) = \frac{e^{-\alpha k}}{\pi} \int_0^\infty e^{-i\theta k} \psi_T(\theta) d\theta .$$

In order to ensure the stability of the term to integrate around 0, one must take α as large as possible (under the constraint mentioned above). One can control the truncation error by noticing that $\psi_T(\theta)$ behaves in $1/\theta^2$ for θ large which implies that $C_T(k) - \frac{e^{-\alpha k}}{\pi} \int_0^{\bar\theta} e^{-i\theta k} \psi_T(\theta) d\theta$ is at most of order of $1/\bar\theta$.

In practice, a *Fast Fourier Transform* algorithm is used to compute this integral for different log-strike k quickly at the same time. This is very useful for calibration procedures, as they require to re-compute these values for different set of parameters. From this perspective, it should be compared to the use of Dupire's equation presented in Sect. 7.2.1.

The first step consists in approximating the integral on a grid over θ:

$$C_T(k) \simeq \frac{e^{-\alpha k}}{\pi} \sum_{n=1}^N e^{-i\theta_n k} \psi_T(\theta_n) \Delta_\theta$$

where $\Delta_\theta := \bar\theta/N$, $N > 1$ and $\theta_n := (n-1)\Delta_\theta$.

Then, another grid on k is constructed: $k_\ell := -\bar k + (\ell-1)\Delta_k$ for $1 \le \ell \le N$, with $\bar k > 0$ and $\Delta_k := 2\bar k/(N-1)$ so that $k_0 = -\bar k$ and $k_N = \bar k$. Then

$$C_T(k_\ell) \simeq \frac{e^{-\alpha k_\ell}}{\pi} \sum_{n=1}^N e^{-i(n-1)(\ell-1)\Delta_\theta \Delta_k} e^{i\bar k\theta_n} \psi_T(\theta_n) \Delta_\theta .$$

Finally the parameters of the grids are chosen so that $\Delta_\theta \Delta_k = 2\pi/N$, which leads to

$$C_T(k_\ell) \simeq \frac{e^{-\alpha k_\ell}}{\pi} \sum_{n=1}^N e^{-i\frac{2\pi}{N}(n-1)(\ell-1)} e^{i\bar k\theta_n} \psi_T(\theta_n) \Delta_\theta .$$

Hence, it remains to perform a computation of the form

$$f_\ell = \sum_{n=1}^N e^{-i\frac{2\pi}{N}(n-1)(\ell-1)} g_n , \quad \ell = 1,\ldots,N ,$$

for which efficient algorithms with a computation cost in $O(N \ln N)$ exist, see e.g. [19].

8.3.3.2   Other Penalty Terms

Note that the multiplication by $e^{-\alpha k}$ is only used to ensure that we face an integrable function and that, in practice, the choice of α may be difficult. It is however possible to use alternative approaches. For instance, one can consider the Fourier transform of $\tilde{c}_T^{\hat{\sigma}_0}(k) = C_T(k) - BS_T^{\hat{\sigma}_0}(k)$ where $BS_T^{\hat{\sigma}_0}(k)$ is the price of a call of maturity T and of strike e^k computed for a level of volatility $\hat{\sigma}_0$ corresponding to the at-the-money implicit volatility.

It is easy to check that $\tilde{c}_T^{\hat{\sigma}_0}(k)$ is square integrable and that there exists $\alpha > 0$ such that $\mathbb{E}\left[S_T^{1+\alpha}\right] < \infty$. Finally, similar computations as above show that

$$\frac{1}{2\pi} \int e^{i\theta k} \tilde{c}_T^{\hat{\sigma}_0}(k) dk = \frac{\phi_T(\theta - i) - \phi_T^{\hat{\sigma}_0}(\theta - i)}{-\theta^2 + i\theta}$$

where

$$\phi_T^{\hat{\sigma}_0}(k) := \frac{1}{2\pi} \int e^{i\theta k} BS_T^{\hat{\sigma}_0}(k) d\theta = e^{-\frac{\sigma^2 T}{2}(\theta^2 + i\theta)}.$$

For the parameters used in practice, the above expression decreases more quickly than any power of θ, when its real part tends to infinity. The convergence of the integral is therefore quick.

8.3.4   The Recalibration Issue

Again, the parameters should be stable to avoid losses purely related to the recalibration of the model. In Heston's type models, the hedging strategy presented in Exercise 7.5 above, which takes the recalibration into account, cannot be used, because the parameters of the models are not assets that can be traded.[2]

It can however be used in Local Stochastic Volatility models, see Exercise 8.5 below, if, given the set of the parameters driving the stochastic part (Y in this exercise), one can calibrate the options prices by solely using the local volatility part. Then, the strategy is the following: (1) Calibrate the whole model at $t = 0$ on liquid options, including possibly exotic ones. (2) Keep the parameter (α, β) driving Y constant but recalibrate at each time t the local volatility function σ on calls and puts. We refer to [7] for more details.

---

[2]In Exercise 7.5, the function $\hat{\sigma}$ is fixed, the only inputs/parameters are the call prices $(C^i)_{i \le I}$.

8.4 Problems

8.1 (Variance swap hedging by using liquid calls) We consider a financial market constituted of a single risky asset S, and risk-free rate equal to 0. The dynamics of S are given by

$$dS_t = S_t \sigma_t dW_t \text{ for } t \in [0, T] ,$$

where W is a Brownian motion under a measure $\mathbb{Q}$, and σ is an adapted process satisfying $\underline{\sigma} \leq \sigma_t \leq \bar{\sigma}$ for all $t \in [0, T]$ $\mathbb{Q}$-a.s. for some constants $0 < \underline{\sigma} \leq \bar{\sigma} < \infty$. The risky asset price at 0 is $S_0 > 0$.

We want to price and hedge the random leg of the variance swap:

$$J_T := \frac{1}{T} \int_0^T \sigma_t^2 dt .$$

1. By using Itô's lemma, show that

$$-\frac{2}{T} \ln(S_T/S_0) + \frac{2}{T} \int_0^T \sigma_t dW_t = J_T .$$

2. Find a function ϕ such that $\frac{2}{T} \int_0^T \sigma_t dW_t = \int_0^T \phi(S_t) dS_t$.
3. Deduce a hedging strategy for $\frac{2}{T} \int_0^T \sigma_t dW_t$.
4. We assume in this question that we can sell at the price p a European contract of payoff $\ln(S_T)$, a log-contract of maturity T. How can we hedge J_T? What is the price of this hedge?

We assume from now on that we can buy and sell dynamically an European call of payoff $[S_{T'} - S_0]^+$ and maturity $T' > T$. Its price at $t < T$ is given by

$$p(t, S_t, \sigma_t) = \mathbb{E}^{\mathbb{Q}}[[S_{T'} - S_0]^+ \mid \mathcal{F}_t]$$

where p is a smooth function. We assume in addition that the dynamics of σ are of the form

$$d\sigma_t = \varphi(S_t, \sigma_t)(dW_t + \eta d\bar{W}_t)$$

with $\sigma_0 > 0$ a constant, φ a continuous and bounded function, $\eta > 0$ a constant and $\bar{W}$ a $\mathbb{Q}$-Brownian motion independent of W.

5. Show by using Itô's lemma that

$$dp(t, S_t, \sigma_t) = (\partial_S p(t, S_t, \sigma_t) S_t \sigma_t + \partial_\sigma p(t, S_t, \sigma_t) \varphi(S_t, \sigma_t)) dW_t$$
$$+ \partial_\sigma p(t, S_t, \sigma_t) \varphi(S_t, \sigma_t) \eta d\bar{W}_t .$$

We now also assume that there exists a smooth function v satisfying

$$v(t, S_t, \sigma_t) = \mathbb{E}^{\mathbb{Q}}[\ln(S_T) \mid \mathcal{F}_t] \text{ for } t \in [0, T] .$$

6. Write down the dynamics of $(v(t, S_t, \sigma_t))_{t \leq T}$ as obtained by using Itô's lemma.
7. Deduce that, in order to hedge dynamically the payoff $\ln(S_T)$, one should hold at each time t

$$\psi_t = \partial_\sigma v(t, S_t, \sigma_t) / \partial_\sigma p(t, S_t, \sigma_t)$$

units of the option of payoff $[S_{T'} - S_0]^+$ and

$$\phi_t S_t \sigma_t = \partial_S v(t, S_t, \sigma_t) S_t \sigma_t + \partial_\sigma v(t, S_t, \sigma_t) \varphi(S_t, \sigma_t)$$
$$- \psi_t \left(\partial_S p(t, S_t, \sigma_t) S_t \sigma_t + \partial_\sigma p(t, S_t, \sigma_t) \varphi(S_t, \sigma_t) \right)$$

units of the risky asset.
8. How can we hedge dynamically J_T by using the risky asset and the call? Provide the price of this hedge in terms of S_0 and $v(0, S_0, \sigma_0)$.
9. Which partial differential equation should the function v solve?

8.2 (Weighted variance swap) Let us consider the framework of Exercise 8.1, except that the random leg is now of the form

$$J_T^w := \int_0^T \sigma_t^2 w(S_t) dt ,$$

where w is a deterministic function, continuous and non-negative. The case $w(x) = x^+$ corresponds to the *variance-gamma swap*.

1. Fix F, a smooth function such that $\partial^2 F(x) = 2w(x)/x^2$ for $x > 0$. Show that

$$J_T^w = F(S_T) - F(S_0) - \int_0^T \partial F(S_t) dS_t.$$

2. How can we build a semi-static hedge for J_T^w without knowing σ?

8.3 (Stochastic correlation and option on log-returns) We continue the Exercise 7.2, and now assume that there exists a liquid option that pays at T the payoff

$$\theta := (\ln(S_T^1/S_0^1) + \frac{1}{2}\sigma_1^2 T)(\ln(S_T^2/S_0^2) + \frac{1}{2}\sigma_2^2 T) .$$

1. Show that there exist two predictable processes $\hat{\phi}^1$ and $\hat{\phi}^2$ such that

$$\theta = \int_0^T \hat{\phi}_t^1 dS_t^1 + \int_0^T \hat{\phi}_t^2 dS_t^2 + \sigma_1\sigma_2 \int_0^T \rho_t dt \ ,$$

and such that $M_t := \int_0^t \hat{\phi}_s^1 dS_s^1 + \int_0^t \hat{\phi}_s^2 dS_s^2$ defines a $\mathbb{P}$-martingale on $[0,T]$.

2. We assume that the market evaluates this product under the measure $\mathbb{P}$. Show that its price at t is given by

$$P_t = \sigma_1\sigma_2\mathbb{E}\left[\int_0^T \rho_s ds \mid \mathcal{F}_t\right] + M_t \ .$$

3. From now on, we assume that f maps $\mathbb{R}$ onto $(-1, 1)$, and denote by φ its inverse. Show that there exist continuous and bounded functions μ_ρ and σ_ρ such that $|\sigma_\rho| > 0$ on $(-1, 1)$ and

$$d\rho_t = \mu_\rho(\rho_t)dt + \sigma_\rho(\rho_t)dW_t^3 \ .$$

4. For the rest of this exercise, we assume that the function

$$v(t, r, c) := \mathbb{E}\left[C_T \mid (\rho_t, C_t) = (r, c)\right]$$

with

$$C_t := \int_0^t \rho_s ds \ ,$$

is $C^{1,2,1}$ on the domain $D := \{(t, r, c) \in [0, T) \times (-1, 1) \times \mathbb{R} \ : \ |c| < t\}$ and continuous on $\bar{D} := \{(t, r, c) \in [0, T] \times (-1, 1) \times \mathbb{R} \ : \ |c| < t\}$. Provide the partial differential equation solved by v.

5. Show that, for $t \leq T$,

$$dP_t = dM_t + \sigma_1\sigma_2\partial_r v(t, \rho_t, C_t)\sigma_\rho(\rho_t)dW_t^3 \ .$$

6. We now admit that $\frac{\partial}{\partial r}v > 0$ on D. Deduce that, for all $\mathcal{F}_T$-measurable and bounded random variable G, there exist predictable processes ϕ^1, ϕ^2 and ϕ^3 such that

$$G = \mathbb{E}[G] + \int_0^T \phi_s^1 dS_s^1 + \int_0^T \phi_s^2 dS_s^2 + + \int_0^T \phi_s^3 dP_s \ .$$

7. Comment.

8. What is the pricing equation for the option with payoff S if the derivative with payoff θ is liquid, with price given by the process P?

8.4 (Robust hedging of a digital barrier option) We want to hedge a digital barrier option that pays 1 at T if the risky asset price S has reached the level B before T. We only assume that S has continuous path. We look for a super-hedging strategy which does not rely on the exact dynamics of S, i.e. that is model independent. We denote by $C(K)$ the payoff of the call of strike $K \geq 0$, and assume that its price at 0, denoted by $c(K)$, is listed on the market. We set $S^* := \max_{t \leq T} S_t$ and $\tau := \inf\{t \geq 0 : S_t \geq B\}$.

1. Show that, for all $0 < K < B$, we can find $\beta(K) \in \mathbb{R}$ such that

$$\mathbf{1}_{\{S^* \geq B\}} \leq \frac{1}{B-K} C(K) + \beta(K)(S_T - S_\tau)\mathbf{1}_{\{\tau < T\}}$$

with equality if $S_T \geq K$ and $S^* \geq B$.
2. To which semi-static strategy does the right-hand side term corresponds?
3. Deduce that the price p of the barrier option with payoff $\mathbf{1}_{\{S^* \geq B\}}$ satisfies[3]

$$p \leq \min_{0 < K < B} \frac{1}{B-K} c(K).$$

4. We assume that there exists a measure $\mathbb{Q}$ such that

$$c(K) = \mathbb{E}^{\mathbb{Q}}[C(K)]$$

for all $K > 0$. We assume that $K > 0 \mapsto c(K)$ is strictly convex, and satisfies $c(0+) = S_0$ and $c(+\infty) = 0$. We also assume that $\mathbb{Q}[S_T = K] = 0$. Show that the value $\hat{K} > 0$ at which the minimum is reached in the above, satisfies $B = \mathbb{E}^{\mathbb{Q}}[S_T \mid S_T \geq \hat{K}]$.

8.5 (Local stochastic volatility models and implied volatility dynamics) Let us consider a model in which $r \equiv 0$ and S is a one dimensional stochastic process satisfying

$$dS_t = \sigma(t, S_t, Y_t)dW_t^1,$$

$$dY_t = \alpha(t, Y_t)dt + \beta(t, Y_t)dZ_t,$$

where

$$dZ_t = \rho(t, S_t, Y_t)dW_t^1 + (1 - \rho(t, S_t, Y_t)^2)^{\frac{1}{2}}dW_t^2.$$

In the above, all coefficients are uniformly Lipschitz with linear growth, $\sigma, \beta > 0$ and ρ takes values in $[-1, 1]$. Moreover, we assume that $S_0 > 0$ and that S remains strictly positive whatever its initial condition $S_0 > 0$ is. We also assume that $\mathbb{P}$ is

---

[3]One can in fact construct a model of price process S such that equality holds.

the unique risk neutral measure, that (S_t, Y_t) admits a density $f(t, \cdot)$ for all t, and that $(t, x, y) \in [0, T] \times (0, \infty) \times \mathbb{R} \mapsto f(t, x, y)$ is C_b^∞. We fix $K > 0$ and denote by $p(t, x, y)$ the price at time t of the call of strike K and maturity T if $(S_t, Y_t) = (x, y)$.

1. Which partial differential equation should p satisfy?
2. We assume that call options of strike K and maturities $T' \leq T$ can be traded dynamically and that the corresponding pricing functions are continuous, and $C^{1,2}$ strictly before their maturity. How can we hedge an option with payoff $G \in \mathbb{L}^1$?
3. Give the above hedging strategy when $G = g(S_T)$, for some bounded measurable map g, in the case where the corresponding pricing function is $C_b^{1,2}$.
4. Let $p_{BS}(t, x, \theta)$ denote the Black and Scholes price of the call option if $S_t = x$ and for the level of volatility $\theta > 0$. What is the partial differential equation satisfied by p_{BS}?
5. We now define $\theta_{\text{imp}}(t, x, y)$ as the solution[4] of $p_{BS}(t, x, \theta_{\text{imp}}(t, x, y)) = p(t, x, y)$. By using the above, show that θ_{imp} solves

$$
0 = \partial_\theta p_{BS} \left(\partial_t \theta_{\text{imp}} + \alpha \partial_y \theta_{\text{imp}} \right) + \sigma^2 \left(\partial_{x\theta}^2 p_{BS} \partial_x \theta_{\text{imp}} + \frac{1}{2} \partial_\theta p_{BS} \partial_{xx}^2 \theta_{\text{imp}} \right)
$$
$$
+ \rho \sigma \beta \partial_{xy}^2 p + \frac{1}{2} \beta^2 \partial_{yy}^2 p,
$$

where

$$
\partial_{xy}^2 p = \partial_{x\theta}^2 p_{BS} \partial_y \theta_{\text{imp}} + \partial_{\theta\theta}^2 p_{BS} \partial_x \theta_{\text{imp}} \partial_y \theta_{\text{imp}} + \partial_\theta p_{BS} \partial_{xy}^2 \theta_{\text{imp}},
$$
$$
\partial_{yy}^2 p = \partial_{\theta\theta}^2 p_{BS} (\partial_y \theta_{\text{imp}})^2 + \partial_\theta p_{BS} \partial_{yy}^2 \theta_{\text{imp}}.
$$

6. Use the Black and Scholes formula[5] to verify that

$$
\partial_{xx}^2 p_{BS}(t, x, \theta) = \frac{1}{\theta(T - t)x^2} \partial_\theta p_{BS}(t, x, \theta),
$$
$$
\partial_{\theta\theta}^2 p_{BS}(t, x, \theta) = \frac{(d_1 d_2)(t, x, \theta)}{\theta} \partial_\theta p_{BS}(t, x, \theta),
$$
$$
\partial_{x\theta}^2 p_{BS}(t, x, \theta) = \frac{-d_2(t, x, \theta)}{\theta x \sqrt{T - t}} \partial_\theta p_{BS}(t, x, \theta),
$$

with d_1 and d_2 given as in the Black and Scholes formula, to be computed.

---

[4] This is the implied volatility associated to our LSV model.
[5] See Exercise 2.4.

7. Verify that $\partial_\theta p_{BS}(t, x, \theta) > 0$ and deduce from the above that

$$\mathcal{D}\theta_{\text{imp}} = \partial_t \theta_{\text{imp}} + \alpha \partial_y \theta_{\text{imp}} + \frac{1}{2}\sigma^2 \partial_{xx}^2 \theta_{\text{imp}} + \frac{1}{2}\beta^2 \partial_{yy}^2 \theta_{\text{imp}} + \rho\sigma\beta \partial_{xy}^2 \theta_{\text{imp}}$$

where

$$\mathcal{D}\theta_{\text{imp}} := \frac{d_2\sigma^2}{\theta x\sqrt{T-t}}\partial_x \theta_{\text{imp}} - \rho\sigma\beta \left(\frac{-d_2}{\theta x\sqrt{T-t}} + \frac{d_1 d_2}{\theta}\partial_x \theta_{\text{imp}} \right) \partial_y \theta_{\text{imp}}$$
$$- \frac{1}{2}\beta^2 \frac{d_1 d_2}{\theta}(\partial_y \theta_{\text{imp}})^2.$$

8. Deduce the dynamics of the implied volatility $(\theta_{\text{imp}}(t, S_t, Y_t))_{t\leq T}$. How is it correlated to S? Does it have a drift?
9. Show that, if we consider the processes $\theta_{\text{imp}}^{\tau,K}$ associated to each maturity τ and strike K, then this model allows to reproduce perfectly at $t = 0$ the prices of the calls available on the market. Can we also use it to calibrate exotic options?
10. What can be the advantages of this model in terms of hedging and tracking of the smile? What are the constraints for it to be performant?

Corrections

8.1

1. Use Itô's lemma.
2. Take $\phi(S_t) = 2S_t/T$.
3. It is an immediate consequence of the preceding question.
4. Sell $2/T$ log-contracts, cost $-(2/T)\mathbb{E}^{\mathbb{Q}}[\ln S_T]$, keep $(2/T)ln(S_0)$ in cash, cost $(2/T)ln(S_0)$, and hedge $\frac{2}{T}\int_0^T \sigma_t dW_t$, cost 0.
5. Apply Itô's lemma and note that $(p(t, S_t, \sigma_t))_{t\leq T}$ is a martingale, so that the dt term is zero.

$$dp(t, S_t, \sigma_t) = (\partial_S p(t, S_t, \sigma_t)S_t\sigma_t + \partial_\sigma p(t, S_t, \sigma_t)\varphi(S_t, \sigma_t)) \, dW_t$$
$$+ \partial_\sigma p(t, S_t, \sigma_t)\varphi(S_t, \sigma_t)\eta d\bar{W}_t .$$

6. Again this is a martingale.
7. Just match the terms.
8. The dynamic hedge of $\ln S_T$ is given in 7. The remaining terms are hedged according to 4.
9. Apply Feynman and Kac's formula.

8.2

1. Apply Itô's lemma to $F(S)$.
2. Hedge dynamically $\int_0^T \partial F(S_t) dS_t$ and hedge statically $F(S_T) - F(S_0)$ by using combinations of calls, puts and cash.

8.3

1. Apply Itô's lemma and check that $\hat{\phi}^1$ and $\hat{\phi}^2$ are square integrable.
2. It is $\mathbb{E}[\theta \mid \mathcal{F}_t]$ and M is a martingale.
3. Apply Itô's lemma.
4. Apply the Feynman and Kac's formula.
5. Just apply Itô's lemma, and note that $(v(t, \rho_t, C_t))_{t \leq T}$ is a martingale.
6. Use the martingale representation which gives a representation in terms of W. To obtain the required representation, solve the corresponding system.
7. and 8. Given question 1. and the last result, if the option of payoff θ is liquid, then any option can be hedged by trading dynamically S^1, S^2 and the option of payoff θ. The existence of this option allows one to take a position in the stochastic correlation.

8.4

1. Take $\beta(K) = -1/(B - K)$.
2. A call position and a buy order when S reaches B.
3. The dynamic hedging part $\beta(K)(S_T - S_\tau)\mathbf{1}_{\{\tau < T\}}$ has cost 0.
4. Use the first order optimality condition and the fact that $\partial_K C(K) = -\mathbb{E}^{\mathbb{Q}}[\mathbf{1}_{S_T \geq K}] = -\mathbb{Q}[S_T \geq K]$ (which follows from the fact that $\mathbb{Q}[S_T = K]) = 0$).

8.5

1. $p(t, x, y) = \mathbb{E}[[S_T - K]^+ \mid (S_t, Y_t) = (x, y)]$. Apply the Feynman and Kac's formula.
2. Trade dynamically the stock and one option. See Section 8.1.
3. Apply the results of Section 8.1.
4. $-\partial_t p_{BS} - \frac{1}{2}\sigma^2 x^2 \partial_{xx}^2 p_{BS} = 0$ on $[0, T) \times (0, \infty)$ and $p_{BS}(T, \cdot) = [\cdot - K]^+$.
5. Use the PDE satisfied by p and write the derivatives of p in terms of p_{BS} and θ_{imp} by using $p_{BS}(t, x, \theta_{imp}(t, x, y)) = p(t, x, y)$.
6. Use the formula in Exercise 2.4 and do the explicit computations.
7. Do the explicit computations of $\partial_\theta p_{BS}(t, x, \theta)0$. Then, use 4. and 6. to simplify the PDE obtained in 5.
8. By Itô's lemma and the preceding question,

$$d\theta_{imp}(t, S_t, Y_t) = \mathcal{D}\theta_{imp}(t, S_t, Y_t) dt$$

$$+ (\partial_x \theta_{imp}\sigma)(t, S_t, Y_t) dW_t^1 + (\partial_y \theta_{imp}\beta)(t, S_t, Y_t) dZ_t.$$

Yes, it has a drift. Its co-quadratic variation with S is given by

$$d\langle\theta_{imp}(\cdot, S, Y), S\rangle_t = \left[\sigma(\partial_x \theta_{imp}\sigma + \rho\partial_y \theta_{imp}\beta)\right](t, S_t, Y_t) dt.$$

9. It suffices to use the initialisation $\theta_{\text{imp}}^{\tau,K} = \sigma_{\text{imp}}(\tau, K)$, in which $\sigma_{\text{imp}}(\tau, K)$ is the implied volatility as observed on the market for the maturity τ and the strike K. Just the local volatility part is enough for calibration (by Dupire's formula). The presence of the extra parameters can be used to calibrate exotic options.

10. It provides dynamics for the smile. If these dynamics match the ones observed on the market, one can expect to avoid the problems of recalibration. If not, then using this model might be problematic. . .

References

1. Alexander, C.O., Nogueira, L.: Stochastic Local Volatility, No. ICMA-dp2008-02. Henley Business School, Reading University (2008)
2. Alfonsi, A.: On the discretisation schemes for CIR (and Bessel squared) processes. Monte Carlo Methods Appl. **11**(4), 355–467 (2006)
3. Avellaneda, M., Friedman,C., Holmes, R., Samperi, D.: Calibrating volatility surfaces via relative-entropy minimization. Appl. Math. Financ. **4**(1), 37–64 (1997)
4. Andersen, L., Piterbarg, V.: Moment explosions in stochastic volatility models. Financ. Stoch. **11**(1), 29–50 (2007)
5. Barles, G., Souganidis, P.E.: Convergence of approximation schemes for fully nonlinear second order equations. Asym. Anal. **4**(3), 271–283 (1991)
6. Bates, D.: Jumps and stochastic volatility: Exchange rate process implicit in Deutschmark options. Rev. Financ. Stud. **9**, 69–107 (1996)
7. Bergomi, L.: Stochastic Volatility Modeling. Chapman and Hall/CRC Financial Mathematics Series. CRC Press, Boca Raton (2016)
8. Bouchard, B., Dang, N.M.: Generalized stochastic target problems for pricing and partial hedging under loss constraints – application in optimal book liquidation. Financ. Stoch. **17**(1), 31–72 (2013)
9. Bouchard, B., Elie, R., Imbert, C.: Optimal control under stochastic target constraints. SIAM J. Control Optim. **48**(5), 3501–3531 (2010)
10. Bouchard, B., Elie, R., Touzi N.: Stochastic target problems with controlled loss. SIAM J. Control Optim. **48**(5), 3123–3150 (2009)
11. Bouchard, B., Moreau L., Nutz, M.: Stochastic target games with controlled loss. Ann. Appl. Probab. **24**(3), 899–934 (2014)
12. Bouchard, B., Touzi, N.: Weak dynamic programming principle for viscosity solutions. SIAM J. Control Optim. **49**(3), 948–962 (2011)
13. Bouchard, B., Vu, T.N.: The American version of the geometric dynamic programming principle: application to the pricing of American options under constraints. Appl. Math. Optim. **61**(2), 235–265 (2010)
14. Bouchard, B., Vu, T.N.: A stochastic target approach for P&L matching problems. Math. Oper. Res. **37**(3), 526–558 (2012)
15. Brézis, H.: Analyse Fonctionnelle. Masson, Paris (1983)
16. Carr, P., Madan, D.: Option valuation using the fast Fourier transform. J. Comput. Financ. **2**, 61–73 (1999)
17. Chassagneux, J.-F., Elie, R., Kharroubi, I.: When terminal facelift enforces Delta constraints. Financ. Stoch. **19**, 329–362 (2015)

© Springer International Publishing Switzerland 2016
B. Bouchard, J.-F. Chassagneux, *Fundamentals and Advanced Techniques in Derivatives Hedging*, Universitext, DOI 10.1007/978-3-319-38990-5

18. Cont, R., Tankov, P.: Financial Modelling with Jump Processes. Chapman and Hall, Boca Raton (2004)
19. Cooley, J.W., Tukey, J.W.: An algorithm for the machine calculation of complex Fourier series. Math. Comput. **19**, 297–301 (1965)
20. Crandall, M.G., Ishii, H., Lions, P.-L.: User's guide to viscosity solutions of second order partial differential equations. Am. Math. Soc. **27**, 1–67 (1992)
21. Crépey, S.: Contribution à des méthodes numériques appliquées à la finance and aux jeux différentiels. Thèse de doctorat, École Polytechnique (2001)
22. Davis, M., Obloj, J.: Market completion using options. Prépublication (2008)
23. Delbaen, F., Schachermayer, W.: A general version of the fundamental theorem of assand pricing. Math. Ann. **300**, 463–520 (1994)
24. Delbaen, F., Schachermayer, W.: The fundamental theorem of asset pricing for unbounded stochastic processes. Math. Ann. **312**(2), 215–250 (1994)
25. Delbaen, F., Shirakawa, H.: A note of option pricing for constant elasticity of variance model. Asia-Pac. Financ. Mark. **9**(2), 85–99 (2002)
26. Dupire, B.: Pricing with a smile. Risk Mag. **7**, 18–20 (1994)
27. Feynman, R.P., Hibbs, A.: Quantum Mechanics and Path Integrals. McGraw-Hill, New York (1965)
28. Föllmer, H., Kabanov, Y.: Optional decomposition and Lagrange multipliers. Financ. Stoch. **2**, 69–81 (1998)
29. Föllmer, H., Leukert, P.: Quantile Hedging. Financ. Stoch. **3**(3), 251–273 (1999)
30. Föllmer, H., Leukert, P.: Efficient hedging: cost versus shortfall risk. Financ. Stoch. **4**, 117–146 (2000)
31. Fournié, E., Lasry, J.-M., Lebuchoux, J., Lions, P.-L.: Applications of Malliavin calculus to Monte Carlo methods in finance II. Financ. Stoch. **5**, 201–236 (2001)
32. Fournié, E., Lasry, J.-M., Lebuchoux, J., Lions, P.-L., Touzi, N.: Applications of Malliavin calculus to Monte Carlo methods in finance. Financ. Stoch. **3**, 391–412 (1999)
33. Glasserman, P.: Monte Carlo Methods in Financial Engineering. Stochastic Modelling and Applied Probability, vol. 53. Springer, New York (2003)
34. Hagan, P.S., Kumar, D., Lesniewski, A.S., Woodward, D.E.: Managing smile risk. In: The Best of Wilmott, vol. 1, pp. 249–296. Jon Wiley & Sons, Chichester (2005)
35. Heston, S.: A closed-form solution for options with stochastic volatility with applications to bond and currency options. Rev. Financ. Stud. **6**(2), 327–343 (1993)
36. Hodges, S.D., Neuberger, A.: Optimal replication of contingent claims under transaction costs. Rev. Futures Mark. **8**, 222–239 (1989)
37. Ishii, H.: On the equivalence of two notions of weak solutions, viscosity solutions and distribution solutions. Funkcial. Ekvac **38**(1), 101–120 (1995)
38. Jourdain, B.: Stochastic flows approach to Dupire's formula. Financ. Stoch. **11**(4), 521–535 (2007)
39. Kabanov, Y., Stricker, C.: A teachers' note on no-arbitrage criteria. In: Séminaire de Probabilités, XXXV. Lecture Notes in Mathematics, vol. 1755, pp. 149–152. Springer, Berlin/London (2001)
40. Kac, M.: On some connections between probability theory and differential and integral equations. In: Proceedings 2nd Berkeley Symposium on Mathematical Statistics and Probability, University of California Press, Berkely/Los Angeles (1951)
41. Karatzas, I., Shreve, S.E.: Brownian Motion and Stochastic Calculus. Springer, New York (1991)
42. Karatzas, I., Shreve, S.E.: Methods of Mathematical Finance. Springer, New York (1998)
43. El Karoui, N.: Les aspects probabilistes du contrôle stochastique. École d'Été de Probabilités de Saint Flour IX. Lecture Notes in Mathematics, vol. 876. Springer (1979)
44. Kramkov, D., Schachermayer, W.: Necessary and sufficient conditions in the problem of optimal investment in incomplete markets. Ann. Appl. Probab. **13**(4), 1504–1516 (2003)
45. Lapeyre, B.J., Sulem, A., Talay, D.: Understanding Numerical Analysis for Financial Models. Cambridge University Press, Cambridge (2003)

46. Musiela, M., Rutkowski, M.: Martingale Methods in Financial Modeling. Stochastic Modelling and Applied Probability, vol. 36. Springer, Berlin/New York (2005)
47. Neveu, J.: Martingales à temps Discret. Masson, Paris (1974)
48. Nualart, D.: The Malliavin Calculus and Related Topics. Springer, Berlin (1995)
49. Overhaus, M., et al.: Equity Hybrid Derivatives. Wiley Finance. Wiley, Hoboken (2007)
50. Pironneau, O.: Dupire-like identities for complex options. Comptes Rendus Mathematique **344**(2), 127–133 (2007)
51. Protter, P.: Stochastic Integration and Differential Equations. Springer, Berlin (1990)
52. Revuz, D., Yor, M.: Continuous Martingales and Brownian Motion. Springer, Berlin (1990)
53. Rockafellar, R.T.: Convex Analysis. Princeton University Press, Princeton (1970)
54. Schachermayer, W.: Optimal Investment in Incomplete Markets when Wealth may Become Negative. Ann. Appl. Probab. **11**(3), 694–734 (2001)
55. Soner, H.M., Touzi, N.: Stochastic targetproblems, dynamic programming and viscosity solutions. SIAM J. Control Optim. **41**, 404–424 (2002)
56. Soner, H.M., Touzi, N.: Dynamic programming for stochastic target problems and geometric flows. J. Eur. Math. Soc. **4**, 201–236 (2002)

Index

adapted, 4, 55
admissible strategy, 58
American option, 23, 71
   callable, 40
   swing, 41
arbitrage opportunity, 6, 58
asian option, 85, 156
attainable, 65

backward Kolmogorov's equation, 229
barrier option, 83, 142, 154
Bermudan option, 155
Black and Scholes model, 80, 227
Black-Scholes-Barenblatt equation, 180, 184, 236
butterfly option, 82
buy-and-hold, 182

calibration, 227
call on spread, 84
CEV model, 239
chooser option, 86
CIR, 247, 260
comparison principle, 135, 137
complete/incomplete markets, 20, 65
Cox-Ingersoll-Ross process, 247, 260

Davis' price, 112
delta, delta hedging, 132, 151
digital option, 82
directed upward, 27
discount factor, 4

discounted value, 5
discounting process, 56
dividend, 154
Doob-Meyer decomposition, 24, 74
dual formulation, 17, 26, 64, 75
Dupire's equation, 230, 231
Dupire's formula, 231
dynamic programming principle, 72, 77, 147
Dynkin operator, 130

essential supremum, 24
European options, 13
exercise date, 23

face-lift, 170, 183
Farkas' Lemma, 59
Fenchel transform, 108, 119
Feynman-Kac Theorem, 130, 138
financial portfolio, 4
first fundamental theorem of asset pricing, 7, 63
Fokker-Planck equation, 229
forward libor rate, 93
forward risk-neutral measure, 90
forward start option, 241
forward swap rate, 94

G2++ model, 92
gamma, 235, 245
geometric dynamic programming principle, 195
Girsanov's theorem, 61

© Springer International Publishing Switzerland 2016
B. Bouchard, J.-F. Chassagneux, *Fundamentals and Advanced Techniques
in Derivatives Hedging*, Universitext, DOI 10.1007/978-3-319-38990-5

Index of Notations

$\mathbf{1}_A$, xi
I_d, 140
$\mathbf{1}_d$, 81

A^c, xi
$\mathcal{A}$, 5, 58
$\mathcal{A}_b$, 62
$\mathcal{A}_K^b$, 74
$\mathcal{A}_K$, 32, 194

β, 4, 56
β^t, 129
B_t, 5
$B_t(T)$, 80

$\|\cdot\|$, xi
$\mathcal{C}_K(\cdot)$, 170
$C^{p,q}(\bar{B})$, xi
$C_b^{p,q}(\bar{B})$, xii

$\delta_K(\cdot)$, 75
∂f, xi
$\partial^2 f$, xi
$\partial^2_{x^i,x^j} f$, xi
$\partial_t f$, xi
$\partial^2_x f$, xi
∇f, xi
$\partial_x S^{t,x}$, 140
$DS^{t,x}$, 139

$\mathbb{E}^{\mathbb{Q}}[\cdot]$, xi

$esssup$, 24

$\tilde{\Gamma}_t$, 8

$\mathcal{H}_b^K(\tilde{S})$, 32

$\hat{K}$, 75
$\hat{\mathcal{K}}_b$, 75, 169

$\mathcal{L}_S$, 130
$\mathbb{L}^p(B, \mathbb{Q}, \mathcal{G})$, xi
L_S, 14
L^τ, 94
$\mathbb{L}_b^0$, 12

$\mathcal{M}(\tilde{S})$, 7, 60
$\mathcal{M}_b(\tilde{S})$, 7
$\mathcal{M}_{\mathrm{loc}}(\tilde{S})$, 60
$\mathbb{M}^{d,n}$, xi

(NA), 6
(NA), 58
(NA_t), 7
$(NFLVR)$, 63

$\bar{\mathcal{O}}$, xii
$\mathcal{O}^c$, xii
$\partial\mathcal{O}$, xii

© Springer International Publishing Switzerland 2016
B. Bouchard, J.-F. Chassagneux, *Fundamentals and Advanced Techniques in Derivatives Hedging*, Universitext, DOI 10.1007/978-3-319-38990-5

Printed in the United States
By Bookmasters